A·N·N·U·A·L E·D·I·T·I·O·N·S

World Politics

03/04

Twenty-fourth Edition

EDITOR

Helen E. Purkitt

U.S. Naval Academy

Dr. Helen Purkitt obtained her Ph.D. in international relations from the University of Southern California. She is professor of political science at the U.S. Naval Academy. Her research and teaching interests include political psychology, African politics, international security, and environment politics. She is currently completing a case study of transnational bioterrorism for use by homeland security first-line defenders and an experimental study of framing effects in budget decision making. Recent publications include a book coauthored with Stephen Burgess, *South Africa and Weapons of Mass Destruction* (forthcoming, 2003: Indiana University Press); "How People Think About Environmental Problems and Political Conflicts," in S. Nagel (ed.) *Environmental Policy and Developing Nations*. McFarland, 2001; "The cognitive basis of foreign policy expertise: Evidence from intuitive analyses of political novices and 'experts' in South Africa," in D. Sylvan and J. F. Voss (eds.) *Problem Representations and Political Decision Making*. Cambridge University Press, 1998.

McGraw-Hill/Dushkin

530 Old Whitfield Street, Guilford, Connecticut 06437

Visit us on the Internet
http://www.dushkin.com

Credits

1. **New World Order**
   Unit photo—United Nations photo.
2. **World Economy**
   Unit photo—© 2003 by PhotoDisc, Inc.
3. **Weapons of Mass Destruction**
   Unit photo—© 2003 by PhotoDisc, Inc.
4. **North America**
   Unit photo—Courtesy of The White House.
5. **Latin America**
   Unit photo—United Nations photo.
6. **Europe**
   Unit photo—Courtesy of British Information Service.
7. **Former Soviet Union**
   Unit photo—United Nations photo.
8. **The Pacific Basin**
   Unit photo—© 2003 by PhotoDisc, Inc.
9. **Middle East and Africa**
   Unit photo—AP/Wide World photo.
10. **International Organizations and Global Issues**
    Unit photo—United Nations photo.

Copyright

Cataloging in Publication Data
Main entry under title: Annual Editions: World Politics. 2003/2004.
1. World Politics—Periodicals. I. Purkitt, Helen E., *comp.* II. Title: World Politics.
ISBN 0–07–283819–1   658'.05   ISSN 1098–0300

Twenty-Fourth Edition

Cover image © 2003 PhotoDisc, Inc.
Printed in the United States of America   1234567890BAHBAH54   Printed on Recycled Paper

Editors/Advisory Board

Members of the Advisory Board are instrumental in the final selection of articles for each edition of ANNUAL EDITIONS. Their review of articles for content, level, currentness, and appropriateness provides critical direction to the editor and staff. We think that you will find their careful consideration well reflected in this volume.

To the Reader

In publishing ANNUAL EDITIONS we recognize the enormous role played by the magazines, newspapers, and journals of the public press in providing current, first-rate educational information in a broad spectrum of interest areas. Many of these articles are appropriate for students, researchers, and professionals seeking accurate, current material to help bridge the gap between principles and theories and the real world. These articles, however, become more useful for study when those of lasting value are carefully collected, organized, indexed, and reproduced in a low-cost format, which provides easy and permanent access when the material is needed. That is the role played by ANNUAL EDITIONS.

Annual Editions: World Politics 03/04 is aimed at filling a void in materials for learning about world politics and foreign policy. The articles are chosen for those who are new to the study of world politics. The goal is to help students learn more about international issues that often seem remote but may have profound consequences for a nation's well-being, security, and survival. This volume has been compiled to convey the dynamic interdependence and complexities of actors and actions in contemporary international relations.

Interdependence means that events in places as far away as Latin America, Asia, the Middle East, and Africa may affect the United States, just as America's actions, and inaction, have significant repercussions for other states. Interdependence also refers to the increased role of nonstate actors such as international corporations, the United Nations, and a rich array of nongovernmental actors such as Cable News Network (CNN) and the terrorist network of Osama bin Laden's al Qaeda.

The September 11, 2001, terrorist attacks on the World Trade towers and the Pentagon tragically underscored the reality that nonstate actors increasingly influence the scope, nature, and pace of events worldwide. The U.S.–led military intervention in Afghanistan and the escalation of ongoing tensions in Iraq and North Korea underscore the fact that nation-state conflicts will also continue as a key feature of international relations.

International events are proceeding at such a rapid pace that what is said about international affairs today may be outdated by tomorrow. It is important, therefore, that readers develop a mental framework or theory of the international system as a complex system of loosely connected and diverse sets of actors who interact around an ever-changing agenda of international issues.

This collection of articles about international events provides up-to-date information, commentaries about the current set of issues on the world agenda, and analyses of the significance of the issues and emerging trends for the structure and functioning of the post–cold war international system.

This twenty-fourth edition of Annual Editions: World Politics is divided into 10 units. The end of the cold war means that we can no longer view international relations through the prism of a bipolar system. Instead, national, regional, subregional, and transnational issues are increasingly important aspects of international relations in the emerging multipolar and multidimensional world system.

There is a growing recognition that new strategies are required to deal with the new problems facing the world. One of the most serious problem areas is the growing recognition of an emerging global imperative to control the spread of old and new infectious diseases. The most serious of these today is the HIV/AIDS pandemic. Even though the HIV/AIDS pandemic has not yet peaked in southern Africa, there are signs that the center of the global pandemic will soon shift to Eurasia. The spread of the disease in Africa is a humanitarian tragedy of epic proportions that has lowered the life expectancy, disrupted the economy, and devastated societies. As the HIV/AIDS pandemic spreads to Russia, India, and China, the global balance of power may be altered. Some experts warn that the effects of the spreading HIV/AIDS pandemic, much like the effects of the black plague in feudal Europe, may help to trigger major changes in power structures within existing nation-states and even the structure of the international system.

Increased numbers of analysts now call the twenty-first century the biological century, in recognition of the dramatic political, economic, and social changes that are likely to occur as the result of major research breakthroughs and biotech innovations. The dark side of the biological revolution is that novel processes and procedures can also be used to create new types of weapons of mass destruction that may be difficult to detect or prevent. As these developments unfold, international politics promises to be a complex mix of old and new trends and actors. A recent UN publication warns that water shortages, global warming, and nitrogen pollution constitute new threats to future world security unless politicians act now to curb conspicuous overconsumption in the world's richer countries. At the same time hunger remains the most serious health threat in the world today.

I would like to thank Theodore Knight and his associates at McGraw-Hill/Dushkin for their help in putting this volume together and previous users of Annual Editions: World Politics, who took the time to contribute articles or comments. Please continue to provide feedback by filling out the postage-paid article rating form on the last page of this book.

Helen Purkitt

Helen E. Purkitt
Editor

Contents

UNIT 1
New World Order

Four articles consider some of the challenges facing the world: the impact of local conflicts on foreign policy, major influences on domestic and international security, and the consequences of globalization.

The concepts in bold italics are developed in the article. For further expansion, please refer to the Topic Guide and the Index.

UNIT 2
World Economy

Four articles examine the global marketplace as politics redefine the rules of the economic game.

UNIT 3
Weapons of Mass Destruction

Four articles discuss nuclear proliferation and the use of toxic weapons.

The concepts in bold italics are developed in the article. For further expansion, please refer to the Topic Guide and the Index.

UNIT 4
North America

These four articles discuss current and future United States and Canadian roles in world policy and international trade.

The concepts in bold italics are developed in the article. For further expansion, please refer to the Topic Guide and the Index.

UNIT 5
Latin America

Three selections consider Latin American relations in the Western Hemisphere with regard to politics, economic reform, and trade.

UNIT 6
Europe

Four selections review some of the historic events that will alter Western and Central Europe. Topics include the European Union's expansion and Central/Eastern Europe's strivings toward democracy.

The concepts in bold italics are developed in the article. For further expansion, please refer to the Topic Guide and the Index.

UNIT 7
Former Soviet Union

Two articles examine the relationship between the Bush administration and Russia and the place of Ukraine between East and West. The country is leaning toward Europe.

UNIT 8
The Pacific Basin

Three articles examine some of the countries instrumental in the economic evolution of the Pacific Basin.

The concepts in bold italics are developed in the article. For further expansion, please refer to the Topic Guide and the Index.

UNIT 9
Middle East and Africa

Five articles review the current state of the Middle East and Africa with regard to conflict, extremism, and democratic trends.

UNIT 10
International Organizations and Global Issues

Three articles discuss international organizations and world peace, UN reform, and the establishment of an international criminal court.

The concepts in bold italics are developed in the article. For further expansion, please refer to the Topic Guide and the Index.

The concepts in bold italics are developed in the article. For further expansion, please refer to the Topic Guide and the Index.

Topic Guide

This topic guide suggests how the selections in this book relate to the subjects covered in your course. You may want to use the topics listed on these pages to search the Web more easily.

On the following pages a number of Web sites have been gathered specifically for this book. They are arranged to reflect the units of this *Annual Edition*. You can link to these sites by going to the DUSHKIN ONLINE support site at *http://www.dushkin.com/online/*.

ALL THE ARTICLES THAT RELATE TO EACH TOPIC ARE LISTED BELOW THE BOLD-FACED TERM.

United Nations

Weapons of mass destruction

World Wide Web Sites

The following World Wide Web sites have been carefully researched and selected to support the articles found in this reader. The easiest way to access these selected sites is to go to our DUSHKIN ONLINE support site at *http://www.dushkin.com/online/*.

AE: World Politics 03/04

The following sites were available at the time of publication. Visit our Web site—we update DUSHKIN ONLINE regularly to reflect any changes.

General Sources

Belfer Center for Science and International Affairs (BCSIA)
http://www.ksg.harvard.edu/bcsia/

BCSIA is a center for research, teaching, and training in international affairs.

Carnegie Endowment for International Peace
http://www.ceip.org

One of the goals of this organization is to stimulate discussion and learning among experts and the public on a wide range of international issues. The site provides links to the journal *Foreign Policy* and to the Moscow Center.

Central Intelligence Agency
http://www.odci.gov

Use this official home page to learn about many facets of the CIA and to get connections to other sites and resources, such as *The CIA Factbook,* which provides extensive statistical information about every country in the world.

Crisisweb: The International Crisis Group (ICG)
http://www.crisisweb.org

ICG is an organization "committed to strengthening the capacity of the international community to anticipate, understand, and act to prevent and contain conflict." Go to this site to view the latest reports and research concerning conflicts around the world.

The Heritage Foundation
http://www.heritage.org

This page offers discussion about and links to many sites of the Heritage Foundation and other organizations having to do with foreign policy and foreign affairs.

World Wide Web Virtual Library: International Affairs Resources
http://www.etown.edu/vl/

Surf this site and its links to learn about specific countries and regions, to research think tanks and organizations, and to study such vital topics as international law, development, the international economy, human rights, and peacekeeping.

UNIT 1: New World Order

Avalon Project at Yale Law School
http://www.yale.edu/lawweb/avalon/terrorism/terror.htm

The Avalon Project web site features documents in the fields of law, history, economics, diplomacy, politics, government, and terrorism.

Human Rights Web
http://www.hrweb.org

YThis useful site offers ideas on how individuls can get involved in helping to protect human rights around the world.

U.S. Air Force Institute for National Security Studies
http://www.usafa.af.mil/inss/occasion.htm

The full-text commissioned peer review reports on a variety of security topics affecting the United States and the world can be found at this Web site, sponsored by the Department of Defense.

UNIT 2: World Economy

International Political Economy Network
http://csf.colorado.edu/ipe/

This premier site for research and scholarship includes electronic archives.

Organization for Economic Cooperation and Development/ FDI Statistics
http://www.oecd.org/daf/investment/

Explore world trade and investment trends and statistics on this site. It provides links to many related topics and addresses global economic issues on a country-by-country basis.

Virtual Seminar in Global Political Economy/Global Cities & Social Movements
http://csf.colorado.edu/gpe/gpe95b/resources.html

This site of Internet resources is rich in links to subjects of interest in regional environmental studies, covering topics such as sustainable cities, megacities, and urban planning.

World Bank
http://www.worldbank.org

News (press releases, summaries of new projects, speeches) and coverage of numerous topics regarding development, countries, and regions are provided at this site. Go to the research and growth section of this site to access specific research and data regarding the world economy.

UNIT 3: Weapons of Mass Destruction

The Bulletin of the Atomic Scientists
http://www.bullatomsci.org

This site allows you to read more about the Doomsday Clock and other issues as well as topics related to nuclear weaponry, arms control, and disarmament.

Federation of American Scientists
http://www.fas.org

This site provides useful information about and links to a variety of topics related to chemical and biological warfare, missiles, conventional arms, and terrorism.

ISN International Relations and Security Network
http://www.isn.ethz.ch

This site, maintained by the Center for Security Studies and Conflict Research, is a clearinghouse for extensive information on international relations and security policy.

The RMA Debate: Terrorism and Counter-terrorism
http://www.comw.org/rma/fulltext/terrorism.html

The RMA Debate is a gateway tp full-text online resources about the Revolution in Military Affairs, information war, and asymmetrical warfare.

Terrorism Research Center
http://www.terrorism.com

The Terrorism Research Center features definitions and research on terrorism, counterterrorism documents. a comprehensive list of Web links, and profiles of terrorist and counterterrorist groups.

UNIT 4: North America

The Henry L. Stimson Center   Peace Operations and Europe
http://www.stimson.org/fopo/?SN=FP20020610372

The Future of Peace Opertions has begun to address specific areas concerning Europe and operations.The site links to useful UN, NATO, and EU documents, and research pieces and news sites.

The North American Institute
http://www.northamericaninstitute.org

NAMI, a trinational public-affairs organization, is concerned with the emerging "regional space" of Canada, the United States, and Mexico and the development of a North American community. It provides links for study of trade, the environment, and institutional developments.

UNIT 5: Latin America

Inter-American Dialogue
http://www.iadialog.org

This is the Web site for IAD, a premier U.S. center for policy analysis, communication, and exchange in Western Hemisphere affairs. The 100-member organization has helped to shape the agenda of issues and choices in hemispheric relations.

UNIT 6: Europe

Central Europe Online
http://www.centraleurope.com

This site contains daily updated information under headings such as news on the Web today, economics, trade, and currency.

Europa: European Union
http://europa.eu.int

This server site of the European Union will lead you to the history of the EU (and its predecessors), descriptions of EU policies, institutions, and goals, and documentation of treaties and other materials.

NATO Integrated Data Service
http://www.nato.int/structur/nids/nids.htm

Check out this Web site to review North Atlantic Treaty Organization documentation, to read *NATO Review,* and to explore key issues in the field of European security and transatlantic cooperation.

Social Science Information Gateway
http://sosig.esrc.bris.ac.uk

A project of the Economic and Social Research Council (ESRC), this is an online catalog of thousands of Internet resources relevant to political education and research.

UNIT 7: Former Soviet Union

Russia Today
http://www.russiatoday.com

This site includes headline news, resources, government, politics, election results, and pressing issues.

Russian and East European Network Information Center, University of Texas at Austin
http://reenic.utexas.edu/reenic.html

This is *the* Web site for information on the former Soviet Union.

UNIT 8: The Pacific Basin

ASEAN Web
http://www.asean.or.id

This site of the Association of South East Asian Nations provides an overview of Asia: Web resources, summits, economic affairs, political foundations, and regional cooperation.

Inside China Today
http://www.insidechina.com

Part of the European Internet Network, this site leads you to information on all of China, including recent news, government, and related sites.

Japan Ministry of Foreign Affairs
http://www.mofa.go.jp

Visit this official site for Japanese foreign policy statements and press releases, archives, and discussions of regional and global relations.

UNIT 9: Middle East and Africa

Africa News Online
http://www.africanews.org

Open this site for up-to-date information on all of Africa, with reports from Africa's leading newspapers, magazines, and news agencies. Coverage is country-by-country and regional. Internet links are among the resource pages.

ArabNet
http://www.arab.net

This page of ArabNet, the online resource for the Arab world in the Middle East and North Africa, presents links to 22 Arab countries. Each country page classfies information using a standardized system.

Columbia International Affairs Online
http://www.ciaonet.org/cbr/cbr00/video/cbr_v/cbr_v.html

At this site find excerpts from al Qaeda's 2-hour videotape used to recruit young Muslims to fight in a holy war. The tape demonstates al Qaeda's use of the Internet and media outlets for propaganda and persuasion purposes.

ei: Electronic Intifada
http://electronicintifada.net/new.shtml

EI is a major Palestinian portal for information about the Palestinian-Israeli conflict from a Palestinian perspective. The site had 600,000 visits in April 2001.

IslamiCity
http://islamicity.com

This is one of the largest Islamic sites on the Web, reaching 50 million people a month. Based in California, it includes public opinion polls, links to television and radio broadcasts, and religious guidance.

Israel Information Center
http://www.accessv.com/~yehuda/

Search the directories in this site for such information as policy speeches. interviews, and briefings; discussions of Israel and the UN; and Web sites of Israel's government.

www.dushkin.com/online/

Mail and Guardian

htp://www.mg.co.za

This daily summary of news articles from a major English-speaking newspaper in South Africa provides a bird's eye view of how major stories in the West and worldwide are covered in Africa.

MEMRI: The Middle East Research Institute

http://www.memri.org/video

Arab satellite channels air recent video clips on topics related to Islamic culture, fundamentalism, and terrorism from this site. For translations of what Arab leaders are telling their followers, go to *http://www.memri.org.*

UNIT 10: International Organizations and Global Issues

Commission on Global Governance

http://www.sovereignty.net/p/gov/gganalysis.htm

This site provides access to *The Report of the Commission on Global Governance,* produced by an international group of leaders who want to find ways in which the global community can better manage its affairs.

Global Trends 2005 Project

http://www.csis.org/gt2005/sumreport.html

The Center for Strategic and International Studies explores the coming global trends and challenges of the new millenium. Read their summary report at this Web site. Also access Enterprises for the Environment, Global Information Infrastructure Commission, and Americas at this site.

InterAction

http://www.interaction.org

InterAction encourages grassroots action, engages policy makers on advocacy issues, and uses this site to inform people on its initiatives to expand international humanitarian relief and development assistance programs.

IRIN

http://www.irinnews.org

The UN Office for the Coordination of Humanitarian Affairs provides free analytical reports, fact sheets, interviews, daily country updates, and weekly summaries through this site and e-mail distribution service.The site is a good source of news for crisis situations as they occur.

The North-South Institute

http://www.nsi-ins.ca/ensi/index.html

Searching this site of the North-South Institute   which works to strengthen international development cooperation and enhance gender and social equity   will help you find information and debates on a variety of global issues.

Uniited Nations Home Page

http://www.un.org

Here is the gateway to information about the United Nations. Also see *http:/www.undp.org/missions/usa/usna/htm* for the U.S. Mission at the UN.

We highly recommend that you review our Web site for expanded information and our other product lines. We are continually updating and adding links to our Web site in order to offer you the most usable and useful information that will support and expand the value of your Annual Editions. You can reach us at: *http://www.dushkin.com/annualeditions/.*

UNIT 1
New World Order

Unit Selections

1. **Clash of Globalizations**, Stanley Hoffmann
2. **Transnational Terrorism and the al Qaeda Model: Confronting New Realities**, Paul J. Smith
3. **Sovereignty**, Stephen D. Krasner
4. **Reconciling Non-Intervention and Human Rights**, Douglas T. Stuart

Key Points to Consider

- Explain why you agree or disagree with Stanley Hoffman's two theses that (a) interstate war is becoming less common and that (b) all states' foreign policies are increasingly shaped by domestic politics, in addition to economic and military power.

- According to Paul Smith, why are we unlikely to witness the demise of the al Qaeda model even if the war on terrorism succeeds in eliminating al Qaeda as a terrorist group?

- Do you agree or disagree with Stephen Krasner's thesis that autonomous, independent nation-states are likely to remain the most important actors in international relations? What types of nonstate actors are likely to increase in influence in the future?

- Review Douglas Stuart's definition and guidelines for determining when humanitarian interventions into another country should be approved, and explain why you believe that these are or are not adequate criteria for nation-states and international organizations to use.

 **Links: www.dushkin.com/online/**
These sites are annotated in the World Wide Web pages.

Avalon Project at Yale Law School
http://www.yale.edu/lawweb/avalon/terrorism/terror.htm

Human Rights Web
http://www.hrweb.org

U.S. Air Force Institute for National Security Studies
http://www.usafa.af.mil/inss/occasion.htm

At the beginning of the twenty-first century, there is a noticeable increase in efforts to predict important changes and to understand new patterns of relationships that may shape international relations. Anyone can engage in this sport as there is little agreement among "futurists" about what to expect regarding causes of tension, types of conflict, or the patterns of interaction that may characterize international relations during the twenty-first century.

With the demise of the cold war, analysts focus more on the political and economic ramifications of the emerging international system. Many cite "globalization" as the dominant characteristic of the new international order. Globalization refers to the increased global interdependence of economic, communication, and transport systems. Globalization also refers to innovations in computer and other high-tech capabilities that are increasingly being used by people in all parts of world society. While many futurologists stress the novel features of the new world order, others emphasize that many of the major changes in trade, communication, and transportation taking place today are similar to the profound remaking of the world that occurred during earlier centuries. During the fifteenth and sixteenth centuries, European nation-states and societies in Africa, the Americas, the Middle East, and to a lesser extent Asia, were tied together in the first modern world economy.

Economic and cultural trends in the modern era of globalization may differ from earlier eras of intense globalization because modern trends are triggering integrative and disintegrative tendencies that are global in scope and unprecedented in the rapidity of the pace of change. The most significant disintegrative global trends include the rise of cultural extremism; increased economic inequality between the developed and developing world; and the diffusion of high-technology weaponry.

For some, the terrorist attacks against the World Trade Center in New York and the Pentagon in Washington D.C., on September 11, 2001, and the anthrax letter attacks the following month highlighted the vulnerabilities of economically developed societies to attacks by disaffected radicals who now pursue their political goals by killing large numbers of civilians using unconventional means. In "Transnational Terrorism and the al Qaeda Model: Confronting New Realities," Paul Smith describes how al Qaeda attacks in various parts of the world illustrate the harsh new reality that terrorist and criminal networks now exist in a transnational milieu that is divorced from state-driven constraints. Other analysts and policy makers interpreted these same incidents as evidence of an increasing clash of cultures or as the start of a new era in international relations.

The juxtaposition of contradictory trends in the international system allows futurologists to make very different predications about the emerging world order. In "Clash of Globalizations," Stanley Hoffmann concludes that the events of September 11, 2001, marked the beginning of a new era, but one in which great-powers rivalries will continue to shape international relations. An important change within this vision is an emphasis on the increasingly important role that domestic politics now plays in the formulation of foreign policies of all states. Stephen Krasner, in "Sovereignty," concludes that those who proclaim the death of sovereignty misread history. Krasner outlines several reasons why nation-states as autonomous, independent entities are likely to be able to adapt to new challenges. In contrast, other analysts predict that the future world order will continue to be a unipolar world dominated by the United States. Still others predict the demise of democratic capitalism and free-market integration and the rise of new empires that will replace the nation-state system during the twenty-first century.

Another distinguishing characteristic of the post–cold war era has been an increase in ethnic and communal conflicts. Throughout the 1990s, as nation-states increasingly intervened in other countries, it became harder to skirt a confrontation between the traditional commitment to state sovereignty and a growing commitment to basic human rights. In "Reconciling Non-Intervention and Human Rights," Douglas Stuart offers a definition of humanitarian intervention and guidelines for determining the circumstances under which such intervention should be authorized and action taken. Such explicit criteria may become increasingly important as nation-states continue to intervene in domestic conflicts with and without the UN's blessing.

Today, governments and the media in developed countries ignore many violent local conflicts in remote areas of the developing world. The remote conflicts that do receive media attention and involvement tend to be ones that are important to major powers due to geographic proximity or strategic value. Some conflicts raise concerns because they may provide sanctuaries or new targets of opportunity for al Qaeda. Other situations gain international prominence due to concerns about the use of weapons of mass destruction by nation-states or other actors as the expertise and materials needed to produce nuclear, chemical, and biological weapons continue to proliferate.

Lasting peace and security requires local, national, regional, and international actors to manage shared resources such as water, while effectively coping with new transnational problems such as the spread of disease. At the same time, local and national authorities must meet the basic needs of growing populations and promote economic growth. How well political regimes cope with environmental and related challenges is receiving attention today because it is recognized that such factors serve as root causes or triggers fueling local conflicts.

Occasionally, problems caused by environmental factors facilitate peace. A devastating drought in Mozambique, in addition to damage caused by prolonged civil war, led warring factions to agree to a negotiated peace settlement in the early 1990s. In a similar fashion, the worsening problems created by drought throughout the Middle East, along with the desires of President Assad of Syria and former prime minister Barak of Israel, prompted the leaders of the two countries to resume peace talks at the end of 1999. However, the breakdown of the peace process in the Middle East, the resumption of the intifada in 2000, the continuing violence, and recent statements by Prime Minister Sharon that Palestinian leader Arafat was "irrelevant to a negotiated settlement" underscore how difficult it will be to construct lasting peace agreements among parties engaged in protracted conflicts.

Clash of Globalizations

Stanley Hoffmann

A NEW PARADIGM?

WHAT IS THE STATE OF international relations today? In the 1990s, specialists concentrated on the partial disintegration of the global order's traditional foundations: states. During that decade, many countries, often those born of decolonization, revealed themselves to be no more than pseudostates, without solid institutions, internal cohesion, or national consciousness. The end of communist coercion in the former Soviet Union and in the former Yugoslavia also revealed long-hidden ethnic tensions. Minorities that were or considered themselves oppressed demanded independence. In Iraq, Sudan, Afghanistan, and Haiti, rulers waged open warfare against their subjects. These wars increased the importance of humanitarian interventions, which came at the expense of the hallowed principles of national sovereignty and nonintervention. Thus the dominant tension of the decade was the clash between the fragmentation of states (and the state system) and the progress of economic, cultural, and political integration—in other words, globalization.

Everybody has understood the events of September 11 as the beginning of a new era. But what does this break mean? In the conventional approach to international relations, war took place among states. But in September, poorly armed individuals suddenly challenged, surprised, and wounded the world's dominant superpower. The attacks also showed that, for all its accomplishments, globalization makes an awful form of violence easily accessible to hopeless fanatics. Terrorism is the bloody link between interstate relations and global society. As countless individuals and groups are becoming global actors along with states, insecurity and vulnerability are rising. To assess today's bleak state of affairs, therefore, several questions are necessary. What concepts help explain the new global order? What is the condition of the interstate part of international relations? And what does the emerging global civil society contribute to world order?

SOUND AND FURY

TWO MODELS made a great deal of noise in the 1990s. The first one—Francis Fukuyama's "End of History" thesis—was not vindicated by events. To be sure, his argument predicted the end of ideological conflicts, not history itself, and the triumph of political and economic liberalism. That point is correct in a narrow sense: the "secular religions" that fought each other so bloodily in the last century are now dead. But Fukuyama failed to note that nationalism remains very much alive. Moreover, he ignored the explosive potential of religious wars that has extended to a large part of the Islamic world.

Fukuyama's academic mentor, the political scientist Samuel Huntington, provided a few years later a gloomier account that saw a very different world. Huntington predicted that violence resulting from international anarchy and the absence of common values and institutions would erupt among civilizations rather than among states or ideologies. But Huntington's conception of what constitutes a civilization was hazy. He failed to take into account sufficiently conflicts within each so-called civilization, and he overestimated the importance of religion in the behavior of non-Western elites, who are often secularized and Westernized. Hence he could not clearly define the link between a civilization and the foreign policies of its member states.

Other, less sensational models still have adherents. The "realist" orthodoxy insists that nothing has changed in international relations since Thucydides and Machiavelli: a state's military and economic power determines its fate; interdependence and international institutions are secondary and fragile phenomena; and states' objectives are imposed by the threats to their survival or security. Such is the world described by Henry Kissinger. Unfortunately, this venerable model has trouble integrating change, especially globalization and the rise of nonstate actors. Moreover, it overlooks the need for international cooperation that results from such new threats as the proliferation of weapons of mass destruction (WMD). And it ignores what the scholar Raymond Aron called the "germ of a universal consciousness": the liberal, promarket norms that developed states have come to hold in common.

Taking Aron's point, many scholars today interpret the world in terms of a triumphant globalization that submerges borders through new means of information and communication. In this universe, a state choosing to stay closed invariably faces decline and growing discontent among its subjects, who are eager for material progress. But if it opens up, it must accept a reduced role that is mainly limited to social protection, physical protection against aggression or civil war, and maintaining national identity. The champion of this epic without heroes is *The New York Times* columnist Thomas Friedman. He contrasts barriers with open vistas, obsolescence with modernity, state control with free markets. He sees in globalization the light of dawn, the "golden straitjacket" that will force contentious publics to understand that the logic of globalization is that of peace (since war would interrupt globalization and therefore progress) and democracy (because new technologies increase individual autonomy and encourage initiative).

BACK TO REALITY

THESE MODELS come up hard against three realities. First, rivalries among great powers (and the capacity of smaller states to exploit such tensions) have most certainly not disappeared. For a while now, however, the existence of nuclear weapons has produced a certain degree of prudence among the powers that have them. The risk of destruction that these weapons hold has moderated the game and turned nuclear arms into instruments of last resort. But the game could heat up as more states seek other WMD as a way of narrowing the gap between the nuclear club and the other powers. The sale of such weapons thus becomes a hugely contentious issue, and efforts to slow down the spread of all WMD, especially to dangerous "rogue" states, can paradoxically become new causes of violence.

Second, if wars between states are becoming less common, wars within them are on the rise—as seen in the former Yugoslavia, Iraq, much of Africa, and Sri Lanka. Uninvolved states first tend to hesitate to get engaged in these complex conflicts, but they then (sometimes) intervene to prevent these conflicts from turning into regional catastrophes. The interveners, in turn, seek the help of the United Nations or regional organizations to rebuild these states, promote stability, and prevent future fragmentation and misery.

Third, states' foreign policies are shaped not only by realist geopolitical factors such as economics and military power but by domestic politics. Even in undemocratic regimes, forces such as xenophobic passions, economic grievances, and transnational ethnic solidarity can make policymaking far more complex and less predictable. Many states—especially the United States—have to grapple with the frequent interplay of competing government branches. And the importance of individual leaders and their personalities is often underestimated in the study of international affairs.

For realists, then, transnational terrorism creates a formidable dilemma. If a state is the victim of private actors such as terrorists, it will try to eliminate these groups by depriving them of sanctuaries and punishing the states that harbor them. The national interest of the attacked state will therefore require either armed interventions against governments supporting terrorists or a course of prudence and discreet pressure on other governments to bring these terrorists to justice. Either option requires a questioning of sovereignty—the holy concept of realist theo-

ries. The classical realist universe of Hans Morgenthau and Aron may therefore still be very much alive in a world of states, but it has increasingly hazy contours and offers only difficult choices when it faces the threat of terrorism.

At the same time, the real universe of globalization does not resemble the one that Friedman celebrates. In fact, globalization has three forms, each with its own problems. First is economic globalization, which results from recent revolutions in technology, information, trade, foreign investment, and international business. The main actors are companies, investors, banks, and private services industries, as well as states and international organizations. This present form of capitalism, ironically foreseen by Karl Marx and Friedrich Engels, poses a central dilemma between efficiency and fairness. The specialization and integration of firms make it possible to increase aggregate wealth, but the logic of pure capitalism does not favor social justice. Economic globalization has thus become a formidable cause of inequality among and within states, and the concern for global competitiveness limits the aptitude of states and other actors to address this problem.

Optimism regarding globalization rests on very fragile foundations.

Next comes cultural globalization. It stems from the technological revolution and economic globalization, which together foster the flow of cultural goods. Here the key choice is between uniformization (often termed "Americanization") and diversity. The result is both a "disenchantment of the world" (in Max Weber's words) and a reaction against uniformity. The latter takes form in a renaissance of local cultures and languages as well as assaults against Western culture, which is denounced as an arrogant bearer of a secular, revolutionary ideology and a mask for U.S. hegemony.

Finally there is political globalization, a product of the other two. It is characterized by the preponderance of the United States and its political institutions and by a vast array of international and regional organizations and transgovernmental networks (specializing in areas such as policing or migration or justice). It is also marked by private institutions that are neither governmental nor purely national—say, Doctors Without Borders or Amnesty International. But many of these agencies lack democratic accountability and are weak in scope, power, and authority. Furthermore, much uncertainty hangs over the fate of American hegemony, which faces significant resistance abroad and is affected by America's own oscillation between the temptations of domination and isolation.

The benefits of globalization are undeniable. But Friedman-like optimism rests on very fragile foundations. For one thing, globalization is neither inevitable nor irresistible. Rather, it is largely an American creation, rooted in the period after World War II and based on U.S. economic might. By extension, then, a deep and protracted economic crisis in the United States could have as devastating an effect on globalization as did the Great Depression.

3

Second, globalization's reach remains limited because it excludes many poor countries, and the states that it does transform react in different ways. This fact stems from the diversity of economic and social conditions at home as well as from partisan politics. The world is far away from a perfect integration of markets, services, and factors of production. Sometimes the simple existence of borders slows down and can even paralyze this integration; at other times it gives integration the flavors and colors of the dominant state (as in the case of the Internet).

Third, international civil society remains embryonic. Many nongovernmental organizations reflect only a tiny segment of the populations of their members' states. They largely represent only modernized countries, or those in which the weight of the state is not too heavy. Often, NGOs have little independence from governments.

Fourth, the individual emancipation so dear to Friedman does not quickly succeed in democratizing regimes, as one can see today in China. Nor does emancipation prevent public institutions such as the International Monetary Fund, the World Bank, or the World Trade Organization from remaining opaque in their activities and often arbitrary and unfair in their rulings.

Fifth, the attractive idea of improving the human condition through the abolition of barriers is dubious. Globalization is in fact only a sum of techniques (audio and videocassettes, the Internet, instantaneous communications) that are at the disposal of states or private actors. Self-interest and ideology, not humanitarian reasons, are what drive these actors. Their behavior is quite different from the vision of globalization as an Enlightenment-based utopia that is simultaneously scientific, rational, and universal. For many reasons—misery, injustice, humiliation, attachment to traditions, aspiration to more than just a better standard of living—this "Enlightenment" stereotype of globalization thus provokes revolt and dissatisfaction.

Another contradiction is also at work. On the one hand, international and transnational cooperation is necessary to ensure that globalization will not be undermined by the inequalities resulting from market fluctuations, weak state-sponsored protections, and the incapacity of many states to improve their fates by themselves. On the other hand, cooperation presupposes that many states and rich private players operate altruistically—which is certainly not the essence of international relations—or practice a remarkably generous conception of their long-term interests. But the fact remains that most rich states still refuse to provide sufficient development aid or to intervene in crisis situations such as the genocide in Rwanda. That reluctance compares poorly with the American enthusiasm to pursue the fight against al Qaeda and the Taliban. What is wrong here is not patriotic enthusiasm as such, but the weakness of the humanitarian impulse when the national interest in saving non-American victims is not self-evident.

IMAGINED COMMUNITIES

Among the many effects of globalization on international politics, three hold particular importance. The first concerns institutions. Contrary to realist predictions, most states are not perpetually at war with each other. Many regions and countries live in peace; in other cases, violence is internal rather than state-to-state. And since no government can do everything by itself, interstate organisms have emerged. The result, which can be termed "global society," seeks to reduce the potentially destructive effects of national regulations on the forces of integration. But it also seeks to ensure fairness in the world market and create international regulatory regimes in such areas as trade, communications, human rights, migration, and refugees. The main obstacle to this effort is the reluctance of states to accept global directives that might constrain the market or further reduce their sovereignty. Thus the UN's powers remain limited and sometimes only purely theoretical. International criminal justice is still only a spotty and contested last resort. In the world economy—where the market, not global governance, has been the main beneficiary of the state's retreat—the network of global institutions is fragmented and incomplete. Foreign investment remains ruled by bilateral agreements. Environmental protection is badly ensured, and issues such as migration and population growth are largely ignored. Institutional networks are not powerful enough to address unfettered short-term capital movements, the lack of international regulation on bankruptcy and competition, and primitive coordination among rich countries. In turn, the global "governance" that does exist is partial and weak at a time when economic globalization deprives many states of independent monetary and fiscal policies, or it obliges them to make cruel choices between economic competitiveness and the preservation of social safety nets. All the while, the United States displays an increasing impatience toward institutions that weigh on American freedom of action. Movement toward a world state looks increasingly unlikely. The more state sovereignty crumbles under the blows of globalization or such recent developments as humanitarian intervention and the fight against terrorism, the more states cling to what is left to them.

Second, globalization has not profoundly challenged the enduring national nature of citizenship. Economic life takes place on a global scale, but human identity remains national—hence the strong resistance to cultural homogenization. Over the centuries, increasingly centralized states have expanded their functions and tried to forge a sense of common identity for their subjects. But no central power in the world can do the same thing today, even in the European Union. There, a single currency and advanced economic coordination have not yet produced a unified economy or strong central institutions endowed with legal autonomy, nor have they resulted in a sense of post-national citizenship. The march from national identity to one that would be both national and European has only just begun. A world very partially unified by technology still has no collective consciousness or collective solidarity. What states are unwilling to do the world market cannot do all by itself, especially in engendering a sense of world citizenship.

Third, there is the relationship between globalization and violence. The traditional state of war, even if it is limited in scope, still persists. There are high risks of regional explosions in the Middle East and in East Asia, and these could seriously affect relations between the major powers. Because of this threat, and because modern arms are increasingly costly, the "anarchical society" of states lacks the resources to correct some of globalization's most flagrant flaws. These very costs, combined with the classic distrust among international actors who prefer to try to preserve their security alone or through traditional alliances, prevent a more satisfactory institutionalization of world politics—for example, an increase of the UN's powers. This step could happen if global society were provided with sufficient forces to prevent a conflict or restore peace—but it is not.

Globalization, far from spreading peace, thus seems to foster conflicts and resentments. The lowering of various barriers celebrated by Friedman, especially the spread of global media, makes it possible for the most deprived or oppressed to compare their fate with that of the free and well-off. These dispossessed then ask for help from others with common resentments, ethnic origin, or religious faith. Insofar as globalization enriches some and uproots many, those who are both poor and uprooted may seek revenge and self-esteem in terrorism.

GLOBALIZATION AND TERROR

Terrorism is the poisoned fruit of several forces. It can be the weapon of the weak in a classic conflict among states or within a state, as in Kashmir or the Palestinian territories. But it can also be seen as a product of globalization. Transnational terrorism is made possible by the vast array of communication tools. Islamic terrorism, for example, is not only based on support for the Palestinian struggle and opposition to an invasive American presence. It is also fueled by a resistance to "unjust" economic globalization and to a Western culture deemed threatening to local religions and cultures.

If globalization often facilitates terrorist violence, the fight against this war without borders is potentially disastrous for both economic development and globalization. Antiterrorist measures restrict mobility and financial flows, while new terrorist attacks could lead the way for an antiglobalist reaction comparable to the chauvinistic paroxysms of the 1930s. Global terrorism is not the simple extension of war among states to nonstates. It is the subversion of traditional ways of war because it does not care about the sovereignty of either its enemies or the allies who shelter them. It provokes its victims to take measures that, in the name of legitimate defense, violate knowingly the sovereignty of those states accused of encouraging terror. (After all, it was not the Taliban's infamous domestic violations of human rights that led the United States into Afghanistan; it was the Taliban's support of Osama bin Laden.)

But all those trespasses against the sacred principles of sovereignty do not constitute progress toward global society, which has yet to agree on a common definition of terrorism or on a common policy against it. Indeed, the beneficiaries of the antiterrorist "war" have been the illiberal, poorer states that have lost so much of their sovereignty of late. Now the crackdown on terror allows them to tighten their controls on their own people, products, and money. They can give themselves new reasons to violate individual rights in the name of common defense against insecurity—and thus stop the slow, hesitant march toward international criminal justice.

Another main beneficiary will be the United States, the only actor capable of carrying the war against terrorism into all corners of the world. Despite its power, however, America cannot fully protect itself against future terrorist acts, nor can it fully overcome its ambivalence toward forms of interstate cooperation that might restrict U.S. freedom of action. Thus terrorism is a global phenomenon that ultimately reinforces the enemy—the state—at the same time as it tries to destroy it. The states that are its targets have no interest in applying the laws of war to their fight against terrorists; they have every interest in treating terrorists as outlaws and pariahs. The champions of globalization have sometimes glimpsed the "jungle" aspects of economic globalization, but few observers foresaw similar aspects in global terrorist and antiterrorist violence.

Finally, the unique position of the United States raises a serious question over the future of world affairs. In the realm of interstate problems, American behavior will determine whether the nonsuperpowers and weak states will continue to look at the United States as a friendly power (or at least a tolerable hegemon), or whether they are provoked by Washington's hubris into coalescing against American preponderance. America may be a hegemon, but combining rhetorical overkill and ill-defined designs is full of risks. Washington has yet to understand that nothing is more dangerous for a "hyperpower" than the temptation of unilateralism. It may well believe that the constraints of international agreements and organizations are not necessary, since U.S. values and power are all that is needed for world order. But in reality, those same international constraints provide far better opportunities for leadership than arrogant demonstrations of contempt for others' views, and they offer useful ways of restraining unilateralist behavior in other states. A hegemon concerned with prolonging its rule should be especially interested in using internationalist methods and institutions, for the gain in influence far exceeds the loss in freedom of action.

In the realm of global society, much will depend on whether the United States will overcome its frequent indifference to the costs that globalization imposes on poorer countries. For now, Washington is too reluctant to make resources available for economic development, and it remains hostile to agencies that monitor and regulate the global market. All too often, the right-leaning tendencies of the American political system push U.S. diplomacy toward an excessive reliance on America's greatest asset—military strength—as well as an excessive reliance on market capitalism and a "sovereigntism" that offends and alienates. That the mighty United States is so afraid of the world's imposing its "inferior" values on Americans is often a source of ridicule and indignation abroad.

ODD MAN OUT

For all these tensions, it is still possible that the American war on terrorism will be contained by prudence, and that other governments will give priority to the many internal problems created by interstate rivalries and the flaws of globalization. But the world risks being squeezed between a new Scylla and Charybdis. The Charybdis is universal intervention, unilaterally decided by American leaders who are convinced that they have found a global mission provided by a colossal threat. Presentable as an epic contest between good and evil, this struggle offers the best way of rallying the population and overcoming domestic divisions. The Scylla is resignation to universal chaos in the form of new attacks by future bin Ladens, fresh humanitarian disasters, or regional wars that risk escalation. Only through wise judgment can the path between them be charted.

We can analyze the present, but we cannot predict the future. We live in a world where a society of uneven and often virtual states overlaps with a global society burdened by weak public institutions and underdeveloped civil society. A single power dominates, but its economy could become unmanageable or distrupted by future terrorist attacks. Thus to predict the future confidently would be highly incautious or naive. To be sure, the world has survived many crises, but it has done so at a very high price, even in times when WMD were not available.

Precisely because the future is neither decipherable nor determined, students of international relations face two missions. They must try to understand what goes on by taking an inventory of current goods and disentangling the threads of present networks. But the fear of confusing the empirical with the normative should not prevent them from writing as political philosophers at a time when many philosophers are extending their conceptions of just society to international relations. How can one make the global house more livable? The answer presupposes a political philosophy that would be both just and acceptable even to those whose values have other foundations. As the late philosopher Judith Shklar did, we can take as a point of departure and as a guiding thread the fate of the victims of violence, oppression, and misery; as a goal, we should seek material and moral emancipation. While taking into account the formidable constraints of the world as it is, it is possible to loosen them.

STANLEY HOFFMANN is Buttenwieser University Professor at Harvard University and a regular book reviewer for *Foreign Affairs*.

Transnational Terrorism and the al Qaeda Model: Confronting New Realities

PAUL J. SMITH

On 6 January 1995, Philippine authorities responded to a fire that had started in room 603 of the Dona Josefa apartment complex in downtown Manila. Although firefighters quickly contained the blaze, which they first attributed to a simple cooking fire, they soon realized that they had stumbled upon something far more sinister. The fire, later investigations revealed, was started by one of the residents, who had mistakenly mixed water with chemicals being prepared for bombs. The incident's timing—coming just one week before the Pope's visit to the Philippines—immediately set off alarm bells within the Philippine security establishment. More alarming was the apartment's location, just minutes away from one of the Pontiff's intended destinations, and the discovery of Roman Catholic vestments that would provide cover for a suicide bomber.

But the most disturbing revelation was found in a laptop computer left in the apartment when the residents fled. Ramzi Ahmed Yousef, one of the residents of the apartment, had reportedly told his roommate, Abdul Hakim Murad, to retrieve the laptop. Murad returned to the apartment but was intercepted by Philippine police. Murad attempted to flee, but he stumbled and was apprehended. Murad then offered large sums of money to the police in an effort to bribe his way out of his predicament, but to no avail. Later, Murad would be subjected to a grueling inquisition—according to reports—about the contents of the computer and his role in the scheme that was code-named "Oplan Bojinka."

Oplan Bojinka, it was later learned, was a complex plan to bomb 11 US airliners over the Pacific Ocean as they traveled from Asia back to the United States. The plot would involve a team of five bombers who would travel on planes for a particular leg of their journey, plant the bomb, and then exit the plane at the next stop. Most of the bombers would later travel on separate routes back to Pakistan, where they would meet. The airplanes, however, would have a very different fate. As the planes journeyed to their next stops—in most cases the United States—the bombs would detonate, destroying the planes in mid-air. More than 4,000 people likely would have died had Oplan Bojinka been completed.

In later trial testimony, it was revealed that the bombing of Philippine Airlines flight 434 from Cebu to Japan on 11 December 1994, in which a Japanese businessman was killed, was a trial run for the larger Bojinka plan. Oplan Bojinka also included airborne suicide attacks with passenger airplanes onto key US targets, including CIA headquarters in Langley, Virginia. When Murad revealed this detail during interrogation, he also admitted attending flying schools in the United States and elsewhere. Subsequent FBI investigations confirmed Murad's attendance in at least two American schools, one in New York and the other in North Carolina.

On 11 September 2001, an analogue of Oplan Bojinka—and some would argue Bojinka itself—was actualized when 19 young men, mostly Saudi Arabian nationals, commandeered four passenger airplanes and rammed three of them into critical US targets, the World Trade Center and the Pentagon. The resulting social and economic impact—some 3,000 lives lost and billions of dollars in economic damage—catapulted terrorism onto an entirely new level of strategic importance. Catastrophic, or mass-casualty terrorism, once a theory, had now become a reality.[1] But the larger issue revolved around the nature of terrorism itself and its emerging modus operandi. Whether the 11 September attacks in the United States were the delayed manifestation of Oplan Bojinka, as some believe, or whether they were an isolated plan, it is clear that terrorism—and particularly that form of terrorism practiced by al Qaeda—has fundamentally changed.[2]

The 11 September attacks on the United States were a bold, calculated transnational attack by an organization that has established and maintained a multinational presence in more than 50 countries, directed by a base located—at least until recently—in Afghanistan. Like many multinational corporations, al Qaeda is both the product and beneficiary of globalization. The organization took advantage of the fruits of globalization and modernization—including satellite technology, accessible air travel, fax machines, the internet, and other modern conveniences—to advance its political agenda. No longer geographically constrained within a particular territory, or financially tied to a particular state, al Qaeda emerged as the ultimate transnational terror organization, relying on an array of legitimate and illicit sources of cash, including international charities that were often based in the West.

In the weeks following the attacks, many politicians, journalists, and pundits pointed to a "massive intelligence failure" that facilitated or allowed the attacks.[3] Some attributed this failure to the lack of human intelligence operations within Afghanistan. However, some experts have argued that the greatest intelligence failure of the 11 September attacks was the inability on the part of intelligence and law enforcement agencies to grasp and understand that al Qaeda represented a different type of terrorism, one less anchored to specific geographic locations or political constituencies and one capable of achieving transglobal strategic reach in its operations.[4]

The 11 September attacks also exposed fundamental weaknesses of modern Western states, including vulnerable borders, inadequate immigration controls, and insufficient internal antiterrorism surveillance. Indeed, investigations conducted following the US terror attacks would reveal an uncomfortable truth about al Qaeda and its affiliate groups. Probably their most important bases of operation—from a financial and logistical perspective—were located not in Afghanistan or Sudan, but rather in Western Europe and North America, including in the United States itself.[5]

The al Qaeda Multi-Cellular Terror Model

Al Qaeda (Arabic for "The Base") traces its roots to Afghanistan and the pan-Islamic resistance to the invasion of Afghanistan by the Soviet Union in 1979. In 1982, Osama bin Laden, then a young Saudi Arabian national, joined the anti-Soviet jihad. He traveled to Afghanistan where, after just a few years, he established his own military camps from which anti-Soviet assaults could be launched. In 1988, bin Laden and others established al Qaeda, not as a terrorist organization, but rather as a reporting infrastructure so that relatives of foreign soldiers who had come to Afghanistan to join the resistance could be properly tracked.[6] Al Qaeda reportedly had the additional function of funneling money to the Afghan resistance.[7] In 1989, the year the Soviets withdrew their last troops from Afghanistan, bin Laden returned to Saudi Arabia, where he began delivering public lectures about topics that were sensitive to the government—including

predictions that Kuwait would soon be invaded by Iraq. When his prediction came true, he became frustrated when the Saudi government ignored his advice (including offers of military assistance), and instead turned to the United States for military help.

Increasingly unhappy with bin Laden's public activities and his militant views, the Saudi government placed him under house arrest. Through his family connections, bin Laden was nevertheless able to secure permission for a business trip to Pakistan. Once in Pakistan, he traveled to Afghanistan and stayed a few months. But soon after, he left for Sudan where he was welcomed by National Islamic Front (NIF) leader, Hassan al-Turabi.[8] Bin Laden's time in Sudan is probably the most important in terms of al Qaeda's development. During this period, al Qaeda forged alliances with militant groups from Egypt, Pakistan, Algeria, and Tunisia, as well as with Palestinian Jihad and Hamas.[9] Also while in Sudan, al Qaeda began to develop its signature transnational modus operandi by engaging in a range of international operations, such as deploying fighters to Chechnya and Tajikistan, establishing satellite offices in Baku, Azerbaijan, and funding affiliates based in Jordan and Eritrea.[10] Under American pressure, however, Sudan forced bin Laden to leave in 1996. He and other members of al Qaeda relocated their operations to Afghanistan where they remained, until recently.

Al Qaeda has traditionally operated with an informal horizontal structure, comprising more than 24 constituent terrorist organizations, combined with a formal vertical structure. Below Osama bin Laden was the "majlis al shura," a consultative council that directed the four key committees (military, religious, finance, and media), members of which were handpicked by senior leadership. The majlis al shura discussed and approved major operations, including terrorist attacks.[11] Bin Laden and his two cohorts, Ayman al-Zawahiri and Mohammed Atef, set general policies and approved large-scale actions. Until the US intervention in Afghanistan, al Qaeda acted in a manner somewhat resembling a large charity organization that funded terrorist projects to be conducted by preexisting or affiliate terrorist groups.

The United States emerged as a central enemy to al Qaeda almost from the beginning of the organization's existence for a variety of reasons, including al Qaeda's unhappiness with US operations in the 1990–91 Gulf War and the 1992–93 Operation Restore Hope in Somalia. Al Qaeda's overarching complaint against the United States has centered on its continued military presence in Saudi Arabia and throughout the Arabian peninsula. To publicize its disdain for the United States, al Qaeda issued various "fatwas" (verdicts based on Islamic law) urging that US forces should be attacked. In 1992 and 1993, the group issued fatwas urging that American forces in Somalia should be attacked. In 1996, the group issued a "Declaration of Jihad on the Americans Occupying the Country of the Two Sacred Places," which urged the expulsion of American forces from the Arabian Peninsula.[12] This was followed by a media interview in 1997 in which bin Laden called for attacks on US soldiers.[13]

The anti-American rhetoric emanating from al Qaeda hit a high pitch in 1998 when the organization essentially fused with Egypt's two main terrorist organizations, al Jihad (Islamic Jihad) and al Gamaa al Islamiya (Islamic Group), both of which were linked to the assassination of former Egyptian President Anwar Sadat. The new campaign would be known as the World Islamic Front for Jihad Against the Jews and the Crusaders, and would also include co-signatories from Pakistan and Bangladesh.[14] Contained in the text that announces the World Islamic Front are calls to attack not only US soldiers, but also US civilians. The proclamation demands that Muslims everywhere should "abide by Allah's order by killing Americans and stealing their money anywhere, anytime, and whenever possible."[15] To understand al Qaeda's evolution, it is especially important to recognize the importance of the Egyptian influence on bin Laden, which dates back to his time in Afghanistan. Currently most of al Qaeda's membership is drawn from these two Egyptian groups. Moreover, one Egyptian in particular, Ayman al-Zawahiri—a former key figure in al Jihad—has had a tremendous intellectual influence on Osama bin Laden and is considered by many to be a candidate to succeed him.[16]

As indicated above, al Qaeda's model has been to establish bases with indigenous groups throughout the world. Early in its existence, al Qaeda developed the ability to penetrate Islamic nongovernmental organizations (NGOs) to the point that it was "inseparably enmeshed with the religious, social, and economic fabric of Muslim communities worldwide."[17] In some cases, al Qaeda pursued a virtual "hands off" policy with its affiliated group. It may have guided or directed the group's operations, but at the same time required it to raise its own funds. Ahmed Ressam, who was intercepted entering the United States in December 1999 as part of the infamous "Millennium Plot," was part of a cell in Montreal, Canada, that survived by engaging in petty theft—including passport theft—and other crimes. However, for certain operations, such as the 11 September attacks in the United States, al Qaeda was much more willing to provide substantial and direct financial support.

Al Qaeda's strength lay in its reliance on a multi-cellular structure, spanning the entire globe, which gave the organization agility and cover. One French terrorism expert recently lamented, "If you have good knowledge of the [al Qaeda] network today, it's not operational tomorrow."[18] He compared its networks to a constantly changing virus that is impossible to totally grasp or destroy. Al Qaeda's multi-cellular international structure provided an ironic backdrop to President George Bush's proclamation that the United States would find terrorists wherever they were located and would consider attacking any nation that harbored terrorists. The uncomfortable reality is that many states—including those allied with the United States—harbored al Qaeda cells, but did nothing to neu-

tralize them, either because they did not know of their presence (or the precise danger they posed) or were unwilling, for political or security reasons, to disrupt their operations. Certain German investigators, for instance, ruefully admit that their lack of aggressive intervention—despite full awareness of al Qaeda's activities in many of its main cities—probably contributed to the 11 September tragedy.

As a truly transnational terrorist organization, al Qaeda has sought to expand beyond the traditional venue of the Middle East, Western Europe, North America, and South Asia. Increasingly the organization has pursued Southeast Asia as a key basing and staging region. Al Qaeda has long cultivated links with groups such as the Philippine-based Abu Sayyaf and Moro Islamic Liberation Front (MILF) and the Indonesian group Laskar Jihad. Al Qaeda is also linked to region-wide organizations, such as Jemaah Islamiah, the mastermind of plots against the US Embassy in Singapore and other critical American and Western targets. In late September 2001, the Philippine military's chief of staff confirmed speculation that al Qaeda was seeking to support the Abu Sayyaf Group with "materiel, leadership, and training support."[19] Similar trends have been detected in Indonesia, where officials suspect growing linkages between al Qaeda and indigenous groups such as Laskar Jihad. In December 2001, the head of Indonesia's intelligence services, Abdullah Hendropriyono, asserted that al Qaeda and other international terrorist organizations were attempting to sow unrest on the Indonesian island of Sulawesi by promoting inter-ethnic violence between Muslims and Christians. He also confirmed that al Qaeda and other international groups had used the territory as a base and training site for international terrorist operations.[20]

Al Qaeda has also established links in Africa and South America. In South America, the "triple border" area (where Brazil, Argentina, and Paraguay meet) is viewed as a base for such Middle Eastern terrorist organizations as Hezbollah, al Gamaa al Islamiya, and Hamas, all al Qaeda constituent or affiliate groups. A 1999 Argentine intelligence report stated that al Qaeda was operating in the region in an attempt to forge links with Hezbollah supporters.[21] The region, and other locations in Brazil, appear to have played a significant role in the planning of the 11 September attacks.[22] Al Qaeda also has established links in various African countries, including Somalia, Sudan, and South Africa.[23] Al Qaeda reportedly has considered moving to Somalia following US military operations in Afghanistan, a possibility that recently prompted a US Naval blockade of the entire Somali coastline.

Al Qaeda has flourished in an environment of weak or quasi-states that are undergoing disruptive political or social change. Vast swaths of political instability in many parts of the world, and particularly in Africa and Asia, have provided a breeding ground for al Qaeda and its analogues. As one French analyst stated, wide expanses of anarchic territory "need no longer be considered a regrettable feature of the postmodern world, but rather a strategic challenge that should be addressed urgently."[24] Such areas are not only hospitable to terrorists, they may also attract transnational crime groups, drug traffickers, and maritime pirates. Despite their isolation, paradoxically, these areas constitute an acute threat to global security.

Al Qaeda's Suicidal Tendencies

In early 2001, Dahmane Abd al-Sattar received what was probably the most important mission of his life. As a member of a Tunisian-dominated al Qaeda cell based in Belgium, he and an unidentified accomplice had been "activated" by the al Qaeda leadership. Their goal would be to conduct a suicide strike on Ahmed Shah Massoud, the legendary leader of the Northern Alliance in Afghanistan, thousands of miles away. With the help of at least 14 European-based co-conspirators, Mr. Sattar, along with his accomplice, began a circuitous journey, posing as European-based Moroccan journalists. They used forged Belgian passports and an apparently forged Pakistani visa.[25] Their journey first involved traveling to the United Kingdom, where they obtained a letter of introduction written by Yasser al-Siri, the head of London's Islamic Observation Center. The letter provided the two assassins with the legitimacy and cover to gain access to Massoud.[26]

The pair next traveled to Pakistan. There, with the al-Siri introduction letter in hand, they were able to obtain visas at the Afghanistan embassy posing as journalists for "Arabic News International."[27] The men then traveled to Kabul, which at that time was firmly controlled by the Taliban. Later they were given permission to cross into the Panjshir Valley, the stronghold of the man whom they would assassinate. After a long series of negotiations, the assassins managed to get approval for their interviews of key Northern Alliance leaders, but they focused particularly on interviewing Massoud. Just before the interview, the cameraman reportedly placed his rigged camera on a low table facing Massoud.[28] Then the interview began. The main journalist, presumably al-Sattar, asked Massoud what he would do with Osama bin Laden if he (Massoud) returned to power. Massoud reportedly laughed at the question, and at that instant the camera exploded. One of the two assassins died immediately in the explosion. The second was shot dead by nearby guards.[29] Massoud, meanwhile, lay on the ground in a pool of blood. He died soon thereafter.

The attack on Commander Massoud is remarkable not simply because of its tactical value for al Qaeda—it took away the Northern Alliance's most capable leader—but also because it highlighted the efficacy of suicide techniques that al Qaeda has increasingly come to rely upon. Suicide terrorism is defined as "the readiness to sacrifice one's life in the process of destroying or attempting to destroy a target to advance a political goal."[30] The difference between a brave combat soldier and a suicide bomber is that the former confronts his fears of death, hoping to avoid its clutches. The suicide bomber, on the other hand, intends to die. If somehow the suicide attacker survived the attack, yet successfully conducted the terrorist operation, he would most likely consider himself a failure.

Before the 11 September attacks, experts generally considered suicide bombers to be usually poor, not particularly well-educated, unmarried, and hungry for revenge.[31] The 11 September suicide attacks, conducted by well-educated and generally prosperous individuals, have shaken that profile. Additionally, Israeli security agents have discovered growing discrepancies to the general suicide bomber profile, such as increased incidence of educated or prosperous attackers.[32] In January 2002, Israel encountered another surprise in the profile of suicide bombers when a 28-year-old Palestinian woman named Wafa Idris blew herself up in a crowded shopping district in downtown Jerusalem. No longer could Israeli security forces concentrate their anti-bombing surveillance exclusively on Palestinian men.

If the 11 September attacks are any guide, al Qaeda or its affiliate groups will increasingly rely on suicide attacks. The method has an array of advantages over more traditional warfare. One Israeli-based analyst enumerates four major points: it is simple and inexpensive; it almost certainly guarantees mass casualties and extensive damage because the bomber can choose the exact time, location, and circumstances of the attack; there are no post-attack fears of interrogation since the attacker will almost certainly die; and it has a powerful effect on the public and the media, due to the widespread horror and sense of helplessness that it cultivates.[33]

Al Qaeda's reliance on suicide attacks has become a key part of its arsenal, particularly in recent years. Prominent al Qaeda suicide attacks have included the bombing of the US Embassies in Kenya and Tanzania in 1998 and the bombing of USS *Cole* in October 2000. Al Qaeda's interest in airborne suicide attacks is significant not only because of the 11 September attacks and Oplan Bojinka, but also because of other attacks that were either disrupted or thwarted while in progress.

In 1994, the Armed Islamic Group (GIA), an al Qaeda affiliate, hijacked an Air France Airbus A300 and flew it to France with the intent of exploding the airplane in the air over Paris. The plane first landed in Marseille for refueling. There French commandos raided the plane, killing all of the hijackers.[34] In 1999, US intelligence officials reportedly received evidence suggesting that Osama bin Laden was planning to blow up at least six airliners at six international airports simultaneously. The 1999 interception of Ahmed Ressam along the US-Canada border by an astute US Customs officer probably thwarted an airport attack in late 1999. Although it is not clear that these planned attacks involved suicide tactics, most likely al Qaeda would have used such tactics if they would have ensured success.

Porous Borders and the Vulnerable State

On 30 November 2001, a US federal judge sentenced a former Mexican immigration inspector, Angel Salvador Molina-Paramo, to 30 months in federal prison for his role in a global human smuggling ring spanning several continents. Molina-Paramo's partner and the chief of the smuggling operation was George Tajirian, an Iraqi-born human smuggler accused of trafficking hundreds of illegal immigrants from the Middle East across the US-Mexican border during the 1990s. US authorities had arrested Tajirian in 1998 and, following a plea agreement, he was sentenced to 13 years in US federal prison. Prosecutors alleged that the ring smuggled Palestinian, Jordanian, Syrian, Iraqi, Yemeni, and other illegal immigrants through Mexico to the United States. The smuggling operation "included smuggling stations in Jordan, Syria, Palestine, and Greece; and staging areas in Greece, Thailand, Cuba, Ecuador, and Mexico."[35]

With the sentencing of Molina-Paramo coming just weeks after the 11 September attacks in the United States, the obvious question surrounding this case was whether any of the migrants—most of whom were smuggled between 1996 and 1998—were possibly terrorists. Mr. Tajirian did not appear to be operating a terrorist-funneling operation; however, it is also clear that Tajirian was not particularly fussy about any criminal or terrorist background of his migrant clients. During his prosecution, authorities introduced evidence that Tajirian had smuggled into the United States "persons with known ties to subversive or terrorist organizations as well as individuals with known criminal histories."[36] If a migrant had a known criminal background, Tajirian simply raised the smuggling fees.[37] Overall, US officials believe that Tajirian and his cohorts smuggled more than 1,000 Middle Eastern residents illegally into the United States.[38]

This immigrant smuggling case might be viewed as simply an oddity, perhaps another indicator of the sinister and depraved international underworld of human smuggling. Most traditional security planners would consider the case—and the issue of human smuggling in general—an interesting social or labor migration phenomenon that, though disturbing, bears little relevance to national or international security. But the case of George Tajirian and the ring he led is also arguably one more example of the vulnerability of US border security, a vulnerability which, in an age of international terrorism where modern terrorists must travel to multiple countries to either raise money, cultivate support, or conduct attacks, cannot simply be dismissed as merely an immigration issue or social policy question.

The reality that few US authorities want to publicly admit is that the notion of border security, particularly within the dark and transient world of transnational crime, is largely fiction. For more than two decades, human smuggling syndicates with links to China, India, Albania, and other countries have developed complex and circuitous pathways into the United States, just as they have in Western Europe and East Asia. In the context of international terrorism, porous borders and the rise of human smuggling—and its attendant side industry of document fraud—pose serious security challenges for states. Just as the human body's lymphatic system provides a stream for the spread of lethal cancer cells, so too can the global stream of human smuggling and illegal migration carry the agents of global terrorism. Ironically, despite the publicity regarding vulnerable borders in the United States following the 11 September attacks, Homeland Security Director Tom Ridge publicly admitted in February 2002 that US borders "remain disturbingly vulnerable to terrorists."[39] It is precisely this concern that has prompted President Bush to consider fusing US Customs with the troubled Immigration and Naturalization Service, the agency nominally in charge of border enforcement.

Identity fraud and illegal migration have emerged as the lifeblood of global terrorism, as critical as any bomb, machine gun, or grenade. Terrorist organizations place a premium on clandestine international mobility, relying on an array of identity fraud techniques. Ayman al-Zawahiri, the physician-terrorist considered to be Osama bin Laden's second-in-command, is known to have carried a "bewildering variety of passports" that, among other things, allowed him to secretly enter the United States in the early 1990s to raise funds from California mosques in order to support terrorist activities of the Egyptian group al Jihad.[40] In another case, Philippine police in early 2002 arrested a 31-year-old Indonesian man, Fathur Rohman al-Ghozi, who was implicated in a plot to bomb the US Embassy in Singapore, after he was discovered carrying at least four fake or forged passports, including ones issued by the governments of Indonesia and the Philippines.[41] Officials determined that the man, a well-known bomb specialist for the terror group Jemaah Islamiah, had relied on these passports—which used various aliases—to travel throughout Southeast Asia to coordinate different cells of the group in preparation for a series of attacks. Similar patterns have emerged with other terrorist and transnational crime organizations around the world.[42]

Just as the illicit transnational migration of people can pose security challenges for the United States or other countries, so too can the flow of commercial cargo. Every year, the United States receives over 5.8 million containers from maritime sources, and over 2.1 million rail cars.[43] Facing the daunting task of inspecting these cargo units is the US Customs Service, which can inspect only about five percent of this international cargo, because of personnel constraints. In one border-crossing area along the US-Canada border, there are only eight primary inspection lanes and customs inspectors typically have only two minutes to inspect each tractor-trailer.[44] Moreover, because cargo can enter the United States and not be inspected for up to 30 days—due to "port of entry" procedures that allow inspection only in the final destination city—US authorities have little basis to verify the identity of the sender or the identity of the contents of thousands of multi-ton containers traveling throughout the United States on trains, trucks, or barges.[45]

In the context of international terrorism, the lack of rigorous cargo inspection procedures makes the United States and other Western countries extremely vulnerable to mass-casualty attacks by terrorist groups and other nonstate actors. Terrorist organizations such as al Qaeda could theoretically attempt to smuggle a nuclear or chemical weapon into the United States within the normal stream of cargo imports in order to conduct a mass-casualty attack. It is well-known that al Qaeda has sought to develop nuclear and chemical weapon capability. It is also documented that al Qaeda has sought to exploit the global container traffic stream in at least one case. In October 2001, Italian authorities discovered an al Qaeda operative locked inside a shipping container destined for Canada. The container was fitted with a bed and bathroom. The Egyptian national who was traveling in the container was also found to be carrying airport maps, airport security passes, and a mechanic's certificate.[46]

The most devastating scenario would involve the smuggling of nuclear weapons via shipping containers. A recent Central Intelligence Agency report suggests that non-missile delivery of a nuclear device by state or nonstate actors is the most likely means through which the United States is likely to suffer a nuclear, chemical, or biological attack. Such groups would likely deploy such devices not through missiles, but through smuggling aboard ships, trucks, airplanes, or other means. The CIA report lists various advantages for such a method, including the ability to deploy the weapon covertly, the ability to mask the source of the weapon (to evade retaliation), the ability to ensure that the weapon was used effectively at the intended location, and the ability to evade missile or other defenses. The report argued that "foreign nonstate actors—including terrorist, insurgent, or extremist groups—have used, possessed, or expressed an interest in CBRN [chemical, biological, radiological, and nuclear] materials."[47] Al Qaeda is among such groups.[48] Clearly, in light of such threats, border security takes on much greater urgency, although it is not clear that this understanding has reached traditional military or security planning circles.

The Dilemma of State Responses to Terrorism

In the early hours of Sunday, 14 October 2001, Abdel Rahman Hamad, having just completed morning prayers, stood on the rooftop of his house and gazed at the surrounding vista. As he leaned against a wall on the edge of his roof, two bullets fired by an Israeli government sharpshooter traveled more than 300 yards and penetrated Hamad's chest, killing him almost instantly.[49] The assassination capped many months of Israeli surveillance of Hamad. Abdel Rahman Hamad was a high-rank-

ing leader of Hamas and was the chief suspect behind the June 2001 suicide bombing at a Tel Aviv disco that had killed 21 people only four months earlier.

The assassination, coming just a few weeks after the 11 September attacks in the United States, elicited an awkward response from Washington. On one hand, the United States, which had publicly discussed its own desire to kill Osama bin Laden, attempted to present a neutral position, cautioning the Israeli government against the employment of its "targeted killing" program. The United States, after all, was in the midst of a debate over its own self-imposed ban against overseas assassinations. Many political leaders were calling for an end to the policy, while others argued that such a policy change was unnecessary since, they reasoned, killing individuals in a lawful use of force—in this case self-defense—is not assassination. Presumably Israel could have relied on the same logic, since Hamad was a known Hamas manager who was preparing suicide bombers for future missions involving the murder of Israeli civilians and widespread destruction of property.

The Hamad assassination also provides another useful insight relevant to state responses to terrorism. If there is anything to be learned from the "war on terror"—whether conducted by the United States, Israel, or France—it should be that it is a form of warfare that is dark, morally ambiguous, and replete with dirty tricks. Spurred by sudden intelligence insights, whether from human or technological sources, states will be presented with time-sensitive opportunities to deploy special forces (or similar types of military or paramilitary units) to inflict deadly results upon individuals or entire organizations. Israel's model of targeted killings—albeit not entirely successful—may emerge as the norm in future warfare, which will likely be quiet, bloody, and murderous.

A state policy of preemptive assassinations, which most likely would be couched in such euphemistic phrases as "permissive termination" or "calculated elimination," may be deemed necessary and politically expedient, and yet would remain anathema to the traditional and honorable soldiering ideal that has bonded military forces, particularly in Western states, for multiple generations. This contradiction will have to be managed head-on; otherwise it may result in diminished morale and civilian misunderstanding. As one analyst has noted: "Western countries have been disinclined to prepare for military action that was considered uncivilized."[50] Civilized or not, states may be required to engage in such actions if the threat of transnational terrorism—with its current predisposition for mass casualties—is to be contained or averted.

Another morally complicated issue involves state use of torture during interrogation. When Abdul Hakim Murad was questioned by Philippine investigators in 1995, he was reportedly subjected to various forms of torture. At one stage in the interrogation, Philippine authorities allegedly deceived Murad into believing that Israeli Mossad agents were behind his interrogation. Whether because of that or for other reasons, Murad eventually confessed the details of Oplan Bojinka. In this case, aggressive interrogation techniques, while arguably morally abhorrent and distasteful, effectively served their purpose. Thousands of innocent civilians did not die in early 1995 because of information gained through these methods.

Traditionally, the United States and certain other Western countries have skirted the torture dilemma by "exporting" their difficult candidates to states and regimes known to engage in the practice—such as Egypt—with full knowledge that interrogation with torture would likely take place.[51] Yet such a practice of "vicarious torture" is imbued with an obvious hypocrisy that prevents the sending state—such as the United States—from having clean hands when it engages in such practices. Moreover, obtaining human intelligence from foreign governments is fraught with its own downside risk: such intelligence, filtered through a foreign government, may contain information tainted by that government's biases or hidden policy objectives.

There is no easy answer to the thorny issue of interrogation involving the application of torture. From a moral and legal perspective, the answer would seem to be a clear-cut no, don't do it. From a practical perspective, moreover, torture can be counterproductive, eliciting a resolve of defiance within the individual being interrogated, and the intelligence so gained may be unreliable. But in the harsh world of transnational terrorism, the reality is far from black and white. As one terrorism expert recently wrote, intelligence is the key weapon against global terrorism, but intelligence does not come cheaply and, moreover, "Americans still do not appreciate the enormously difficult—and morally complex— problem that the imperative to gather 'good intelligence' entails."[52] In a conversation between that writer and a Sri Lankan army officer credited with thwarting attacks by the ruthless terrorist group the Liberation Tigers of Tamil Eelam (LTTE), the officer noted that terrorism could not be fought with laws or moral dicta, but only by thoroughly "terrorizing" the terrorists—in other words, "inflicting on them the same pain that they inflict on the innocent."[53] In these murky aspects of dealing with terrorists, there are no easy answers.

Conclusion

Al Qaeda's attack on the United States on 11 September 2001 was a major turning point in the evolution of international terrorism. In this case, the United States was attacked not by a fellow state, but a non-state terrorist organization. Al Qaeda represents the worst that globalization has to offer. Its transnational tentacles reach into every corner of the globe. Its ability to penetrate countries with passport fraud and other illegal immigration techniques is unparalleled, and its virus-like ability to infect indigenous groups—even those with originally benign goals—is now well-documented.

The lesson to be learned from al Qaeda is that terrorist groups can now exist in a transnational milieu, divorced from state-driven constraints. Even if we witness the demise of al Qaeda, we are not likely to witness the demise of its model. Terrorist groups can thrive in the dark pockets of anarchy that pervade the globe. But they can also coexist alongside their targets by planting cells in Western Europe and North America. The question thus becomes: Have we learned the lesson, and, moreover, are we prepared for the next attack?

NOTES

1. Regarding warnings about the rise of "catastrophic terrorism," see Ashton Carter et al., "Catastrophic Terrorism: Tacking the New Danger," *Foreign Affairs*, 77 (November/December 1998), 80.
2. Investigations conducted in Southeast Asia after the 11 September attacks in the United States provide linkages between Bojinka and the attacks on New York City and Washington, D.C., suggesting that the former was possibly a blueprint for the latter. Specifically, certain key leaders of a newly discovered group, Jemaah Islamiah, have reported links to both plots. See Richard C. Paddock, "Southeast Asian Terror Exhibits Al Qaeda Traits," *Los Angeles Times*, 3 March 2002, p. A1.
3. This specific phrase was uttered by US Senator Richard Shelby on the CBS News Show *Face the Nation* on 16 September 2001 (file accessed through Lexis-Nexis).
4. Ed Blanche, "Al-Qaeda Recruitment," *Jane's Intelligence Review*, 14 (January 2002), 27–28.
5. Ed Blanche, "What the Investigation Reveals," *Jane's Intelligence Review*, 13 (November 2001), 16–17; see also article by John Mintz and Rom Jackman, "Finances Prompted Raids on Muslims," *The Washington Post*, 24 March 2001, p. A01.
6. *A Biography of Osama bin Laden*, background article for PBS documentary series *Frontline*, internet, www.pbs.org/wgbh/pages/frontline/shows/binladen/who/bio.html, accessed 24 March 2002.
7. *Osama bin Laden: A Chronology of His Political Life*, background article for PBS documentary series *Frontline*, internet, www.pbs.org/wgbh/pages/frontline/shows/binladen/etc/cron.html, accessed 24 March 2002.
8. Peter L. Bergen, *Holy War: Inside the Secret World of Osama bin Laden* (New York: Free Press, 2001), p. 79.
9. Ibid., p. 85.
10. Ibid., p. 86.
11. Indictment of Zacarias Moussaoui, United States District Court for the Eastern District of Virginia, Alexandria Division (December 2001).
12. Bergen, p. 93.
13. Ibid., p. 94.
14. Ibid., p. 95.
15. Ibid., p. 96.
16. Ed Blanche, "The Egyptians Around Bin Laden," *Jane's Intelligence Review*, 13 (December 2001), 19–21.
17. Phil Hirschkorn, et al., "Blowback," *Jane's Intelligence Review*, 13 (August 2001), 42–45.
18. Steven Erlanger and Chris Hedges, "Terror Cells Slip Through Europe's Grasp," *The New York Times*, 28 December 2001, p. 1.

19. "Abu Sayyaf Gets Arms, Training from Bin Laden: Philippine Military," *Agence France Presse*, 28 September 2001.

20. Fabiola Desy Unidjaja, "International Terrorists Train in Poso," *Jakarta Post*, 13 December 2001.

21. Mario Daniel Montoya, "War on Terrorism Reaches Paraguay's Triple Border," *Jane's Intelligence Review*, 13 (December 2001), 12–15.

22. John C. K. Daly, "Moroccan has 'much to tell,'" *Jane's Intelligence Review*, 13 (December 2001), 13.

23. See testimony of Dr. J. Stephen Morrison before the House International Relations Committee, Subcommittee on Africa, Hearing on "Africa and the War on Global Terrorism" *Federal Document Clearing House Congressional Testimony*, 15 November 2001.

24. Therese Delpech, "The Imbalance of Terror," *Washington Quarterly*, 25 (Winter 2002), 31.

25. Dan Bilefsky and John Carreyrou, "Arrests Are Made in Probe of Alliance Leader's Death," *The Wall Street Journal*, 27 November 2001.

26. Dan Bilefsky, "Belgian Terrorist Cell Cited in Massoud Killing—Murder of Guerrilla Leader Linked to bin Laden," *The Wall Street Journal*, 10 December 2001.

27. Mohamad Bazzi, "America Strikes Back: The London Connection," *Newsday*, 9 October 2001.

28. Sylvain Cypel and Erich Inciyan, "On the Trail of Massoud's Killers," *Manchester Guardian Weekly*, 24 October 2001, p. 30.

29. Alan Cullison and Andrew Higgins, "Files Found: A Computer in Kabul Yields a Chilling Array of al Qaeda Memos," *The Wall Street Journal*, 31 December 2001.

30. Rohan Gunaratna, "Suicide Terrorism: a Global Threat," *Jane's Intelligence Review*, 12 (April 2000), 52–55.

31. David Eshel, "Israel Reviews Profile of Suicide Bombers," *Jane's Intelligence Review*, 13 (November 2001), 20–21.

32. Ibid.

33. Ehud Sprinzak, "Rational Fanatics," *Foreign Policy*, No. 120 (September/October 2000), 66.

34. Blanche, "The Egyptians Around Bin Laden."

35. *Former Mexican Immigration Official Sentenced to Federal Prison for Role in Alien Smuggling Organization*, US Attorney's Office—Western District of Texas—Press Release, internet, http://www.usdoj.gov/usao/txw/Molina.htm, accessed 14 January 2002..

36. Ibid.

37. Sam Dillon, "Iraqi Accused of Smuggling Hundreds in Mideast to U.S.," *The New York Times*, 26 October 2001, p. A18.

38. Ibid.

39. Eric Pianin and Bill Miller, "U.S. Borders Remain Vulnerable Despite New Measures, Ridge Says," *The Washington Post*, 12 February 2002, p. A23.

40. Blanche, "The Egyptians Around Bin Laden."

41. Blontank Poer, "Fathur Used Fake ID to Obtain RI Passport," *Jakarta Post*, 28 January 2002; Raymond Bonner, "A National Challenged: Asia Arena; A Terrorism Suspect is Questioned about Manila Bombings," *The New York Times*, 28 February 2002.

42. Paul J. Smith, "The Terrorists and Crime Bosses Behind the Fake Passport Trade," *Jane's Intelligence Review*, 13 (July 2001), 42–44.

43. Prepared testimony of Governor James. S. Gilmore before the State Committee on Governmental Affairs, *Federal News Service*, 21 September 2001.

44. Stephen E. Flynn, "America the Vulnerable," *Foreign Affairs*, 81 (January/February 2002), 60.

45. Ibid.

46. "Remarks by Customs Service Commissioner Robert Bonner at the Center for Strategic and International Studies" forum titled, "Pushing Borders Outward: Rethinking Customs Border Enforcement," reported in *Federal News Service*, 17 January 2002.

47. *Foreign Missile Developments and the Ballistic Missile Threat* (Washington: Central Intelligence Agency, 2002), internet, http://www.cia.gov/nic/pubs/other_products/Unclassified ballisticmissilefinal.htm, accessed 10 January 2002.

48. Alan Cullison and Andrew Higgins, "Forgotten Computer Reveals Thinking Behind Four Years of al Qaeda Doings," *The Wall Street Journal*, 31 December 2001, p. 1.

49. Arieh O'Sullivan et al., "Israel Kills Hamas Planner of Tel Aviv Disco Bombing," *The Jerusalem Post*, 15 October 2001; James Bennet, "Israel Kills a Hamas Leader, and Eases West Bank Restrictions," *The New York Times*, 15 October 2001, p. A8.

50. Rob de Wijk, "The Limits of Military Power," *The Washington Quarterly*, 25 (Winter 2002), 75.

51. Bob Woodward, "50 Countries Detain 360 Suspects," *The Washington Post*, 22 November 2001, p. A01.

52. Bruce Hoffman, "A Nasty Business," *The Atlantic Monthly*, January 2002.

53. Ibid.

Paul J. Smith is an assistant professor at the Asia-Pacific Center for Security Studies, Honolulu, Hawaii. He earned his B.A. from Washington and Lee University, his M.A. from the University of London, and his J.D. from the University of Hawaii. A previous article, "Transnational Security Threats and State Survival: A Role for the Military?" appeared in the Autumn 2000 issue of *Parameters*.

SOVEREIGNTY

The idea of states as autonomous, independent entities is collapsing under the combined onslaught of monetary unions, CNN, the Internet, and nongovernmental organizations. But those who proclaim the death of sovereignty misread history. The nation-state has a keen instinct for survival and has so far adapted to new challenges—even the challenge of globalization.

By Stephen D. Krasner

The Sovereign State Is Just About Dead

Very wrong. Sovereignty was never quite as vibrant as many contemporary observers suggest. The conventional norms of sovereignty have always been challenged. A few states, most notably the United States, have had autonomy, control, and recognition for most of their existence, but most others have not. The polities of many weaker states have been persistently penetrated, and stronger nations have not been immune to external influence. China was occupied. The constitutional arrangements of Japan and Germany were directed by the United States after World War II. The United Kingdom, despite its rejection of the euro, is part of the European Union.

Even for weaker states—whose domestic structures have been influenced by outside actors, and whose leaders have very little control over transborder movements or even activities within their own country—sovereignty remains attractive. Although sovereignty might provide little more than international recognition, that recognition guarantees access to international organizations and sometimes to international finance. It offers status to individual leaders. While the great powers of Europe have eschewed many elements of sovereignty, the United States, China, and Japan have neither the interest nor the inclination to abandon their usually effective claims to domestic autonomy.

In various parts of the world, national borders still represent the fault lines of conflict, whether it is Israelis and Palestinians fighting over the status of Jerusalem, Indians and Pakistanis threatening to go nuclear over Kashmir, or Ethiopia and Eritrea clashing over disputed territories. Yet commentators nowadays are mostly concerned about the erosion of national borders as a consequence of globalization. Governments and activists alike complain that multilateral institutions such as the United Na-

tions, the World Trade Organization, and the International Monetary Fund overstep their authority by promoting universal standards for everything from human rights and the environment to monetary policy and immigration. However, the most important impact of economic globalization and transnational norms will be to alter the scope of state authority rather than to generate some fundamentally new way to organize political life.

Sovereignty Means Final Authority

Not anymore, if ever. When philosophers Jean Bodin and Thomas Hobbes first elaborated the notion of sovereignty in the 16th and 17th centuries, they were concerned with establishing the legitimacy of a single hierarchy of domestic authority. Although Bodin and Hobbes accepted the existence of divine and natural law, they both (especially Hobbes) believed the word of the sovereign was law. Subjects had no right to revolt. Bodin and Hobbes realized that imbuing the sovereign with such overweening power invited tyranny, but they were predominately concerned with maintaining domestic order, without which they believed there could be no justice. Both were writing in a world riven by sectarian strife. Bodin was almost killed in religious riots in France in 1572. Hobbes published his seminal work, *Leviathan*, only a few years after parliament (composed of Britain's emerging wealthy middle class) had executed Charles I in a civil war that had sought to wrest state control from the monarchy.

This idea of supreme power was compelling, but irrelevant in practice. By the end of the 17th century, political authority in Britain was divided between king and parliament. In the United States, the Founding Fathers established a constitutional structure of checks and balances and multiple sovereignties distributed among local and national interests that were inconsistent with hierarchy and supremacy. The principles of justice, and especially order, so valued by Bodin and Hobbes, have best been provided by modern democratic states whose organizing principles are antithetical to the idea that sovereignty means uncontrolled domestic power.

If sovereignty does not mean a domestic order with a single hierarchy of authority, what does it mean? In the contemporary world, sovereignty primarily has been linked with the idea that states are autonomous and independent from each other. Within their own boundaries, the members of a polity are free to choose their own form of government. A necessary corollary of this claim is the principle of nonintervention: One state does not have a right to intervene in the internal affairs of another.

More recently, sovereignty has come to be associated with the idea of control over transborder movements. When contemporary observers assert that the sovereign state is just about dead, they do not mean that constitutional structures are about to disappear. Instead, they mean that technological change has made it very difficult, or perhaps impossible, for states to control movements across their borders of all kinds of material things (from coffee to cocaine) and not-so-material things (from Hollywood movies to capital flows).

Finally, sovereignty has meant that political authorities can enter into international agreements. They are free to endorse any contract they find attractive. Any treaty among states is legitimate provided that it has not been coerced.

The Peace of Westphalia Produced the Modern Sovereign State

No, it came later. Contemporary pundits often cite the 1648 Peace of Westphalia (actually two separate treaties, Münster and Osnabrück) as the political big bang that created the modern system of autonomous states. Westphalia—which ended the Thirty Years' War against the hegemonic power of the Holy Roman Empire—delegitimized the already waning transnational role of the Catholic Church and validated the idea that international relations should be driven by balance-of-power considerations rather than the ideals of Christendom. But Westphalia was first and foremost a new constitution for the Holy Roman Empire. The preexisting right of the principalities in the empire to make treaties was affirmed, but the Treaty of Münster stated that "such Alliances be not against the Emperor, and the Empire, nor against the Publick Peace, and this Treaty, and without prejudice to the Oath by which every one is bound to the Emperor and the Empire." The domestic political structures of the principalities remained embedded in the Holy Roman Empire. The Duke of Saxony, the Margrave of Brandenburg, the Count of Palatine, and the Duke of Bavaria were affirmed as electors who (along with the archbishops of Mainz, Trier, and Cologne) chose the emperor. They did not become or claim to be kings in their own right.

Perhaps most important, Westphalia established rules for religious tolerance in Germany. The treaties gave lip service to the principle (*cuius regio, eius religio*) that the prince could set the religion of his territory—and then went on to violate this very principle through many specific provisions. The signatories agreed that the religious rules already in effect would stay in place. Catholics and Protestants in German cities with mixed populations would share offices. Religious issues had to be settled by a majority of both Catholics and Protestants in the diet and courts of the empire. None of the major political leaders in Europe endorsed religious toleration in principle, but they recognized that religious conflicts were so volatile that it was essential to contain rather than repress sectarian differences. All in all, Westphalia is a pretty medieval document, and its biggest explicit innovation—provisions that undermined the power of princes to control religious affairs within their territories—was antithetical to the ideas of national sovereignty that later became associated with the so-called Westphalian system.

Universal Human Rights Are an Unprecedented Challenge to Sovereignty

Wrong. The struggle to establish international rules that compel leaders to treat their subjects in a certain way has been going on for a long time. Over the centuries the emphasis has shifted from religious toleration, to minority rights (often focusing on specific ethnic groups in specific countries), to human rights (emphasizing rights enjoyed by all or broad classes of individuals). In a few instances states have voluntarily embraced international supervision, but generally the weak have acceded to the preferences of the strong: The Vienna settlement following the Napoleonic wars guaranteed religious toleration for Catholics in the Netherlands. All of the successor states of the Ottoman Empire, beginning with Greece in 1832 and ending with Albania in 1913, had to accept provisions for civic and political equality for religious minorities as a condition for international recognition. The peace settlements following World War I included extensive provisions for the protection of minorities. Poland, for instance, agreed to refrain from holding elections on Saturday because such balloting would have violated the Jewish Sabbath. Individuals could bring complaints against governments through a minority rights bureau established within the League of Nations.

But as the Holocaust tragically demonstrated, interwar efforts at international constraints on domestic practices failed dismally. After World War II, human, rather than minority, rights became the focus of attention. The United Nations Charter endorsed both human rights and the classic sovereignty principle of nonintervention. The 20-plus human rights accords that have been signed during the last half century cover a wide range of issues including genocide, torture, slavery, refugees, stateless persons, women's rights, racial discrimination, children's rights, and forced labor. These U.N. agreements, however, have few enforcement mechanisms, and even their provisions for reporting violations are often ineffective.

The tragic and bloody disintegration of Yugoslavia in the 1990s revived earlier concerns with ethnic rights. International recognition of the Yugoslav successor states was conditional upon their acceptance of constitutional provisions guaranteeing minority rights. The Dayton accords established externally controlled authority structures in Bosnia, including a Human Rights Commission (a majority of whose members were appointed by the Western European states). NATO created a de facto protectorate in Kosovo.

The motivations for such interventions—humanitarianism and security—have hardly changed. Indeed, the considerations that brought the great powers into the Balkans following the wars of the 1870s were hardly different from those that engaged NATO and Russia in the 1990s.

Globalization Undermines State Control

No. State control could never be taken for granted. Technological changes over the last 200 years have increased the flow of people, goods, capital, and ideas—but the problems posed by such movements are not new. In many ways, states are better able to respond now than they were in the past.

The impact of the global media on political authority (the so-called CNN effect) pales in comparison to the havoc that followed the invention of the printing press. Within a decade after Martin Luther purportedly nailed his 95 theses to the Wittenberg church door, his ideas had circulated throughout Europe. Some political leaders seized upon the principles of the Protestant Reformation as a way to legitimize secular political authority. No sovereign monarch could contain the spread of these concepts, and some lost not only their lands but also their heads. The sectarian controversies of the 16th and 17th centuries were perhaps more politically consequential than any subsequent transnational flow of ideas.

In some ways, international capital movements were more significant in earlier periods than they are now. During the 19th century, Latin American states (and to a lesser extent Canada, the United States, and Europe) were beset by boom-and-bust cycles associated with global financial crises. The Great Depression, which had a powerful effect on the domestic politics of all major states, was precipitated by an international collapse of credit. The Asian financial crisis of the late 1990s was not nearly as devastating. Indeed, the speed with which countries recovered from the Asian flu reflects how a better working knowledge of economic theories and more effective central banks have made it easier for states to secure the advantages (while at the same time minimizing the risks) of being enmeshed in global financial markets.

In addition to attempting to control the flows of capital and ideas, states have long struggled to manage the impact of international trade. The opening of long-distance trade for bulk commodities in the 19th century created fundamental cleavages in all of the major states. Depression and plummeting grain prices made it possible for German Chancellor Otto von Bismarck to prod the landholding aristocracy into a protectionist alliance with urban heavy industry (this coalition of "iron and rye" dominated German politics for decades). The tariff question was a basic divide in U.S. politics for much of the last half of the 19th and first half of the 20th centuries. But, despite growing levels of imports and exports since 1950, the political salience of trade has receded because national governments have developed social welfare strategies that cushion the impact of international competition, and workers with higher skill levels are better able to adjust to changing international conditions. It has become easier, not harder, for states to manage the flow of goods and services.

Globalization Is Changing the Scope of State Control

Yes. The reach of the state has increased in some areas but contracted in others. Rulers have recognized that their effective control can be enhanced by walking away from issues they cannot resolve. For instance, beginning with the Peace of Westphalia, leaders chose to surrender their control over religion because it proved too volatile. Keeping religion within the scope of state authority undermined, rather than strengthened, political stability.

Monetary policy is an area where state control expanded and then ultimately contracted. Before the 20th century, states had neither the administrative competence nor the inclination to conduct independent monetary policies. The mid-20th-century effort to control monetary affairs, which was associated with Keynesian economics, has now been reversed due to the magnitude of short-term capital flows and the inability of some states to control inflation. With the exception of Great Britain, the major European states have established a single monetary authority. Confronting recurrent hyperinflation, Ecuador adopted the U.S. dollar as its currency in 2000.

Along with the erosion of national currencies, we now see the erosion of national citizenship—the notion that an individual should be a citizen of one and only one country, and that the state has exclusive claims to that person's loyalty. For many states, there is no longer a sharp distinction between citizens and noncitizens. Permanent residents, guest workers, refugees, and undocumented immigrants are entitled to some bundle of rights even if they cannot vote. The ease of travel and the desire of many countries to attract either capital or skilled workers have increased incentives to make citizenship a more flexible category.

Although government involvement in religion, monetary affairs, and claims to loyalty has declined, overall government activity, as reflected in taxation and government expenditures, has increased as a percentage of national income since the 1950s among the most economically advanced states. The extent of a country's social welfare programs tends to go hand in hand with its level of integration within the global economy. Crises of authority and control have been most pronounced in the states that have been the most isolated, with sub-Saharan Africa offering the largest number of unhappy examples.

NGOs Are Nibbling at National Sovereignty

To some extent. Transnational nongovernmental organizations (NGOs) have been around for quite awhile, especially if you include corporations. In the 18th century, the East India Company possessed political power (and even an expeditionary military force) that rivaled many national governments. Throughout the 19th century, there were transnational movements to abolish slavery, promote the rights of women, and improve conditions for workers.

The number of transnational NGOs, however, has grown tremendously, from around 200 in 1909 to over 17,000 today. The availability of inexpensive and very fast communications technology has made it easier for such groups to organize and make an impact on public policy and international law—the international agreement banning land mines being a recent case in point. Such groups prompt questions about sovereignty because they appear to threaten the integrity of domestic decision making. Activists who lose on their home territory can pressure foreign governments, which may in turn influence decision makers in the activists' own nation.

But for all of the talk of growing NGO influence, their power to affect a country's domestic affairs has been limited when compared to governments, international organizations, and multinational corporations. The United Fruit Company had more influence in Central America in the early part of the 20th century than any NGO could hope to have anywhere in the contemporary world. The International Monetary Fund and other multilateral financial institutions now routinely negotiate conditionality agreements that involve not only specific economic targets but also domestic institutional changes, such as pledges to crack down on corruption and break up cartels.

Smaller, weaker states are the most frequent targets of external efforts to alter domestic institutions, but more powerful states are not immune. The openness of the U.S. political system means that not only NGOs, but also foreign governments, can play some role in political decisions. (The Mexican government, for instance, lobbied heavily for the passage of the North American Free Trade Agreement.) In fact, the permeability of the American polity makes the United States a less threatening partner; nations are more willing to sign on to U.S.-sponsored international arrangements because they have some confidence that they can play a role in U.S. decision making.

Sovereignty Blocks Conflict Resolution

Yes, sometimes. Rulers as well as their constituents have some reasonably clear notion of what sovereignty means—exclusive control within a given territory—even if this norm has been challenged frequently by inconsistent principles (such as universal human rights) and violated in practice (the U.S.- and British-enforced no-fly zones over Iraq). In fact, the political importance of conventional sovereignty rules has made it harder to solve some problems. There is, for instance, no conventional sovereignty solution for Jerusalem, but it doesn't require much imagination to think of alternatives: Divide the city into small pieces; divide the Temple Mount vertically with the Palestinians controlling the top and the Israelis the bottom; establish some kind of international authority; divide control over different issues (religious practices versus taxation, for instance) among different authorities. Any one of these solutions would be better for most Israelis and Palestinians than an ongoing stalemate, but political leaders on both sides have had trouble delivering a settlement because they are subject to attacks by counterelites who can wave the sovereignty flag.

Conventional rules have also been problematic for Tibet. Both the Chinese and the Tibetans might be better off if Tibet could regain some of the autonomy it had as a tributary state within the traditional Chinese empire. Tibet had extensive local control, but symbolically (and sometimes through tribute payments) recognized the supremacy of the emperor. Today, few on either side would even know what a tributary state is, and even if the leaders of Tibet worked out some kind of settlement that would give their country more self-government, there would be no guarantee that they could gain the support of their own constituents.

If, however, leaders can reach mutual agreements, bring along their constituents, or are willing to use coercion, sovereignty rules can be violated in inventive ways. The Chinese, for instance, made Hong Kong a special administrative region after the transfer from British rule, allowed a foreign judge to sit on the Court of Final Appeal, and secured acceptance by other states not only for Hong Kong's participation in a number of international organizations but also for separate visa agreements and recognition of a distinct Hong Kong passport. All of these measures violate conventional sovereignty rules since Hong Kong does not have juridical independence. Only by inventing a unique status for Hong Kong, which involved the acquiescence of other states, could China claim sovereignty while simultaneously preserving the confidence of the business community.

The European Union Is a New Model for Supranational Governance

Yes, but only for the Europeans. The European Union (EU) really is a new thing, far more interesting in terms of sovereignty than Hong Kong. It is not a conventional international organization because its member states are now so intimately linked with one another that withdrawal is not a viable option. It is not likely to become a "United States of Europe"—a large federal state that might look something like the United States of America—because the interests, cultures, economies, and domestic institutional arrangements of its members are too diverse. Widening the EU to include the former communist states of Central Europe would further complicate any efforts to move toward a political organization that looks like a conventional sovereign state.

The EU is inconsistent with conventional sovereignty rules. Its member states have created supranational institutions (the European Court of Justice, the European Commission, and the Council of Ministers) that can make decisions opposed by some member states. The rulings of the court have direct effect and supremacy within national judicial systems, even though these doctrines were never explicitly endorsed in any treaty. The European Monetary Union created a central bank that now controls monetary affairs for three of the union's four largest states. The Single European Act and the Maastricht Treaty provide for majority or qualified majority, but not unanimous, voting in some issue areas. In one sense, the European Union is a product of state sovereignty because it has been created through voluntary agreements among its member states. But, in another sense, it fundamentally contradicts conventional understandings of sovereignty because these same agreements have undermined the juridical autonomy of its individual members.

The European Union, however, is not a model that other parts of the world can imitate. The initial moves toward integration could not have taken place without the political and economic support of the United States, which was, in the early years of the Cold War, much more interested in creating a strong alliance that could effectively oppose the Soviet Union than it was in any potential European challenge to U.S. leadership. Germany, one of the largest states in the European Union, has been the most consistent supporter of an institutional structure that would limit Berlin's own freedom of action, a reflection of the lessons of two devastating wars and the attractiveness of a European identity for a country still grappling with the sins of the Nazi era. It is hard to imagine that other regional powers such as China, Japan, or Brazil, much less the United States, would have any interest in tying their own hands in similar ways. (Regional trading agreements such as Mercosur and NAFTA have very limited supranational provisions and show few signs of evolving into broader monetary or political unions.) The EU is a new and unique institutional structure, but it will coexist with, not displace, the sovereign-state model.

Stephen D. Krasner is Graham H. Stuart professor of international relations at Stanford University and is currently on leave as a fellow at the Wissenschaftskolleg of Berlin.

Reconciling Non-intervention and Human Rights

By Douglas T. Stuart

"State sovereignty, in its most basic sense, is being re-defined.... At the same time, individual sovereignty... has been enhanced by a renewed and spreading consciousness of individual rights."
—Secretary-General Kofi Annan, 1999

"**W**hen you come to a fork in the road, take it." Yogi Berra's well-known aphorism is an apt introduction to the topic of this essay. Over the last decade, the international community has been engaged in a protracted debate over issues raised by humanitarian interventions in defense of human rights. To date, world leaders have attempted to finesse these issues by committing themselves to the human rights path without abandoning the well-trodden path of non-intervention. When Governments have authorized interventions in support of human rights, they have been able to avoid a direct clash with the principle of non-intervention by citing established legal principles such as "threats to international peace and security" and "failed States". As the precedents accumulate, however, it is becoming harder to skirt a confrontation between the traditional commitment to state sovereignty and the growing commitment to the protection of basic human rights. Some choices have to be made, but they will not be easy or without costs.

The debate between the proponents of non-intervention and the supporters of intervention in defense of human rights is sometimes presented as a struggle between a cadre of reactionary lawyers and authoritarian rulers on the one hand, and a community of enlightened humanitarians on the other. This is both misleading and unfair. In fact, it is better understood as a choice between two important value clusters. The first is associated with a 400-year tradition of international law based on the principle of state sovereignty. This principle, which is usually traced to the Treaty of Westphalia of 1648, has been the basis for a rich body of rules and practices associated with non-recourse to force, the legal equality of States and respect for differing cultural traditions within countries. The second value cluster can also be traced back four centuries as a principle in international law—to Hugo Grotius' assertion of a "right vested in human society" to intervene in the event that a tyrant "should inflict upon his subjects such treatment as no one is warranted to inflict". In diplomatic practice, however, this principle was almost completely eclipsed by the commitment to state sovereignty prior to the 1990s. Viewed in this historical context, the changes that have taken place since the end of the cold war can only be described as revolutionary. Today, it would be irresponsible to publish a textbook on international relations which did not accord a substantial section to human rights and humanitarian intervention.

The international community in general, and the United Nations in particular, now recognize a clear mandate to become involved in situations of large-scale, imminent or ongoing human rights violations. *The Economist* magazine, on 6 January 2001, contended that this international consensus has only developed since 1999 and is the result of successes achieved in Kosovo and East Timor. If this were true, we would have little reason to predict that humanitarian intervention will become established as a permanent and influential principle in international law and diplomacy. In fact, international interest in humanitarian intervention has been developing for more than a decade, in reaction to such events as the 1989 Tiananmen Square incident, the Iraqi repression of its Kurdish minorities, mass killings and population displacements in Rwanda, and the violent collapse of the former Yugoslavia. The Kosovo and East Timor operations represent stages on this continuum, but they also gave the international community new reasons to be concerned about the way in which humanitarian interventions are being initiated and managed. Kosovo raised questions about the appropriate procedures for authorizing humanitarian intervention and about the appropriate use of force. East Timor highlighted the costs associated with the United Nations Security Council's enduring concern for state sovereignty.

In spite of the mixed results achieved to date, the growth of world public support for humanitarian intervention seems inevitable, since it is driven by transformative changes in communications and transportation technology. This places a great burden on scholars and policy makers. They will have to act quickly and assertively to establish new rules for humanitarian intervention if they hope to stay ahead of this trend. But they will not succeed if their actions are interpreted as a frontal assault on the Westphalian order. The resistance will be especially fierce among Third World Governments, which will fear that an ambitious mandate for humanitarian intervention will provide great powers with a convenient fig leaf for interference in their legitimate affairs.

The first step in establishing new rules for humanitarian intervention is to agree upon terminology. For purposes of discussion, I define humanitarian intervention as a coercive action by an outside Government or an authorized agent directed toward or within another State, in order to alleviate or avoid a mass humanitarian crisis. Coercive action is meant to subsume not only military force but also a range of punitive options, including economic and arms embargoes, international sanctions and diplomatic pressure.

It is tempting to distinguish between military and non-military coercive options in terms of "hard" and "easy" choices. In the wake of the air campaign in Kosovo, however, this distinction has broken down. As Michael Ignatieff in *Virtual War: Kosovo and Beyond* has argued, the impunity with which this conflict was fought, in the name of human rights, raises some new and cautionary ethical questions. My definition also establishes humanitarian intervention as a subset of humanitarian action to alleviate human suffering. This broader category includes not only coercive situations but also non-coercive relief operations in response to natural or man-made emergencies. The line between these two types of humanitarian action is often hard to establish and even harder to maintain. All too often, over the last decade, we have seen situations in which relief missions have escalated (or more appropriately collapsed) into intervention situations. It is worth noting that most of these situations have involved a reversal of the traditional relationship between *jus ad bellum* and *jus in bello*. According to the traditional just-war model, a government typically considers the moral and legal justification for going to war, and then once the conflict has begun, monitors the behaviour of its troops to assess the legitimacy of their actions in combat. Recently, however, the norm has been for humanitarian relief workers to enter combat situations as neutrals, and either witness gross abuses of the principles of *jus in bello* or be subjected themselves to such abuses. This has forced outside Governments and international agencies to confront *jus ad bellum* issues relating to coercive intervention.

The following seven guidelines are offered to stimulate debate about the circumstances under which humanitarian intervention should be authorized, and the actions that should be taken.

- *The international community has the right and the obligation to intervene in massive humanitarian crises, even if such intervention constitutes an infringement on state sovereignty.*

- *Security Council authorization should be sought in all cases of humanitarian intervention. In emergency situations where such authorization is not available, unilateral or multilateral action by Governments may be necessary. But Governments which undertake such actions incur a special burden to justify their intervention, and to cease and desist if instructed to do so by the Council.*

- *The United Nations should have primary responsibility for designating a situation as a massive humanitarian crisis. All States should accept an obligation to facilitate UN fact-finding to accomplish this task.*

- *Recourse to military force should be the last resort in humanitarian intervention. When military force is deemed to be necessary, rules of proportionality and discrimination must be strictly adhered to.*

- *Decisions on where and when to intervene should be impartial, guided only by concern for the alleviation of human suffering.*

- *Governments which undertake a humanitarian intervention incur a special responsibility for reconstruction and rehabilitation once the immediate goal of halting the humanitarian crisis has been accomplished.*

- *The international community must provide the United Nations with the resources necessary for fact-finding, protection of relief workers and, in extreme circumstances, coercive military action. These forces must be available to the Secretary-General for immediate deployment. Procedures must also be established for transferring responsibility from these UN "fire brigades" to authorized national or multinational forces.*

The issue of capabilities, which is addressed in the last guideline, is arguably the most important and most immediate matter for action by the world community. [This issue is also central to the August 2000 report by the Panel on United Nations Peace Operations, the so-called Brahimi Report, which is summarized and discussed in the *United Nations Chronicle*, Issue No. 3, 2000.] The United Nations is the only international organization which stands a chance of legitimizing humanitarian interventions in the eyes of the world's population. But it can only do so if it is provided with the resources to respond effectively to humanitarian crises when they occur. In the words of the International Institute for Strategic Studies' *Strategic Survey, 1999–2000*: "Success is a powerful impetus to evolution, as much in international relations as in biology."

UNIT 2
World Economy

Unit Selections

5. **Terrorism's Financial Lifeline: Can It Be Severed?** Kimberley L. Thachuk
6. **Measuring Globalization**, *Foreign Policy*
7. **The Rich Should Not Forget the ROW (Rest of the World)**, Jose Ramos-Horta
8. **Prisoners of Geography**, Ricardo Hausmann

Key Points to Consider

- Explain why you agree or disagree with Kimberley Thachuk's conclusion that more concerted efforts by the G-8 nation-states to implement transparent banking practices and reduce corrupt practices will reduce the flow of illegal capital worldwide.

- After reviewing the indicators used in the Globalization Index, explain which ones are the most useful for monitoring global trends and interdependencies among nation-states in the modern world economy.

- Which countries in the world may be able to greatly expand their market share of service exports in the future? Will this growth translate into increased international political influence or power?

- Explain why you agree or disagree with Ricardo Hausmann that geography rather than poor policies is the most serious problem facing tropical, landlocked countries.

- Given the important role of the Internet in promoting technological innovation as well as communications worldwide, do you believe that the World Wide Web should be a "commons" resource or a resource that should be available mainly for commercial development? Defend your answer.

 **Links: www.dushkin.com/online/**
These sites are annotated in the World Wide Web pages.

International Political Economy Network
http://csf.colorado.edu/ipe/

Organization for Economic Cooperation and Development/FDI Statistics
http://www.oecd.org/daf/investment/

Virtual Seminar in Global Political Economy/Global Cities & Social Movements
http://csf.colorado.edu/gpe/gpe95b/resources.html

World Bank
http://www.worldbank.org

The September 11, 2001, terrorist attacks in the United States highlighted the importance of rogue capital for terrorist and criminal groups. The U.S. war on terrorism has documented in detail the continuing importance of unreported international economic exchanges of millions of dollars throughout the world. The United States is unlikely to ever know all the details of how Osama bin Laden and other terrorists, in addition to unknown numbers of other individuals, moved money across national borders. This is because the ancient system for transferring money between countries, the *hawala* that is widespread in the Middle East and Africa, leaves no electronic or physical transfers of funds. Instead, informal mechanisms for moving money across national borders rely on trust among members of interpersonal networks in different countries.

Recent war-on-terrorism investigations have also documented that the al Qaeda network shipped large amounts of gold out of Afghanistan through an untraceable ancient *hawala* money transfer system. How much and where

the gold and other illegal transfers went may never be known because terrorist groups and criminal syndicates worldwide have grown extremely adept in recent years at finding new ways to generate revenue and move financial flows across national borders. Kimberley L. Thachuk, in "Terrorism's Financial Lifeline: Can It Be Severed?" describes the importance of both traditional and more sophisticated techniques used by terrorists and criminals to build multinational empires. While it is clear that more concerted efforts are needed to implement G-8 standards for transparent banking practices, building the capacity to implement these standards and curbing corrupt practices worldwide will probably only be able to reduce the size of the illegal flows. Few observers seriously believe that it is possible to eliminate illegal transnational capital and resource transfers, as these changes are now an integral feature of the modern world economy.

Governments around the world increasingly focus on issues that relate to how to promote market shares for national industries in world markets as competition becomes more intense. In the increasingly integrated global economy, national governments must scramble to ensure that deep-seated political disagreements do not interfere with the pursuit of commercial gain by industry.

China's entrance into the World Trade Organization (WTO) in 2001 was a historic moment in the free-trade movement. The United States also negotiated several new bilateral free-trade agreements and launched trade talks with Australia and countries in Central America and Africa. A new round of talks that were designed to eliminate restrictions to global trade also opened in Doha. However, the momentum of the Doha talks stalled during 2002 after both the United States and the European community implemented new protectionist measures.

Many governments in Europe and worldwide remain upset about the Bush administration's decisions to impose tariffs on imported steel and support $180 billion in subsidies for American farmers over the next 8 years.

In an effort to jump-start the global trade negotiations, the Bush administration proposed a new plan designed to eliminate all tariffs on industrial and consumer goods by 2015. The proposal included eliminating tariffs on large industrial products such as cars and machinery, in additional to labor-intensive consumer goods like clothing, textiles, and leather articles that are still fairly heavily protected in the United States. The European community proposed a more modest set of proposals that would reduce but not eliminate all tariffs. Both proposals face objections during the 2003 round of global trade talks from countries in Latin America and Asia. Representatives of many countries in the developing world argue that their domestic industries would have to endure severe competitive jolts on the industrial side while still facing steep barriers to the American markets for agriculture.

The pattern of recent trade policies of the United States and European community illustrates how powerful nation-states increasingly use free-trade policies to promote national economic prosperity. The United States and other economically mature states vigorously support trade liberalization because leaders share a belief that increased trade and greater globalization promotes prosperity at home and stability worldwide. Several debates about who benefits and who loses from globalization is at the heart of many contemporary discussions of the world political economy. Typically, the debates rely more on anecdotal evidence than empirical facts.

The article "Measuring Globalization" describes the results from the second round of rankings of nation-states based on a

Globalization Index developed by A. T. Kearney and *Foreign Policy* magazine. The index attempts to measure the complex forces driving the integration of ideas, people, and economics in a standardized composite measure of the vague concept of globalization. The result of the data collection exercise confirms the trend of a dramatic downturn in international travel and tourism after the September 11, 2001, terrorist attacks. However, the slowdown, which was already in progress at the time of the terrorist attacks, quickly dissipated. For a second year in a row, Singapore and Ireland ranked at the top of the index. At the same time, technological innovation and integration grew exponentially and the longer-term trend of increasing income inequality worldwide continued.

One reason for the slowing of globalization is that the three main engines of world economic growth, the United States, Japan, and the countries of Europe, all slowed down and moved towards their first collective contraction since the mid-1970s. The contraction of most other national economies followed. The recent trend followed on the heels of the ripple effects of the Asian financial crisis. The trend dramatically underscored how increased trade interdependencies and investment integration may result in more national economies contracting and recovering at the same time. Argentina's recent economic crisis fits a general pattern whereby developing nations that enthusiastically follow neoclassical Western economic advice and kept their markets opens and their currencies freely convertible suffered more during the global contraction than countries such as China who are less well integrated into the modern world economy.

Recent economic trends highlight the fact that the gap between the rich and poor in the world continues to grow. Jose Ramos-Horta, the Nobel Peace Prize winner of 1996, in "The Rich Should Not Forget the ROW (Rest of the World)," proposes a future agenda to stop the growing disparity. The agenda includes debt cancellation for nation-states with per capita incomes of less than $1,000, increased development aid, improved market access for developing countries, and an anti-poverty coalition to represent the interests of the developing countries. The proposals reflect a shared consensus of critics of neoclassical policy prescriptions supported by the G-8, the IMF, and the World Bank.

In "Prisoners of Geography," Ricardo Hausmann disputes the conventional wisdom among economic development experts who say that with the correct mix of pro-market policies, poor countries will eventually prosper. Instead, Hausmann discusses the provocative thesis that geography rather than policy is the root problem. The implication of this thesis is that tropical, land-locked nations may never enjoy access to the markets and new technologies that they need to flourish in the modern global economy.

The highly vocal and disruptive protesters that now gather at World Trade Organization meetings and other G-8 summit meetings are another manifestation of the global divide about the causes of the growing gap between the wealthy and poor in the world. Many critics of the modern world system argue that recent trade agreements do not reflect the concerns of most people in developing countries. In most developing countries, trade issues are linked to larger problems that relate to poverty. As the number of poor and unemployed people in developing countries increase, developing nation-states increasingly complain that their citizens, who represent three-fourths of humanity, are not benefiting from recent trade agreements.

A growing number of analysts and some policy makers in the developed world now link several of the problems evident in poor countries to complex market failures. The linkages are being discussed because recent efforts to provide partial debt relief to Highly Indebted Poor Countries (HIPCs), to close the gap between the richest and poorest countries, failed. The private movement, Jubilee 2000, continues to advocate more debt relief for more countries. In one of several efforts designed to address the concerns of critics and real world trends, the World Bank developed a new approach, the "New Poverty Reduction Strategy," that emphasizes country ownership, transparency, accountability, and greater civil society involvement in debt relief efforts.

Another expression of protest against economic trends associated with globalism was a recent World Wide Web petition drafted by a group of young radical economists that declared war on neoclassical economics. The dissident economists' problem with neoclassical economics is with its narrow focus on theories of the firm and little consideration for geography, national origin, or how to regulate the behavior of multinational firms. For this group, neoclassical economics is the root cause rather than the theory to use to solve many contemporary political-economic problems.

Opposition to traditional economic prescriptions is more widespread among practitioners and analysts in the developing world. The growing world protest movement is facilitated by the existence of the Internet that offers equal access to all. However, many corporations, with the support of national governments, are currently engaged in attempts to wall off portions of cyberspace. The fear is that in so doing, the Internet's potential to foster democracy and economic growth worldwide will be destroyed.

Terrorism's Financial Lifeline: Can It be Severed?

by Kimberley L. Thachuk

Key Points

The attacks of September 11 and the global campaign against terrorism have put the spotlight on *rogue capital*, a growing problem for law enforcement and the financial sector. Using traditional and sophisticated techniques, terrorist and criminal groups have extended their reach beyond states to build multinational empires with pervasive, well-funded subsidiaries. Weak states, lax banking regulations, persistent corruption, and shadow financial systems compound the problem.

Terrorists and criminals generate, manipulate, and launder funds in different ways and for different ends, but the links between them are growing stronger. As direct state sponsorship declines, terrorists have shifted increasingly toward illicit moneymaking. The question is how best to expose this money lifeline, render it vulnerable, and ultimately sever it.

America has led the global effort to cut off the sources of money to these groups. It has worked through the Group of 8 and bilateral partners to strengthen initiatives against money laundering and manipulation. Yet compliance with such measures alone will not guarantee success. Economic assistance is needed to ease the domestic impact associated with the loss of illicit sources of foreign exchange in weak states. Helping such states to fight this problem is essential. It will entail strengthening justice systems and making bureaucracies more transparent to offset the increasing resort to corruption by those determined to circumvent stricter regulations.

To operate effectively, transnational terrorists and criminals need ready access to money and the ability to maneuver it quickly and secretly across borders. On a large scale, such money maneuvers can ripple across entire regions, embroiling global markets and threatening vital American economic interests as well as destabilizing other countries politically. The ability to move vast quantities of wealth rapidly and anonymously across the globe—sometimes combining modern-day wire transfers, faxes, and Internet connections with centuries-old practices, such as the *hawala*, of personal connections and a handshake—gives terrorist and criminal networks a strategic advantage over many states. Yet it also might be their vulnerability.

Terrorist manipulation and laundering of money has received partic-ular attention since the attacks of September 11.[1] While some of the methods used by terrorists differ from those employed by organized crime groups, most are similar. As state sponsorship for terrorist groups has steadily declined in recent years, terrorists increasingly have resorted to crime to sustain activities. Like organized crime groups, terrorists are engaged in moneymaking schemes that are illegal in most states. Such activities may include drug trafficking; extortion and kidnapping; robbery; fraud; gambling; smuggling and trafficking counterfeit goods, humans, and weapons; soliciting both direct sponsorship and contributions and donations from states; selling publications (legal and illegal); and deriving funds from legitimate business enterprises.

While impossible to quantify accurately, it is estimated that illicit financial transactions account for between 2 and 5 percent of the world's gross domestic product (approximately $600 billion to $1.5 trillion). Drug trafficking alone nets between $300 billion and $500 billion, with trafficking in humans and small arms, counterfeiting ($150 billion to $470 billion), and computer crimes ($100 billion) constituting the remainder of the "gross criminal profit." More significant are the direct threats to national security when

25

terrorists and organized crime groups launder money to underwrite and strengthen operations and global reach. The key question for policymakers is how to expose the money trail and render it vulnerable; terrorists and international criminal groups simply cannot function without ready access to money and an ability to move it efficiently.

Moving Cash Illicitly

Under U.S. law, *money laundering* is defined as the "movement of illicit cash or cash equivalent proceeds into, out of, or through United States financial institutions."[2] While engaging in both licit and illicit transactions, terrorists and criminal groups must be able to obscure the movements of cash, especially as these movements pertain to the funding of ongoing illegal operations. This is important regardless of whether the money is being laundered to disguise its origins or to distribute it clandestinely to network cells of operatives, as is the case with terrorist organizations. The process of laundering involves three stages: placement, layering, and integration.

Placement typically involves a person who is adept at exploiting loopholes in financial regulations to move money quickly through the international banking system. Such an individual might charge a third of the total amount being laundered. Offshore banking centers that are largely unregulated or lack transparency are often the preferred vehicles. At approximately a dozen locations in the Caribbean, Southeast Asia, and Europe, for example, money may be deposited with the assurance of secrecy and tax exemption. The United Nations International Drug Control Programme estimates that approximately one-half of the world's money flows through offshore banks. Not only do terrorist groups and criminal networks benefit from the comparative lack of scrutiny at these offshore banks, but they often can obtain a higher rate of return from banking in safe havens.

Layering is the process of redistributing funds to obscure their origins and give them the appearance of being legitimate. Such layering techniques as *smurfing*, or conducting multiple cash deposits or wire transfers for amounts under the standard $10,000 reporting requirement for a Currency Transaction Report, are the most popular devices. Another technique is to create multiple accounts under different names and then move the money back and forth through these accounts to complicate detecting their origins. Because of heightened monitoring of bank accounts since the September 11 terrorist attacks, layering has become exceedingly difficult to accomplish, at least in the United States and close partner countries.

Integration involves the use of funds for other transactions, such as payments to persons or front companies involved in the laundering conspiracy. The money is further invested in legitimate businesses, real estate, or money-generating activities to hide its origins and to increase profits. In recent years, up to $7 billion is estimated to have moved illegally from Russia through the Bank of New York alone. Over time, such sums can facilitate the control by terrorist and criminal groups over a number of major banks and private businesses. With such significant economic power, terrorist and criminal groups are able to suborn public officials and legislators to obstruct unfavorable legislation or to gain preferential treatment in a number of sectors. It also can contribute to instability in the form of rampant inflation, an exponentially increased dependence on a false economy, widespread corruption, and increasing violence and lawlessness in weak states.

Effects of Globalization

Traditionally, illicit money movements involved black market currency exchanges and parallel remittances. In the case of black market money exchanges, the proceeds

of criminal activity would be generated in one currency, sold to a currency broker, then exchanged for a different currency. To thwart illicit transactions, most initial money laundering legislation focused on large cash deposits in excess of $10,000. Given the physical bulk of cash proceeds raised by criminal activity at the retail level (for example, drug sales), disclosure of large deposits was an important step forward. As recently as 1999, 70 to 80 percent (approximately $200 million) of the U.S. Treasury Forfeiture Fund came from currency seizures, as most illegal money maneuvers continue to be in cash. To make the tracking of funds difficult, however, the manipulators of money shifted tactics and began to buy jewelry, real estate, stocks and bonds, vehicles, furniture, antiques, and other expensive items to be resold at a later date.

increasing reliance by terrorists on laundered and manipulated money means that what may appear to be a law enforcement problem is in fact a national security threat

As law enforcement has refined its ability to detect and interdict illegal money movements, criminal methods have become significantly more complex.[3] Intricate financial trading schemes, at times involving "U-turn" movements of funds through various shell companies in several countries, are more representative of money laundering today. Yet the more transactions that occur, the longer the paper trail becomes—and thus the greater possibility for detection, tracking, and successful prosecution.

It is clear that financial globalization generally has been a boon for money manipulators. With advancements in banking techniques and information-age technology, criminals more easily can find and penetrate states whose laws (or lack thereof) make them susceptible to criminal financial transactions. We are witnessing the increasing criminal exploitation of on-line banking, automated teller machines, and Internet casinos; the misuse of trusts and other noncorporate vehicles (such as credit card fraud) to launder and obscure money; the employment of lawyers, accountants, and other professionals to act as legitimate cloaks for transactions; the greater use of methods other than cash in laundering and manipulation schemes; and an upsurge in suspicious wire transfer activity involving shell and front companies. All this would be daunting enough were it not for the fact that the increasing reliance by terrorists on laundered and manipulated money means that what may appear to be a law enforcement problem is in fact a national security threat with criminal elements.

Among all these trends, the importance of offshore banking cannot be exaggerated. While proximity to major financial centers was once paramount, now anonymity and remote locations in the world are more beneficial for anyone wishing to obscure the source of money. For small countries that have few or no natural resources or industry, banking is an attractive way to become incorporated into the global economy. For example, with 570 banks, 2,240 mutual funds, 500 insurance companies, and 45,000 offshore businesses, the Cayman Islands (with just 350,000 inhabitants) have benefited from assets exceeding $670 billion. Other safe havens—such as Antigua and Barbuda, Bahrain, Dominica, Grenada, Liechtenstein, Nauru, Niue, the Seychelles, Tuvalu, and Vanuatu—quickly have realized the opportunities for healthy profits in the offshore banking and business market.

The patterns of troublesome offshore banking practices are familiar. The states in question enact laws that establish strict bank secrecy, crimi-

nalize the release of customer information, prohibit cooperation with international law enforcement, and allow the licensing of banks that have neither personnel nor a physical presence in the country. Further, the creation of anonymous companies, asset-protection trusts, numerous tax advantages, and even the sale of economic citizenship provide advantages to terrorists and criminals who need to conceal and manipulate money efficiently. Trusts, in particular, are generally the source of operating funds for charities, some of which have recently been revealed as raising and manipulating money for terrorist organizations. Not only have these countries cut Faustian bargains to make quick profits, but they also have used their sovereign status to establish relations with banks in other states. Hence, they are moving rogue capital into the international financial system via legal state channels. In addition to the manipulation and laundering of money, many safe havens also are acting as transit zones for illegal arms, drugs, and smuggled and trafficked humans.

Contrasting Money Flows

Criminals and terrorists manipulate money in somewhat different ways. Organized criminal activity is motivated by simple profit—amassing staggering sums either legally or illegally. For terrorist groups, though, the money (however amassed) is a means to other ends.

Generally, organized criminal activity, such as drug trafficking, generates such great quantities of small bills that traffickers are known to weigh money rather than count it. The retail transactions of drugs are mostly in 5-, 10- and 20-dollar bills such that some distributors accumulate 1,000 to 3,000 pounds of bills on a monthly basis, which means that it must be carried in suitcases, on the persons of human carriers known as *mules*, or shipped in cargo containers.[4] The money is often a liability, as it necessitates a near-constant search

for safe storage, discreet bankers to help invest it legitimately, and any number of schemes to obscure its origins. This situation leaves the operations of international criminals vulnerable to disruption by law enforcement and other officials.

Terrorists, however, are interested in sustaining an interlocking global cell structure and finding ways to distribute money discreetly. Logistically, the biggest obstacle that terrorists face is bank reporting requirements. Groups such as Al Qaeda must find ways to disaggregate and distribute significant amounts of money into smaller denominations to sustain or expand a network of comparatively small cells (typically 3 to 4 operatives) that do the recruitment and conduct operations. Additionally, if the money is derived from criminal activity, such as narcotics trafficking, it must first be amassed, laundered, and *then* redistributed. Again, the more transactions that occur, the more vulnerable these groups are to detection.

A Shadow System

The *hawala* (or trust) system is an alternate method by which Islamic terrorists and organized crime groups distribute money.[5] The system works on an honor scheme, with transaction records being kept only until the money is delivered, at which time they are destroyed. Anyone can go to a *hawala* dealer in thousands of cities around the world and have any quantity of cash transferred to any place on the globe in a matter of hours. In this way, no cash moves across a border or through an electronic or financial transfer system. The sender typically identifies neither himself nor the recipient. Rather, a code word is used that will allow the recipient to collect the cash from a trusted associate of the originating *hawala* dealer (or *hawaladar*). Over time, the accounts between dealers are balanced through future money transfers or exchanges between the two *hawaladars* or a number of *hawaladars* in their circle of associates. The accounts may be reconciled by mutual agreement as an exchange of money or goods.

The ancient Chinese used a similar system called *fei qian*, or flying money, which Arab traders adopted to avoid robbery on the Silk Road. Millions of hardworking Pakistanis, Indians, Filipinos, and others living abroad use the system to send remittances home to family members. *Hawala* merchants typically charge a 1 percent commission for a completed transaction; their main profit comes from currency fluctuations and extra fees for moving large amounts of money for big clients, who at times happen to be drug traffickers and smugglers of other illicit goods.

a significant revenue source for terrorist networks is the trafficking and smuggling of narcotics, arms, and people

In September 2001, President George W. Bush listed 27 terrorist organizations and individuals whose assets were to be blocked in American financial institutions. Since then, more than 202 entities and individuals have been identified for punitive financial action worldwide. The principals behind the Al Qaeda financing network reportedly are Al Barakaat and Al Taqwa/Nada Management Group. Al Barakaat is a Somali-based international financial conglomerate with operations in over 40 countries, including the United States. The organization's founder, Shaykh Ahmed Nur Jimale, reportedly is closely linked to Osama bin Laden and has used Al Taqwa/Nada Group to facilitate the financing and operations of Al Qaeda and other terrorist organizations, such as Hamas. Before its U.S. operations were closed down, Al Barakaat reportedly wired at least $500 million in annual worldwide profits to the company's central

money-exchange office in the United Arab Emirates. Al Qaeda allegedly received a flat 5 percent cut of that money, amounting to approximately $25 million a year.

Sources of Funds

Abuse of charitable organizations poses another problem. A number of Islamic organizations allegedly have been penetrated and manipulated by terrorist groups such as Al Qaeda. In other instances, terrorists have formed charitable organizations as fronts for distributing money to international networks. In both cases, it is likely that numerous innocent citizens have been contributing to what they think are charities for causes such as refugee resettlement and the nourishment of children in Palestine. One of the largest such charitable organizations in the United States, the Holy Land Foundation, recently had its assets frozen by the U.S. Government for its alleged funding of Hamas activities. The Holy Land Foundation raised a reported $13 million from Americans last year. Other moneymaking schemes include raising money from wealthy patrons and providing seed funding to start-up terrorist cells. It further involves shake-downs and extortion, protection rackets, racketeering, and credit card fraud.

A significant revenue source for terrorist networks is the trafficking and smuggling of narcotics, arms, and people. Al Qaeda has been supported by the trafficking of heroin—if not directly, at least indirectly via the Taliban in Afghanistan. Heroin destined for European markets originates primarily in the Golden Crescent region of Southwest Asia, which includes parts of southern Afghanistan, northern Pakistan, and eastern Iran, as well as Central Asia. In 2000, Afghanistan was estimated to have produced between 3,276 and 3,656 metric tons of opium. This represents about 70 percent of the world's supply and is valued between $190 and $212 million. The Taliban allegedly replaced smaller criminal traf-

fickers and exercised a virtual monopoly for drugs exported from that country.

not surprisingly, Osama bin Laden excels at amassing and distributing large sums of money to support his terrorist schemes

The illicit sale of gold, diamonds, and other precious gems is another method Al Qaeda uses to generate and hide revenues. Diamonds purchased illegally and below fair market value from rebels in Sierra Leone are then resold in Europe at a significant profit. Gems are easy to hide, generally maintain their value, and are virtually untraceable. Gold is often used by *hawaladars* to balance the accounts. Its origins also are untraceable because it can be smelted, made into jewelry, and reconstituted in a variety of forms. Gold is probably more easily manipulated than gems because it can be deposited on account without a transaction report being required. Dubai is one of the world's least regulated gold markets. Situated as it is at the center of the Gulf, Africa, and South Asia, it is reportedly one financial hub for militant groups that use gold for illicit purposes.

Bin Laden's Terror Capital

Since September 11, $34 million in terrorist assets, including $27 million belonging to Al Qaeda and bin Laden, have been frozen in the United States. A total of 161 nations have blocked the assets of known terrorist organizations, amounting to another $70 million. Action also is being taken to disrupt severely the misuse of the *hawala* system and other underground remittance systems used by bin Laden, Al Qaeda, and other terrorist organizations.

Not surprisingly, Osama bin Laden excels at amassing and distributing large sums of money to support his terrorist schemes. His main sources for financial support include his personal wealth, estimated between $280 million and $300 million, funds siphoned from overt Muslim charities, and wealthy well-wishers, especially in the Gulf States. Allegedly, a wide variety of international banks in the Gulf are used to manipulate and move funds using business front organizations owned by bin Laden. Mohammad Jamal Khalifa, bin Laden's brother-in-law, is responsible for managing parts of the financial network that deal with major investments in Malaysia, Mauritius, the Philippines, and Singapore.[6] Reportedly, bin Laden has funded a number of network cell operating expenses, including accommodations, safe houses, cars, and payments to operatives for the recruitment of new members. His contributions have further purchased explosives and key components for explosive devices. At least $5,000 is known to have been transferred from bin Laden holdings to operatives in Yemen to fund the attack against the U.S.S. *Cole* in 2000. The investment for bin Laden to mount the September 11 attacks is estimated to have been approximately $500,000, while the total costs to the United States for cleanup, property losses, and Federal Government bailouts will exceed $135 billion.

Addressing the Problem

Although financial manipulation and money laundering have long been stigmatized in various resolutions by the United Nations and other intergovernmental organizations, broad-based declarations generally have lacked muscle. The focal point for more serious practical cooperation has been the Group of 8 (G-8), which in 1989 established the Financial Action Task Force (FATF) on Money Laundering. While this task force does not have universal membership and operates on consensus, the effect of peer pressure by

this multidisciplinary, intergovernmental organization on states that do not comply with its mandates is nevertheless significant. It is composed of 29 member states, as well as international organizations, such as the United Nations, European Union, and World Bank.

The most important FATF contribution is setting standards. It maintains a list of 40 recommendations for combating money laundering in addition to publishing a bulletin of noncooperative states maintained by the FATF Non-Cooperative Countries and Territories Initiative. These measures have become the de facto international anti-money-laundering standard. They cover financial systems and their regulation, criminal justice systems, including law enforcement, and international cooperation. The recommendations allow flexibility in implementation according to particular circumstances and existing laws and represent general principles of action to which states have made a political commitment to adhere.

Working alongside the FATF is the Financial Stability Forum (FSF) Offshore Working Group, which focuses almost exclusively on offshore financial centers. Convened in 1999 at the request of the Group of 8 finance ministers, the FSF promotes international financial stability through information exchange and international cooperation in financial supervision. The Working Group on Offshore Financial Centers is charged with considering the impact and significance of offshore financial centers in relation to global financial stability. Jurisdictions are categorized in terms of their financial supervision, cross-border cooperation, and transparency. The weaker states—such as Antigua and Barbuda, Aruba, the Bahamas, Cayman Islands, Turks and Caicos, and Vanuatu—have been identified as undermining efforts to strengthen the global financial system through their lack of cooperation and low-quality supervision of financial transactions.

Following the attacks of September 11, the FATF expanded its mission beyond money laundering to include an effort to combat terrorist financing. Eight special recommendations with regard to terrorist financing were implemented in October 2001, coupled with an appeal to all FATF members for immediate compliance. The measures range from requiring members to provide closer scrutiny of transactions with noncooperative countries to prohibiting financial transactions with these countries.

Yet these measures are not entirely new. In October 1995, President William Clinton signed Presidential Decision Directive 42, which explicitly recognized that international criminal activity poses a threat to U.S. national security. Specifically, President Clinton noted the number of egregious overseas money laundering centers and ordered the Departments of Justice, State, and Treasury, the Coast Guard, the National Security Council, the intelligence community, and other Federal agencies to increase and integrate efforts against international criminal activity and money laundering. Interagency teams began to negotiate with known money laundering safe havens. More than a dozen countries were identified as vulnerable to money laundering, and these were targeted with a two-pronged approach: increased bilateral law enforcement cooperation using the critical *nowhere to hide* principle and *warnings* about consequences of failing to take action.

The consequences of failing to take action continue to include the effective use of International Emergency Economic Powers Act authority to block foreign business and individual assets, as well as prohibiting American entities from dealing with them. The Office of Foreign Assets Control oversees a sanctions program that includes prohibitions against trading with identified enemies of the United States as set forth in a variety of lists generated by different agencies in the U.S. Government. These include not only the names of known money launderers and terrorists but also members of narcotics cartels and individuals identified in the Foreign Narcotics Kingpin Designation Act.

President Bush has emphasized that the fight against terrorists cannot be won without attacking the money that supports them

Particularly successful has been cooperation between and among financial intelligence units (FIUs). These bodies are the vehicles for much informal cooperation between law enforcement agencies, especially in the area of investigative information exchanges. FIUs receive international suspicious activity reports (required under their respective domestic laws), analyze financial information, disseminate information to domestic agencies, and generally exchange information globally. The U.S. Financial Crimes Enforcement Network (FinCEN) is in charge of enhancing the international exchange of financial intelligence through the FIUs. In 2000, FinCEN arranged 159 information exchanges between FIUs.[7]

Post-9/11 Initiatives

The events of September 11 pushed money laundering and the financing of terrorism to the forefront of domestic and foreign policy concerns. President Bush has emphasized that the fight against terrorists cannot be won without attacking the money that supports them: "the first strike in the war against terror targeted the terrorists' financial support." As a result, Congress approved the USA Patriot Act (2001).[8] This act builds on and strengthens many of the previous money laundering acts.[9] Among other things, it amends Federal law

governing a range of illegal monetary transactions, including money being manipulated for the purposes of corruption of officials. It further prescribes guidelines under which the Secretary of the Treasury may require domestic financial institutions and agencies to take specified measures if reasonable grounds exist for concluding that jurisdictions, financial institutions, types of accounts, or transactions operating outside or within the United States are of primary money laundering concern.

Along with these initiatives, asset forfeiture laws have been toughened substantially. Funds that are found to have been laundered are subject to seizure from interbank accounts.[10] Legislation has closed off any possible loophole protecting legal money used by terrorists by stating, inter alia, that any assets used or intended to be used in the commission of a terrorist act or the proceeds of such an offense, foreign or domestic, also are subject to asset forfeiture.

In an effort to dismember Al Qaeda and other similar groups, the U.S. Customs Service is leading an interagency task force called Operation *Green Quest*. Its main target is abuse of the *hawala* system and the exploitation by Al Qaeda operatives of safe havens, such as Dubai, Hong Kong, and Malaysia. In the trial of the four terrorists convicted for their role in the 1998 terrorist bombings of the American Embassies in Tanzania and Kenya, testimony was given that pointed to Dubai, Hong Kong, London, and Malaysia, as the end sources for the finances for the attacks. As a result of this and other evidence, in 1999 and early 2000, approximately $225 million of assets linked to Al Qaeda through the Taliban were blocked in U.S. financial institutions.

To strengthen law enforcement overseas, the United States has entered into numerous mutual legal assistance treaties (MLATs), which provide for the exchange of financial information and evidence in criminal and related matters. MLATs are particularly helpful in asset forfei-

ture proceedings. To facilitate cooperation in joint investigations, the United States shares the profit of the seized assets with the state with which it cooperated and has urged foreign governments to share forfeited assets to improve cooperative international efforts.

The United States also has concluded a number of customs mutual assistance agreements (CMAAs), which use the World Customs Organization model. Using the CMAAs, the U.S. Customs Service assists in gathering information and evidence for trade fraud, smuggling, violations of export control laws, money laundering, and narcotics trafficking. Using agreements such as these, it is possible to attack the elusive problem of money manipulation, albeit more circuitously than enacting legislation. Unfortunately, if no prior agreement exists with a state, the United States is left in the difficult position of painstaking negotiation on even the most trivial details.

Shortly after the September 11 attacks, the U.S. Treasury Department established the Foreign Terrorist Asset Tracking Center that aims to disrupt terrorists' ability to manipulate money in the international financial system. Financial data will be used to track and target terrorist financing worldwide and to dismantle terrorist organizations by attacking their financial structures. The Center is further mandated to uncover all links between terrorist groups and legitimate business and financial institutions.

a more plausible policy goal would be to press for near-universal adoption of G-8 standards for transparent banking practices

Yet exclusive focus on overseas financial operations is inadequate.

Many terrorist organizations are believed to have financial support within the United States. In July 2000, U.S. authorities arrested 18 Hizballah members operating a cigarette smuggling cell in Charlotte, North Carolina. A number of these people were later indicted for immigration fraud, bribery and related conspiracies, and money laundering. Seven of the 18 are believed to have been providing support and resources directly to Hizballah. In addition to Al Barakaat, which is linked to Al Qaeda, another group—al-Itihaad al-Islamiya—has operated a fundraising organization in Minneapolis-St. Paul, Minnesota. These Minnesotan Somalis reportedly have wired more than $75 million in donations to East Africa in recent years, mainly to impoverished relatives, along with smaller sums for guns and clan-based militia operations.[11]

Next Steps

Internationally, the United States and its G-8 partners set the standard for regulations designed to expose money manipulation. There is, of course, no perfect way to suppress or interdict every conceivable financial transaction that might pose risks, even within the G-8. A more plausible, though still ambitious, policy goal would be to press for near-universal adoption of G-8 standards for transparent banking practices, utilizing a mix of inducements and pressures to obtain such adherence. The aim would be to increase dramatically the costs and risks of timely exposure that criminals and terrorists would face when they seek to move their funds into, out of, or through the publicly regulated banking system in any country.

This objective may be accomplished in a number of mutually reinforcing ways: first, through capacity-building measures; second, through measures to offset a surge in corruption that will result as financial regulations become stricter; and third, by securing compliance with current multilateral money launder-

ing agreements and enforcing the measures against delinquent states as set forth in the extraordinary FATF session in late 2001.

Capacity-Building. Building capacity will entail a dedicated effort to assist countries to find alternate ways to generate foreign exchange. It also will entail training law enforcement and justice officials to conduct anti-money-laundering and manipulation investigations and prosecutions. The decrepit condition of some criminal justice systems constitutes an essential obstacle to attacking money laundering and manipulation. The criminal justice systems of some states will require reforms so that there is a better strategic fit with other state systems.

Much frustration, confusion, and uncertainty often accompany the reform of justice agencies. Two issues are critical. The first concerns the efficiency and effectiveness of criminal justice generally. Frequently, a change in the focus of jurisdictional responsibility requires a corresponding reassessment of the management and administrative structures. Systems that are marred by mismanagement of scarce resources, administrative duplication, poor communication, inefficiency, and ineffectiveness inevitably find it difficult to sustain the confidence of the publics that they serve, let alone be effective international partners. Second, justice agencies must aspire to the greatest possible openness in their work, making it subject to external audit and review. This helps justice institutions to garner respect both symbolically and practically, but, as with any bureaucratic entity, openness does not come easily.

Curbing Corruption. The relationship between money manipulation and corruption is often overlooked in combating terrorist and criminal activity; corruption is a critical enabler of terrorists and organized crime. Not only does corruption minimize opportunities for state control over the activities of international organized criminals and terrorists, but also it inevitably

prevents real sovereignty from being exercised. Secret networks that operate fluidly across the frontiers of several states require impunity from detection and capture. The subornation of public officials through bribery, graft, collusion, or extortion is the vehicle by which to secure that exemption. Moreover, resort to such methods likely will increase in vulnerable states as international pressures begin to produce stricter financial controls at the national level.

Any effort to suppress corrupt practices must necessarily involve greater transparency as well as administrative constraints within the official institutions. Furthermore, mechanisms that detect and punish corruption and limit opportunities for corrupt behavior will assist in the cleanup effort. The media can be put to good use in such anticorruption campaigns; publicizing and scrutinizing the behavior of public servants regularly acts as a deterrent to would-be abusers of power.

Multilateral Agreements. Internationally, wider reporting of suspicious transactions to the FATF of issues related to terrorism is a logical complement to effective mutual law enforcement assistance on a bilateral basis. Such reporting, symbolically, is quite significant as it demonstrates that countries are not solely responsible for the activities of transnational actors within their sovereign borders. Hand-in-hand with wider reporting, foreign countries must be urged to impose anti-money-laundering laws on alternate remittance systems and ensure that nonprofit organizations, such as charities, cannot be used to finance terrorism. This would close the door on groups that are using underground methods to obscure international money laundering and manipulation schemes.

It is evident that terrorists and criminals are more adroit than many states at adapting to the realities of a globalizing world. Governments not only must catch up to the methods being employed by such groups, but they also must surpass them by responding creatively, consistently, and quickly to new challenges. The sooner the source of funding for organized crime and terrorists can be halted significantly, the sooner their operations will be restricted. A number of multilateral and bilateral vehicles which the United States can use to elicit cooperation already exist. What is needed is the enforcement of existing methods and practices for the suppression of money laundering and manipulation. Yet more work needs to be done to understand why this problem exists and what the inherent constraints are in achieving positive results.

Inherent Limitations

While tightening the noose around illicit financial transactions is clearly necessary, the scale of the challenge remains forbidding. To start with, policymakers face continued challenges posed by poor or uneven sharing of information between various states. Classified information cannot easily be shared with foreign nationals, complicating prosecution or extradition proceedings. A concerted effort must be made to ensure that information that is unclassified remains so in order to guarantee that it may be shared with other countries. At the same time, more effective information-sharing necessarily requires greater numbers of analysts with expertise on the international financial system, as well as greater numbers of people in international law enforcement with fluency in foreign languages.

economic assistance needs to be factored in as a way to cushion the domestic impact in weak states of compliance with anti-money-laundering regulations

Beyond the problems of information-sharing, policymakers would be ill-advised to ignore the factors that impel developing and transitional economies to tolerate—indeed often to *encourage*—illicit money transactions in the first place. The so-called losers of globalization are finding it difficult to survive in an increasingly competitive and interconnected world. Hence, simply condemning such states as being uncooperative, setting standards of best practices, or increasing scrutiny by FATF member states will do little for their balance of payments deficits. Many of the states currently operating as safe havens are democratizing. Their leaders now have to respond to citizen demands for prosperity. If looking the other way means vast amounts of foreign exchange will flow into the country, governments may find themselves bending to their need for political survival first and international public opinion second.

For these reasons, economic assistance needs to be factored in as a way to cushion the impact on weak states of complying with internationally recognized anti-money-laundering regulations. After all, if financial systems were brought to the verge of bankruptcy, the countries themselves would be destabilized. Organized crime and terrorist groups could more easily infiltrate such regions, using the dual strategies of corruption and extortion of officials. Consequently, to close loopholes for money laundering and manipulation, countries in the money laundering business need to find alternate sources of foreign exchange while simultaneously detecting and preventing official state corruption.

Policymakers also need to anticipate that increasingly successful interdiction of illegal monetary transactions by law enforcement will tend to drive terrorists and criminal groups further underground or toward more sophisticated methods of evasion. Thus, to the greatest extent possible, anti-money-laundering/manipulation operations must address entire networks rather than

component parts; partial quick fixes will only generate new problems. If only some members of a terrorist network are apprehended and their assets forfeited, for example, other members will simply fill the void.

The pressures facing countries to make the global financial system more transparent are bound to increase over time, especially in the face of ongoing threats of catastrophic terrorism. The question is how to maintain the right set of incentives to elicit the collaboration of countries.

For the moment, the main obstacle that confronts cooperation on money laundering and manipulation remains the same one that plagues the detection and prosecution of terrorists and criminals more generally. Sovereign powers interpret public international law, treaties, and agreements discretely. Terrorists and criminals move through jurisdictions with relative ease, while state and law enforcement officials remain behind to negotiate effective instruments that transcend borders. The exigencies of political survival for some national leaders may also constrain international cooperation, if that means ultimately to surrender one's nationals to foreign law enforcement. Ironically, domestic law enforcement authorities may find that their interests are more in line with their counterparts in other states than they are with other agencies in their own governments. While none of these factors should dissuade aggressive law enforcement action against financial activity that underwrites terrorists and criminals, they do need to be considered in devising a long-term strategy for combating the problem.

Clearly, the exploitation of global financial systems by terrorists and criminals poses threats to international security that cannot be ignored. Money laundering and manipulation are problems that know no boundaries and yet are within the means of states to resolve. Meeting these challenges head-on must be part of an overall campaign plan if the global war on terrorism is ultimately to be won.

Notes

A version of this paper with additional citations can be found at <http://www.ndu.edu/insshp.html>

1. *Money manipulation* connotes the ability to move and maneuver sums of money without detection, while *money laundering* denotes the steps by which the proceeds of a crime must be obscured and reintroduced into the financial system.

2. "The Money Laundering and Financial Crimes Strategy Act of 1998" Pub.L. 105–310, U.S.C. 5340(2)(A), October 1998.

3. Cash transactions comprise between two-thirds and three-quarters of all suspicious transaction reports in the Organization for Economic Cooperation and Development countries, which is surprising as the reliance on cash by the general public has decreased rapidly in favor of a reliance on credit and debit cards.

4. For example, 25 pounds of $20 bills may be carried in a briefcase, but this only amounts to $20,000. $1.8 million in $20 bills weighs 184 pounds and would fit into a suitcase. Twenty million dollars weighs 2,000 pounds, and $300 million weighs 30,000 pounds, which would require a C–130 or a cargo container to transport. One billion dollars would require a container or airplane such as a 747 with a payload of 100,000 pounds.

5. I am indebted to Esther Bacon for her background research on the *hawala* system, bin Laden's financial holdings, and Somali operations in the United States.

6. One of the four main committees that report directly to bin Laden. The others are military, religio-legal, and media.

7. See, for example, *The 2001 National Money Laundering Strategy* (Washington, DC: U.S. Department of Treasury, Office of Enforcement in Consultation with the U.S. Department of Justice, September 2001).

8. Otherwise known as the Uniting and Strengthening America by Providing Appropriate Tools Required to Intercept and Obstruct Terrorism (USA PATRIOT Act) Act of 2001.

9. See especially Title III of *International Money Laundering Abatement and Anti-Terrorist Financing Act of 2001*.

10. For the purposes of this act, foreign banks that have a correspondent banking relationship with banks in the United States are defined as being, for the purposes of summons and subpoena power, the same as a bank in the United States.

11. Al-Itihaad al-Islamiya is thought to have formed in Somalia following the 1991 overthrow of Muhammad Siad Barre. It initially served as a fundraising apparatus for bin Laden's fledgling terrorist network. Prominent businessmen associated with Al-Itahhad and acted as conduits for the money. Through its separate business operations, al-Itihaad al-Islamiya is one of the most financially influential groups in Somalia. Al-Itihaad members control telecommunications firms and money-transfer companies. The group also profits from the trade of *khat*, a leaf that many Somalis chew.

Kimberley L. Thachuk is a senior research fellow in the Institute for National Strategic Studies at the National Defense University. Dr. Thachuk may be reached by phone at (202) 685–2377 or by e-mail at thachukk@ndu.edu.

From *Strategic Forum*, May 2002. © 2002 by Strategic Forum.

Measuring Globalization

*Everyone talks about globalization, but no one has tried to measure its extent… at least not until now. The A.T. Kearney/*Foreign Policy *Magazine Globalization Index™ dissects the complex forces driving the integration of ideas, people, and economies worldwide. Which countries have become the most global? Are they more unequal? Or more corrupt?*

When you can measure what you are speaking about, and express it in numbers, you know something about it," the British physicist Lord Kelvin once observed. "But when you cannot measure it, when you cannot express it in numbers, your knowledge is of a meagre and unsatisfactory kind."

"Unsatisfactory" is the word that best describes the contemporary debate over globalization. There seems to be a consensus that globalization—whether economic, political, cultural, or environmental—is defined by increasing levels of interdependence over vast distances. But few people have undertaken the task of actually trying to measure those levels of interdependence. For instance, how do we determine the extent to which a country has become embedded within the global economy? How do we demonstrate that globalization is racing ahead, rather than just limping along? And how do we know just how worldwide the World Wide Web has become?

Like the physical universe that Lord Kelvin sought to understand, globalization may be too vast a concept to be fully captured by today's still limited set of statistical measurements. But that same challenge has not deterred physicists from their relentless pursuit to measure with ever greater accuracy the forces that hold the universe together. Nor should it deter those who seek a deeper understanding of globalization and its impact on the contemporary world. Without some means to quantify the extent of globalization, any meaningful evaluation of its effects will remain elusive.

With this challenge in mind, we present the A.T. Kearney/Foreign Policy Magazine Globalization Index™, which offers a comprehensive guide to globalization in 50 developed countries and key emerging markets worldwide. The Globalization Index "reverse-engineers" globalization and breaks it down into its most important component parts. On a country-by-country basis, it quantifies the level of personal contact across national borders by combining data on international travel, international phone calls, and cross-border remittances and other transfers. It charts the World Wide Web by assessing not only its growing number of users, but also the number of Internet hosts and secure servers through which they communicate, find information, and conduct business transactions.

The Globalization Index also measures economic integration. It tracks the movements of goods and services by examining the changing share of international trade in each country's economy, and it measures the permeability of national borders through the convergence of domestic and international prices. The index also tracks the movements of money by tabulating inward- and outward-directed foreign investment and portfolio capital flows, as well as income payments and receipts.

Given the unprecedented range of factors that the Globalization Index encompasses, we believe that it is a unique and powerful tool for understanding the forces shaping today's world. And the results of this year's index prove startling. Much of the conventional wisdom cherished by both champions and critics of globalization collapses under the weight of hard data, ranging from the pace and scale of global integration and the characteristics of the "digital divide" to the impact of globalization on income inequality, democratization, and corruption.

Is Globalization Slowing Down?

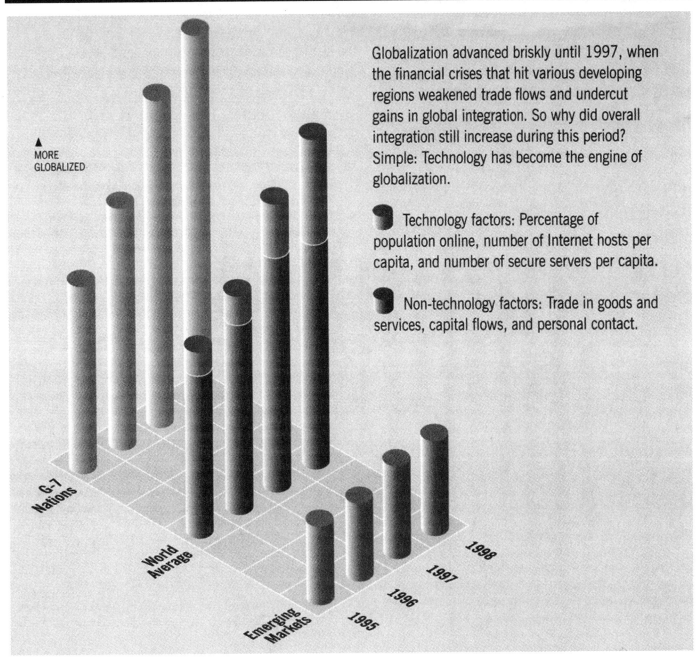

Globalization advanced briskly until 1997, when the financial crises that hit various developing regions weakened trade flows and undercut gains in global integration. So why did overall integration still increase during this period? Simple: Technology has become the engine of globalization.

▮ Technology factors: Percentage of population online, number of Internet hosts per capita, and number of secure servers per capita.

▮ Non-technology factors: Trade in goods and services, capital flows, and personal contact.

▲
MORE
GLOBALIZED

CHARTS BY AGNEW MOYER SMITH

The A.T. Kearney/FOREIGN POLICY Magazine Globalization Index™ may not settle the question of whether globalization does more good than harm. But the index provides an objective starting point for a debate that has typically relied more on anecdotal evidence than empirical facts.

LEADERS OF THE PACK

In recent years, indicators of global integration have shown remarkable growth. The number of international travelers and tourists has risen, now averaging almost three million people daily—up from only one million per day in 1980. The latest data from the United Nations Conference on Trade and Development show that foreign direct investment jumped 27 percent in 1999 to reach an all-time high of U.S. $865 billion, while total cross-border flows of short- and long-term investments have more than doubled between 1995 and 1999. Due to the falling cost of international telephone calls and the rising levels of cross-border activity, the traffic on international switchboards topped 100 billion minutes for the first time in 2000. And with an online population estimated at more

The Global Top 20

Singapore leads the rankings as the most global nation in the index, due in large part to its high trade levels, heavy international telephone traffic, and steady stream of international travelers. European nations round out the rest of the top five countries. Despite high levels of integration on various technological measures, the United States remains less integrated in economic terms, leaving it twelfth in the index.

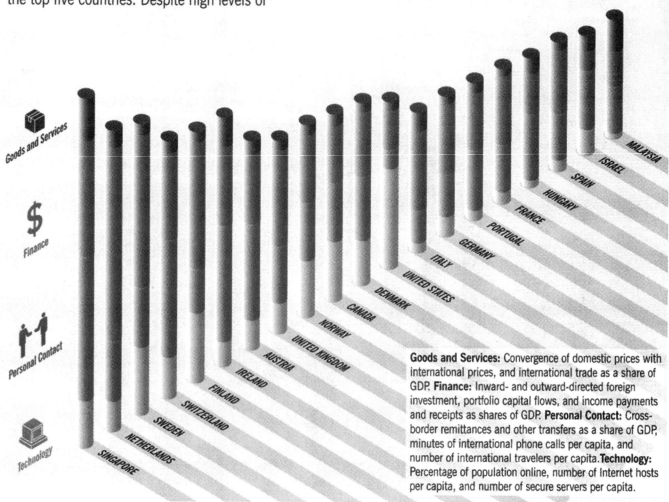

Goods and Services: Convergence of domestic prices with international prices, and international trade as a share of GDP. **Finance:** Inward- and outward-directed foreign investment, portfolio capital flows, and income payments and receipts as shares of GDP. **Personal Contact:** Cross-border remittances and other transfers as a share of GDP, minutes of international phone calls per capita, and number of international travelers per capita. **Technology:** Percentage of population online, number of Internet hosts per capita, and number of secure servers per capita.

CHARTS BY AGNEW MOYER SMITH

than 250 million and growing, more people in more distant places have the opportunity for direct communication than ever before.

The expansion of information technologies adds to globalization in ways other than facilitating communication. Some nations fear that the Internet is an engine driving U.S. cultural hegemony. Others see the Internet as a catalyst for creating global cultural communities, from Moroccan sports enthusiasts rooting for their favorite Canadian ice hockey team to antiglobalization protestors mobilizing against the World Trade Organization and the International Monetary Fund. The Internet is also an unprecedented means for disseminating ideology to a glo-bal audience, whether it is pro-democracy activists in Serbia rerouting dissident radio broadcasts to the World Wide Web or Chechen rebels maintaining their own on-line news service.

The full impact of information technologies on political and social life is not easily measured. But it is possible to gauge their effects on the economic sector. Information technologies make it possible for nations to sustain deeper levels of economic integration with one another. Nowhere is this integration more evident than in financial markets, which use advanced information technologies to move U.S. $1.5 trillion around the world every day. For the United States, cross-border flows of bonds and equities alone are 54 times higher now than they were in 1970. Such flows have multiplied by 55 times for Japan and 60 times for Germany.

Why does globalization remain sluggish even as indicators of technological integration continue to grow exponentially?

At first glance, these trends lend credence to the popular notion that globalization is fast creating a world that, as former Citicorp Chairman Walter Wriston put it, is "tied together in a single electronic market moving at the speed of light." But a closer look reveals that global integration appears to be growing no more rapidly now than it has been for years, and its pace may even be slowing.

Why does globalization remain sluggish even as indicators of technological integration—the number of Internet hosts, online users, and secure servers—continue to grow exponentially? The data from our broad spectrum of developed and developing markets suggest that global economic integration has wound down to something of a crawl. The drop in total trade to and from the 50 countries surveyed weighs particularly heavy in this slowdown. The chief culprit was the series of financial crises that rippled through Southeast Asia, Latin America, and Russia in the late 1990s. Strong growth in portfolio investments and foreign direct investment helped to moderate these declines, and the value of world trade has rebounded since 1999. As a result, we see a situation in which economic globalization slowed even as technological globalization continued at a rapid clip [see chart, "Is Globalization Slowing Down?"].

Some nations have pursued integration with the rest of the world more aggressively than others. The most globalized countries are small nations for which openness allows access to goods, services, and capital that cannot be produced at home. In some cases, geography has played an important role in sustaining integrated markets. The Netherlands, for instance, benefits from (among many other factors) its position at the head of the Rhine, which knits together countries that account for almost three quarters of total Dutch trade. In other cases, such as Sweden and Switzerland, relatively small domestic markets and highly educated workers have given rise to truly global companies capable of competing anywhere in the world. And a host of other factors has contributed to the globalization of other small states. Austria, for example, benefits from heavy travel and tourism, while remittances from large populations living abroad contribute to Ireland's integration with the outside world.

Tiny Singapore stands out clearly as the world's most global country [see chart, "The Global Top 20"]. The country far outdistances its nearest rivals in terms of cross-border contact between people, with per capita international outgoing telephone traffic totaling nearly 390 minutes per year. Singapore also boasts a steady stream of international travelers, equal to three times its total population. In contrast, the United States hosts only one sixth that level of international tourists and travelers and can claim less than one fourth the per capita outgoing international telephone traffic.

Yet in recent years, Singapore has struggled to maintain high levels of trade, foreign investment, and portfolio investment, which help support its globalization lead. The Asian flu is partly to blame, since the financial crisis undermined the entire region's economic performance. But Singapore's slow progress in privatizing state industries, its failure to win endorsement for a regional free-trade agreement, and its tight controls over Internet development have also slowed its integration with other countries.

Another country that ranks high on the Globalization Index is the Netherlands. But here, the story is largely economic. Within only a few short years, the Dutch have both invested heavily in other countries and seen foreign participation in their own economy rise to levels that few other nations have been willing or able to sustain. In the wake of aggressive reforms that have stripped regulations and enhanced labor flexibility, foreign investment increased from 8 percent of gross domestic product (GDP) in 1995 to more than 19 percent of GDP in 1998. Likewise, portfolio investments grew from only 5 percent to more than 30 percent over the same period, the highest levels in the world—more than double those in France and Germany and five times higher than those in the United Kingdom.

With Sweden and Finland riding the wave of Internet development to similar gains in integration with the rest of the world, the current globalization rankings may well be in flux. Singapore could slip from the lead in the coming years, as countries that are better positioned to benefit from global communications technologies or that are more aggressive about reforms to attract foreign trade and investment develop stronger ties with their neighbors.

Yet despite signs of greater openness among these few leading countries, many others remain stalled at much lower levels of integration, with little indication of imminent change. Thus, there is reason to believe that the countries at the top of the rankings are only running further and further away from the pack.

THE DIGITAL ABYSS

Not all countries around the world have participated equally in the transition to the new global economy. As the chart below indicates, the digital divide between developed and emerging-market countries is now more like a digital abyss. On many relevant measures—from the diffusion of Internet users to the number of Internet hosts—the vast majority of economic activity related to information and communications technologies is concentrated in the industrialized world.

Digital Divides

An overwhelming majority of economic activity linked to information and communications technology is concentrated in the industrialized world, with developing nations lagging far behind. But there is also a divide within the divide, with the United States, Canada, and the Scandinavian countries far outpacing most West European economies.

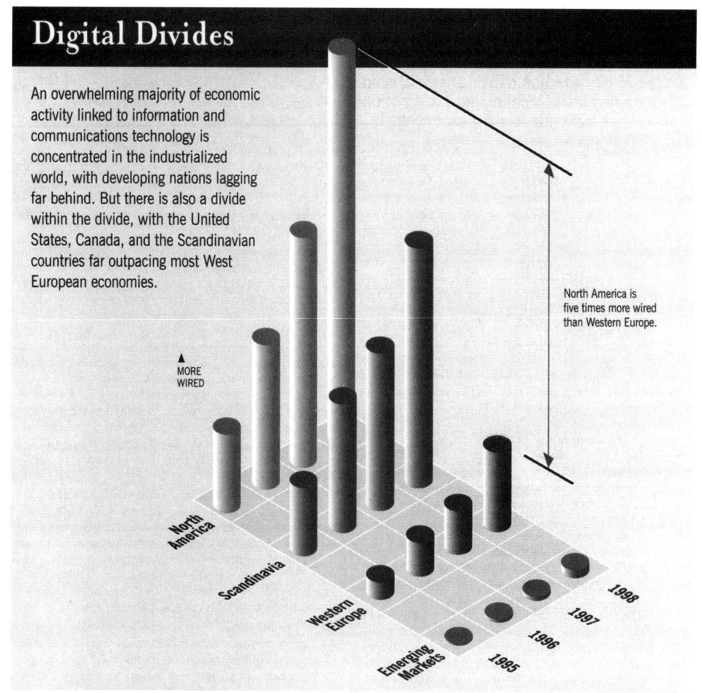

North America is five times more wired than Western Europe.

MORE WIRED

CHARTS BY AGNEW MOYER SMITH

But among industrialized countries, another digital divide exists. The Internet has penetrated deeply in the United States, with neighboring Canada not far behind. In both countries, over 25 percent of the population enjoyed Internet access by 1998 (the last year for which data are available for all countries in the survey). More recent estimates put that number above 40 percent in both countries. Perhaps more important, the United States and Canada lead the world in secure servers suitable for electronic commerce, signifying that their well-developed Internet networks can be used effectively to enhance commercial activities as well as personal communication.

In addition to the United States and Canada, Scandinavian countries also rank among the world's most wired nations. Thirty-nine percent of Sweden's population was online in 1998, growing to 44 percent in more recent surveys. Finland and Norway led in Internet hosts, each with more than 70 servers per 1,000 inhabitants connected directly to the World Wide Web.

Indeed, if any region of the world exemplifies the changing face of global integration, that region is Scandinavia, where Sweden, Finland, and Norway have turned their traditional engineering and manufacturing

Globalization, Freedom, and Corruption

Freedom House, a U.S. nonpartisan organization, each year rates the levels of political rights and civil liberties in countries worldwide. A clear correlation exists between the Freedom House ratings and the rankings in the A.T. Kearney/FOREIGN POLICY Magazine Globalization Index™. More globalized countries (such as the Netherlands and Finland) tend to have more civil liberties and political rights, while less globalized countries (such as China and Kenya) score poorly in these categories. There are some important exceptions: Singapore, for instance, is the world's most globalized economy, but it ranks poorly in the Freedom House index compared with other countries at similar levels of development.

But if Singaporean officials are somewhat authoritarian, at least they are honest. The strong correlation between the Globalization Index's country rankings and levels of perceived corruption, as measured by the international nongovernmental organization Transparency International, suggests a clear relationship between globalization and clean government. Indeed, investors perceive public officials and politicians as less corrupt in more globalized countries such as Singapore, Finland, and Sweden but more underhanded in closed countries such as Indonesia and Nigeria.

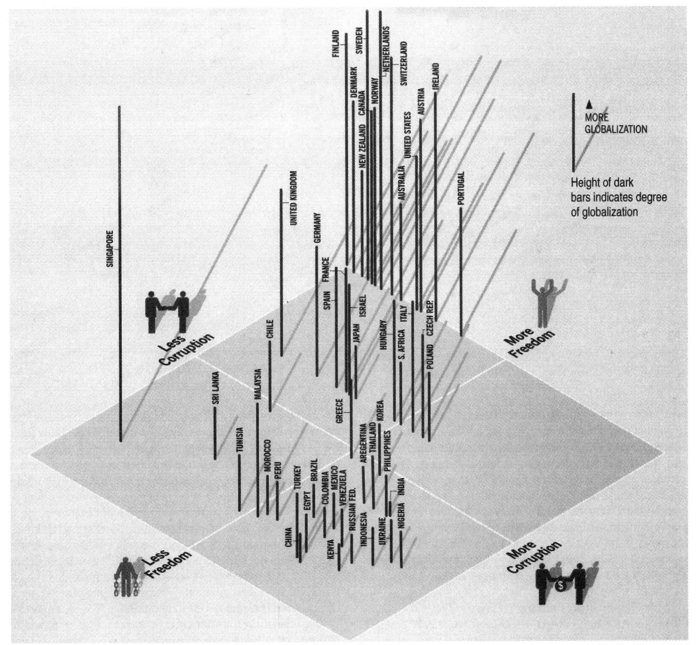

CHARTS BY AGNEW MOYER SMITH

Globalization and Inequality

Are more globalized societies also more unequal? Not necessarily. With some exceptions, countries scoring high in the Globalization Index enjoy more egalitarian income patterns, while nations that are less integrated with the rest of the world display more skewed distributions of income.*

*Inequality data are not available for all countries in the survey.

(continued on next page)

CHARTS BY AGNEW MOYER SMITH

prowess to work in the information technology boom while further opening their countries to trade and investment flows.

Scandinavia's technological takeoff should come as little surprise. In the last century, Sweden was among the first countries to realize the full potential of the telephone. It offered a means of mitigating distance in often sparsely populated lands. Thirty years ago, Sweden's leading technology company, Ericsson, was among the pioneers in mobile telephony, and this decade the country has embraced Internet technologies far ahead of the curve. Stockholm, with nearly 60 percent of its population online, is perhaps the most wired city in the world.

In similar ways, neighboring Finland suggests the possibilities of this Internet-led revolution. In 1995, Finland topped all others in terms of Internet access. Information

technology made it possible for Finnish companies to respond to competitive pressures by diversifying both their export markets and their workforce. Recent studies show that over one quarter of Finnish exports now go to countries beyond Europe, up from less than one fifth in 1990. And nearly half the staff of Finland's 30 largest companies now operate overseas, as compared to only 15 percent in 1983. Although other countries have since pulled ahead in levels of Internet penetration, Finland has witnessed rising levels of trade and investment that have pushed it into the fifth position overall in the Globalization Index, much higher than it would have placed only a few years ago. One other symbol of success: The market capitalization of Nokia, Finland's global telecommunications giant, is now higher than the country's gross domestic product.

Globalization and Inequality *(continued)*

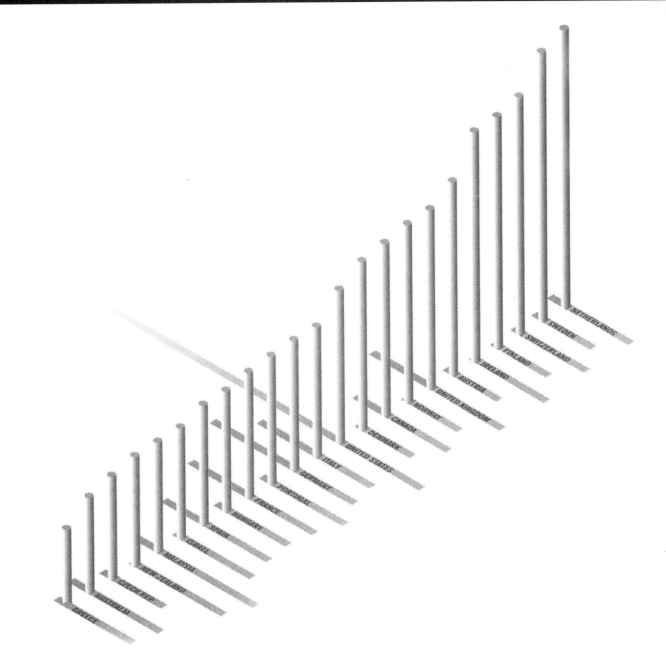

CHARTS BY AGNEW MOYER SMITH

The general pattern of higher globalization and greater income equality holds for most countries, both in mature economies and emerging markets.

The fact that Sweden, Finland, and the rest of Scandinavia have been able to nurture fast-moving technological developments with their traditionally lumbering regulatory and tax regimes offers an unexpected contra-

diction, confusing traditional assumptions about how high levels of regulation impede globalization. But what about areas of relatively high regulation where no technological takeoff has yet been achieved? Look no further than continental Europe to see the negative effects of an unfavorable business climate on integration. Indeed, most of the countries in the euro zone, weighed down by their relatively low scores in Internet development, rank at the bottom of the top 20 globalized countries.

Concerns about the disparities between industrialized and developing countries, especially with respect to Internet access and use, have touched off a worldwide debate about the global digital divide. Rather than a division between developed and developing countries, however, the divide at this moment reflects the vast technological advances in North America and the Scandinavian countries compared with the rest of the world. Together, those two regions stand on one side of a gaping digital chasm that appears to have left much of the remaining world behind.

If this "digital abyss" is to be bridged, developing nations have the most ground to cover. But deciding how to use their limited resources poses a difficult dilemma. Malaysia offers but one example of the perverse choices that can ensue. In an effort to attract investment and develop its high-technology capabilities, Malaysia has spent more than U.S. $3.6 billion on its Multimedia Super Corridor. At the same time, over 70 percent of the nation's primary schools lack computer facilities, and almost 10 percent lack proper connections for water and electricity. The result is an impressive infrastructure not sufficiently supported by human capital.

For other countries, Internet development cannot proceed unless more fundamental concerns about infrastructure are addressed. In Chile, one of the most prosperous emerging markets, 57 percent of the fixed telephone lines and 58 percent of the mobile-phone subscribers are located in the capital city, leaving most of the country without Internet access. And Africa's underdeveloped telecommunications sector has left much of that continent without reliable connections to the World Wide Web. For instance, the Democratic Republic of the Congo still has no direct link to the Internet, and a large number of African countries can count no more than a few hundred active Internet users.

MORE EQUAL THAN OTHERS

Antiglobalization critics frequently claim that globalization increases income inequality. This assertion is elegant in its simplicity, but it ignores a host of other important factors. The level of income disparity in an economy might have more to do with history, economic growth, price and wage controls, welfare programs, and education policies than it does with globalization or trade liberalization.

Moreover, the empirical evidence suggests a very different story about income disparity and globalization [see chart, "Globalization and Inequality"]. Emerging-market countries that are highly globalized (such as Poland, Is-

rael, the Czech Republic, and Hungary) exhibit a much more egalitarian distribution of income than emerging-market nations that rank near the bottom of the Globalization Index (such as Russia, China, and Argentina). There are some exceptions: Malaysia, for instance, is more globalized but less equal than Poland. But the general pattern of higher globalization and greater income equality holds for most countries, both in mature economies and emerging markets.

These findings should reinvigorate the debate over whether countries are poor and unequal because of globalization, or because they are not globalized enough. Moreover, efforts to redress global inequality should be tempered with the recognition that many countries with skewed income distribution patterns, including Brazil and Nigeria, also have large populations. That only underscores the difficulty of pulling the mass of humanity out of poverty.

A CAT SCAN OF GLOBALIZATION

Trade, foreign direct investment, international telephone calls, Internet servers—considered individually, statistics on each of these phenomena are accurate, albeit insufficient, measures of global interdependence. Yet, just as a CAT scan creates a three-dimensional image of the human anatomy from a series of two-dimensional images, the A.T. Kearney/FOREIGN POLICY Magazine Globalization Index™ provides a comprehensive view of global integration through an analysis of its component parts.

There is, of course, an irony associated with trying to measure globalization on a nation-by-nation basis. Even the least integrated countries are being drawn together by new forces beyond their ability to control, whether it is global warming, the spread of infectious diseases, or the rise of transnational crime. And some of the most significant aspects of globalization—the spread of culture and ideas—cannot be easily quantified. These and other challenges highlight the need for a closer and more refined examination of the forces driving global integration, not to mention further refinement of the tools used to measure it.

The Rich Should Not Forget the ROW (Rest of the World)

JOSE RAMOS-HORTA WON THE NOBEL PEACE PRIZE IN 1996.

DILI, EAST TIMOR—The world's economy is many times larger than it was only 50 years ago. Particularly in the Northern countries, which have the greatest concentration of personal wealth, the quality of life has improved dramatically. Mind-boggling advances have occurred in vast fields of human endeavor, from genetics to computers. Human beings have walked on the moon, Mars is being studied at ever closer range.

Yet the same human intelligence that has produced such advances seems so far unable to eliminate extreme poverty or tropical diseases such as malaria and cannot provide clean water to hundreds of millions in Africa, Asia and Latin America.

> ## Child labor, prostitution and sex slavery are rampant in the impoverished but aspiring societies.

And the gap between rich and poor has grown, not diminished. Hundreds of millions survive on less than $1 a day, children walk miles to go to school, if at all, or to fetch firewood and water for the household. Child labor, prostitution and sex slavery are rampant in the impoverished but aspiring societies.

In their pursuit of even greater prosperity, weapons-producing countries aggressively push arms in developing countries that cannot even afford to provide clean water to most of their people, fueling local, often ethnic, conflicts.

Do we have some answers to these challenges from the dark side of the human condition?

There is no dispute that abject poverty, child labor and prostitution are indictments of all humanity. However, poverty should not only touch our conscience: It is also a matter of peace and security because it destabilizes entire countries and regions. That in turn threatens the integration of the global economy that is vital if the rich are to stay rich or if the poor are to move up, if only an inch.

One need not be original to propose some key elements of a solution to these problems, as those more enlightened than I have.

Drawing on the ideas already in circulation, here is the agenda I propose:

DEBT CANCELLATION

The G-8, European Union and World Bank should lead the initiative in writing off the entire public-sector debt of all non-oil-producing countries whose per capita income is less than $1,000.

In addition, a special fund should be developed by the World Bank and the UN Development Agency (UNDP) to assist these countries in improving governance and generating employment for the poorest.

Other highly indebted countries (for instance, Indonesia, Nigeria, Brazil and Mexico) should also benefit from a special debt-relief package of up to half of their public-sector debt if the proceeds are aimed at poverty reduction and education.

> ## The United States, Canada and Japan should open up their markets for goods from the HIC (highly indebted countries) and ease some of the stringent quality-control and quarantine rules that make it impossible for the poorest countries to export their goods and commodities.

For all cases there must be strict conditions involving reduction in defense expenditures, democratic reforms (including of the security forces), good governance and accountability, and allocation of saved resources to eradicating poverty.

Debt cancellation or relief can be phased in tandem with the reform policies being adopted and implemented by the targeted country.

INCREASE OVERSEAS DEVELOPMENT ASSISTANCE

All rich countries should increase the percentage of their overseas development assistance within the next 10 years to the UN-recommended 0.8 percent of GDP. Perhaps on a dollar-for-dollar basis where applicable, such aid could match reduction in military expenditures associated with debt relief.

IMPROVE MARKET ACCESS

Following the example set by the European Union, the United States, Canada and Japan should open up their markets for goods from the HIC (highly indebted countries) and ease some of the stringent quality-control and quarantine rules that make it impossible for the poorest countries to export their goods and commodities.

BUILD AN ANTI-POVERTY COALITION

The violent demonstrations that greet every gathering of world leaders from Seattle to Prague and Davos to Gothenberg and Genoa reflect the justified frustration of those who are genuinely concerned about the effects of globalization on the poorest of the world.

However, one can also see the opportunistic manipulation of these people by Communist-era hard liners who, seeing their world revolution agenda discredited with the collapse of the Soviet Union, now try to hijack what is otherwise a genuine anti-poverty movement.

This past year I attended the World Economic Forum in Davos, Switzerland, during the last week of January. Looking around me, I saw the richest and the most powerful people in the world and realized that I was the poorest among that lot. Yet I did not see, hear or read any complicated plot by the rich to rule the world.

From that modern-day Robin Hood, George Soros, to Bill Gates and James Wolfensohn, the World Bank president, I heard genuine concern and motivation to help the poor.

An ocean away in Porto Alegre, Brazil, thousands were meeting in defiance of the rich running the world from Davos. From that snowy perch in the Alps, as one of the poorest among the richest, I concluded there was enough good will on both sides of the divide to meet halfway.

The poor will not see their lot improved if we opt for the arrogant and discredited Marxist dogma still trotted out by far too many as a solution to the world's ills. The rich will not be able to continue to reap the profits of their investment in globalization if they do not seriously address the issues of poverty on a world scale.

The poor will not see their lot improved if we opt for the arrogant and discredited Marxist dogma still trotted out by far too many as a solution to the world's ills. The rich will not be able to continue to reap the profits of their investment in globalization if they do not seriously address the issues of poverty on a world scale.

To the end of establishing this middle position, I propose a world summit bringing together representatives from the G-8, World Bank, Group of 77 (developing nations' group in the United Nations), development and human rights NGOs (nongovernmental organizations) as well as the corporate world to debate and fashion a global strategy. The ultimate aim should be to boost the poorest with a sort of global Marshall Plan that involves debt cancellation or relief as well as proactive programs to reduce poverty.

Globalization has tied the G-8 to the ROW (rest of the world). To sink or to swim is the choice they now have to make together.

Prisoners of Geography

Economic-development experts promise that with the correct mix of pro-market policies, poor countries will eventually prosper. But policy isn't the problem—geography is. Tropical, landlocked nations may never enjoy access to the markets and new technologies they need to flourish in the global economy.

By Ricardo Hausmann

So you are a Scorpio. Then you must be passionate. So the barometer says that the atmospheric pressure is declining. Then it is going to rain. So your latitude is less than 20 degrees. Then your country must be poor.

There may be some debate about which of these statements is true, but only one is truly offensive—the last one. Indeed, the notion that a country's geography determines its level of economic development is fraught with controversy. People take offense at such a connection because it smacks of racism and undermines the notion of equal opportunity among nations and individuals. It is also paralyzing and defeatist: What can policymakers and politicians do or promise if nothing can overcome geography? From World War II through the mid-1980s, these sentiments prompted a backlash against the study of economic geography in much of the academic world. Today, however, new theories of economic growth coupled with empirical research have brought economic geography back to the forefront of the development debate. Speaking at the United Nations Conference on Women and Development in June 2000, U.S. Treasury Secretary Lawrence Summers decried "the tyranny of geography," particularly in African countries, and warned against concluding that "the economic failures of isolated, tropical nations with poor soil, an erratic climate and vulnerability to infectious disease can be traced simply to the failure of governments to put in place the right enabling environment." The prevailing development paradigm—according to which market-oriented economic policies and the rule of law alone suffice to make all countries rich—appears to be losing credibility. What if geography gets in the way of the Promised Land?

LOCATION, LOCATION, LOCATION

Closing the income gap between rich and poor countries has been a stated objective of the international community for the last 50 years. This commitment spawned the creation or redesign of institutions such as the World Bank, specialized United Nations agencies such as the United Nations Development Programme and the United Nations Conference on Trade and Development, regional development banks such as the Inter-American Development Bank (IDB), bilateral aid agencies in the governments of the most advanced economies, and innumerable foundations, research centers, and other non-governmental organizations.

But the global gap between rich and poor countries has not closed. Instead, it has widened. Economist Angus Maddison estimates that, in 1820, Western Europe was 2.9 times richer than Africa. By 1992, this gap had risen to 13.2 times. The trend continues—albeit less dramatically—in South Asia, the Middle East, Eastern Europe, and Latin America. In 1997, the richest 20 percent of the world's population enjoyed 74 times the income of the lowest 20 percent, compared to 30 times in 1960.

Tropical diseases do not "merit" the sort of R&D investments that a cure for baldness attracts in Western markets.

The countries left behind have distinguishing geographical characteristics: They tend to be located in tropical regions or, because of their location, face large transportation costs in accessing world markets—or both.

In 1995, tropical countries had an average income equivalent to roughly one third of the income of temperate-zone countries. Of the 24 countries classified as "industrial," not one lies between the Tropics of Cancer and Capricorn, except for the northern part of Australia and most of the Hawaiian Islands. Among the richest 30 economies in the world, only Brunei, Hong Kong, and Singapore are in tropical zones, and their geographical locations leave them ideally suited for growth through trade. Tropical nations tend to have annual rates of economic growth that are between one half and a full percentage point lower than temperate countries. A recent IDB study found that after considering the quality of institutions and economic policies, geography explained about a quarter of the income difference between industrialized and Latin American countries in 1995. Tropical countries also have poorer health conditions than their nontropical counterparts. After considering income levels and female education, life expect ancy in tropical regions is seven years lower than in temperate zones. Nations in tropical areas often display especially skewed income distributions. In Africa and Latin America, the richest 5 percent of the population earn nearly 25 percent of the national income, while in industrial countries they earn only 13 percent. Latitude alone can explain half of this difference. Even within regions of the same country, living standards are strongly linked to geography. For example, in Mexico, the southern states of Chiapas, Oaxaca, and Guerrero have twice the infant mortality rate and half the educational attainment of the country's northern states.

Nations with populations far from a coastline also tend to be poorer and show lower rates of economic growth than coastal countries. A country whose population is farther than 100 kilometers from the sea grows 0.6 percent slower per year than nations in which the entire population is within 100 kilometers of the coast. That means, for example, that the post-Soviet republics will experience as

much difficulty battling their geographical disadvantages as they will overcoming the aftereffects of communism. Countries that are tropical, far from the coast, and landlocked have three geographical strikes against them. Many countries in Africa are handicapped by one or all of these factors.

There is still much we do not understand about the links between geography and economic growth. But what we do know suggests that the challenges of economic development must be examined from a very new perspective. Denying the impact of geography will only lead to misguided policies and wasted effort. Geography may pose severe constraints on economic growth, but it need not be destiny.

LATITUDE PROBLEMS

To understand why geography can matter so much for economic development, consider what economists regard as the main engines of growth: Access to markets (based on the work of Scottish economist Adam Smith) and technological progress (drawn from the writings of U.S. economist Joseph Schumpeter).

For Adam Smith, productivity gains achieved through specialization are the secret to the wealth of nations. But for these gains to materialize, producers must have access to markets where they can sell their specialized output and buy other goods. The larger the market, the greater the scope for specialization. In today's global marketplace, most industrial products require inputs from various locations around the world. Therefore, if transportation costs are high, local companies will be at a disadvantage in accessing the imported inputs they need and in getting their own goods to foreign markets.

Unfortunately, transportation costs are often determined by a country's geography. A recent study found that shipping goods over 1 additional kilometer of land costs as much as shipping them over 7 extra kilometers of sea. Maritime shipping is particularly suited to the bulky, low-value-added goods that developing nations tend to produce; therefore, countries lacking cheap access to the sea will be shut out of many potential markets. Moreover, if countries far from the sea do not enjoy the physical infrastructure (the system of roads, railways, and ports) needed for access to navigable rivers or the sea, they will not develop the very industries that could help maintain such an infrastructure.

Land transportation is especially costly for landlocked countries whose products need to cross borders, which are a much more costly hurdle than previously thought. Studies on trade between U.S. states and Canadian provinces find that simply crossing the U.S.-Canadian border is equivalent to adding from 4,000 to 16,000 kilometers worth of transportation costs. Little wonder, then, that the median landlocked country pays up to 50 percent more in transportation costs than the median coastal na-

Living Between the Poverty Lines

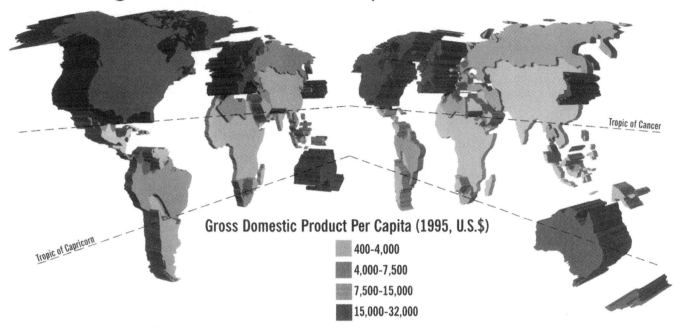

Gross Domestic Product Per Capita (1995, U.S.$)

- 400-4,000
- 4,000-7,500
- 7,500-15,000
- 15,000-32,000

SOURCE: John Luke Gallup, Jeffery D. Sachs, and Andrew D. Mellinger "Geography and Economic Development" (National Bureau of Economic Research Working Paper No. W6849, December 1998)

tion. In practical terms, these differences can be enormous: Shipping a standard container from Baltimore to the Ivory Coast costs about $3,000, while sending that same container to the landlocked Central African Republic costs $13,000.

Governments in landlocked countries face the additional challenge of coordinating infrastructure expenditures with neighboring countries. Sometimes, political or commercial problems inhibit passage to the sea. For example, the agricultural potential of the upper Parana River basin in landlocked Paraguay remained dormant until a Mercosur agreement in the mid-1990s facilitated barge transportation through Brazil and Argentina. Jordan's access to the Mediterranean requires crossing the Israeli border or those of Syria and Lebanon. These instances illustrate why landlocked nations suffer from sluggish economic growth. Countries and territories like Hong Kong, Taiwan, and Singapore have an advantageous geographical position, but much of inland Africa, China, and India remains far from markets and maritime trade.

Geography harms developing countries in other ways. Joseph Schumpeter showed that technological innovations, through research and development (R&D), are powerful engines of economic growth. (This notion is what Schumpeter had in mind when he coined his famous term, "creative destruction.") R&D displays increasing returns: The more people who use and pay for a new idea, the greater its market value. (For example, a new computer program or novel may cost a lot to pro-

duce, but subsequent copies are extremely cheap.) In order to recoup their initial costs, R&D investors will tend to focus on innovations for which potential customers abound. Unsurprisingly, rich countries with large, middleclass populations are more lucrative markets than poor nations with little purchasing power.

Even though innovations such as computers or cellular phones work in many geographical conditions and are therefore easily adopted by developing countries, technologies in other sectors often require research that is very location-specific. Many technologies are not universally applicable; their effectiveness depends on the geographical or climatic conditions in which they are used.

Consider agriculture. The divergence in agricultural productivity between the developed and developing world is grounded in dramatically different R&D capabilities. Governments in advanced economies spend up to five times more (as a percentage of total agricultural production) on agriculture-related R&D than their counterparts in developing countries. Rich nations also benefit from the expenditures of private agricultural producers—a source of funding that is virtually nonexistent in developing nations. Geography aggravates this disparity. Plant varieties need to be adapted to the local climate, meaning that R&D geared toward rich, temperate-zone agriculture is of little use in tropical areas. Countries like Argentina, Chile, Australia, New Zealand, and South Africa can enjoy thriving export sectors in fruit, wine, cereals, oilseeds, and salmon thanks to the technologies developed for these products in temperate zones in the

Northern Hemisphere. But the tropical countries—with their production of coffee, cocoa, sugar cane, and cassava—are left out of the modern-technology club. The result is that the agricultural sector is much less dynamic in tropical areas than in temperate zones. Since unproductive agricultural workers can produce little more than what they require for personal subsistence (and therefore cannot support large urban populations), rural areas remain sparsely populated, have small, poor markets, and suffer from high transportation costs—all of which hamper economic growth.

Climate differences and economies of scale have long played a powerful role in the development of agriculture in different geographical zones. In his Pulitzer Prize-winning book *Guns, Germs, and Steel*, physiologist Jared Diamond explains how Eurasia's east-west geographical layout and the north-south layout in Africa and the Americas determined these regions' historical patterns of economic growth. Since climate changes little with longitude but quite rapidly with latitude, the Eurasian landmass enjoyed fairly uniform climatic conditions. Hence, agricultural innovations developed in one region could travel long distances and be shared by many people, resulting in a large set of plant and animal varieties available throughout the region. By contrast, new varieties developed in the Americas or in Africa could not migrate very far since climates change swiftly, limiting the technological opportunities available to these regions and stunting economic growth.

Of course, agricultural productivity and transportation cost advantages do not necessarily go together. As historian David S. Landes points out in *The Wealth and Poverty of Nations*, the ancient civilizations of Mesopotamia and Egypt had their most fertile lands along rivers. This location—far removed from the seashore—limited their ability to expand their economies through trade. Their power eventually waned and they were supplanted by the seafaring Phoenicians, Greeks, and Romans. More recently, in India and China, agricultural conditions encouraged large populations to cluster along riverbeds far away from the sea, hurting the countries' long-term prospects for economic growth and development through trade.

Investments in health research and technology are also very sensitive to geography. Diseases such as malaria, hookworm, schistosomiasis, river blindness, and yellow fever are hard to control in tropical regions because the lack of seasons makes the reproduction of mosquitoes and other disease transmitters rather constant throughout the year. Since the afflicted countries tend to be poor, tropical diseases do not "merit" the sort of R&D investments that a cure for baldness or erectile dysfunction can attract in Western markets. (Of the aforementioned tropical diseases, only yellow fever has been controlled through an effective vaccine.) Technological development is skewed away from the needs of geographically disadvantaged countries. Thus, children in tropical regions often die of gastrointestinal and other infectious diseases, while many nations still suffer from endemic tropical ailments. Economists John Luke Gallup and Jeffrey Sachs estimate that per capita economic growth in countries with severe malaria is more than a full percentage point lower than in nations where this illness is not prevalent, and that a 10 percent reduction in the incidence of malaria is associated with 0.3 percent higher growth.

The enormous divide in agricultural productivity ensures that standards of living in tropical areas are likely to remain stagnant.

The costs of not dealing with disease in tropical countries go far beyond higher healthcare expenses and reduced worker productivity. Disease can no longer be considered a mere public health problem, but a socioeconomic development issue that affects everything from trade flows to migration patterns. The 1991 cholera outbreak in Peru cost the country's fishing sector nearly $800 million in lost revenues because of a temporary ban on seafood exports. The 1994 plague outbreak in Surat, India, prompted 500,000 people to move from the region and led to work stoppages across several industries, as well as new restrictions on international trade. Estimates of the cost India bore for this plague reach $2 billion.

BORDERING ON POVERTY

The dominant development paradigm these days holds that market-oriented economic policies and the rule of law are all that matter for economic progress. In other words, Mozambique could become Singapore if it would only get its institutions and policies in order; in the meantime, we could alleviate poverty through targeted social spending for the poor, such as the financing of education for girls. But this mantra vastly oversimplifies the challenges of development. If a region is poor because its geography undermines agricultural productivity, impedes market access, and facilitates endemic disease, then good domestic policies will hardly suffice to foster growth. Poverty will not disappear because of expanded nutrition programs or improvements in the teaching materials available in schools. (At best, better trained students simply will migrate to more prosperous regions.)

From this perspective, it may be more important to devote time and resources to transportation infrastructure, which lowers the costs of trading, new technologies for agriculture and public health, and economic integration projects than to focus solely on areas like health, education, and the rule of law.

Infrastructure Development If small, rural communities in developing countries are to experience economic

Locational Correctness

Economic geography offends people because it seems to imply an immutable destiny—if you live in one area, you are poor; if you live in another, you are rich. When the Inter-American Development Bank dared highlight economic geography in its *Economic and Social Progress in Latin America* report in 2000, the Brazilian media attacked the institution for reviving racist and determinist theses. "Ideas from Another Century" screamed the headline in *Gazeta Mercantil*, Brazil's leading business newspaper.

This virulent reaction was not lacking in irony, particularly since income differences in Brazil are closely related to latitude, with the tropical northeast being very poor while the more temperate south is much richer. But these attacks should not be surprising. Since the Enlightenment, economic geography has been a matter of great debate and controversy among scholars and political leaders throughout the world. Their interpretations of the issue have ranged from sensible to silly to outright dangerous: Adam Smith regarded ports, navigable rivers, and canals as essential for industrialization—assets that Great Britain possesses but that

places like Africa and Siberia lack. Montesquieu saw a close relationship between geography and politics, concluding that democracy was fine for Switzerland because of its low agricultural productivity, but that wealthier nations such as France needed a monarchy. During the European imperialist expansion of the 19th century, and under the impact of social Darwinism, geography became a way to justify notions of white racial supremacy. The "fittest" race had become so because, among other reasons, the temperate climate where it developed helped forge populations more prone to thoughtfulness and responsibility than to ebullient pleasure seeking.

Such racially charged views became increasingly unacceptable after the rise of the Nazi regime and the horror of the Holocaust. The reputations of 20th-century geographers such as the famed Ellsworth Huntington of Yale University, author of the landmark 1915 work *Civilization and Climate*, suffered greatly (and unfairly) by association. Historian David S. Landes attributes this reaction not so much to weaknesses in geographers' analyses, of

which there were plenty, but to their pessimistic message that nature, like life, is unfair. Victimized by this backlash, the geography departments at Harvard, Michigan, Northwestern, Chicago, and Columbia universities were shut down in short order following World War II. As a result, several generations of academics disregarded geography as a key factor in socioeconomic development.

In recent years, however, geography has slowly made its way back into mainstream economic thinking; new theories and techniques for studying trade, growth, and the environment have contributed to this resurgence. And interest in geography as a discipline is also rising: In the United States alone, the number of bachelor's degrees awarded in geography rose from about 3,000 in 1985–86 to nearly 4,300 in 1994–95. In the academic arena, economic geography is no longer taboo. It is only a matter of time before the discipline becomes acceptable in broader circles—maybe even among Brazilian editorialists.

—R.H

growth, it is crucial to connect them with the rest of their country and the world through investments in roads and other transportation infrastructure. Many of these investments must be made outside of the particular countries in question. For example, for Rwandan and Ugandan goods to reach new markets, the Kenyan rail system must be improved. This complication poses severe coordination and political challenges; it is not clear, for instance, that such an improvement should be a priority for Kenyan authorities. Unfortunately, the major regional development banks operate with this same narrow focus, granting loans to national governments on the basis of perceived national priorities. Important region-oriented projects remain chronically underfunded. To overcome this problem, bilateral or multilateral organizations should

provide financial incentives to national governments to encourage them to cofinance investment projects that benefit themselves as well as neighboring countries.

Technological Development Although it is fashionable (and accurate) to decry the "digital divide" between advanced and developing economies, this information-technology gap need not be a major concern for poor countries since they benefit from global innovations in these arenas. For instance, Latin American countries soon will have more cellular phones than regular telephone lines, allowing for a major expansion in the region's telecommunications system by skipping the need to install underground cables. By contrast, the dramatic difference between rich and poor countries in agricultural and pharmaceutical R&D ensures that standards of living in tropical areas are likely to remain low and stagnant. Governments in developing nations lack sufficient resources to address this problem by themselves, and the world's private sector allocates very little financing to agricultural R&D for developing nations. Although the well-known difficulties in enforcing intellectual property rights create a significant disincentive for this sort of investment, there may be ways to enlist the knowledge and research capabilities of corporations such as Pfizer and Archur Daniels Midland. Economists Michael Kremer and Jeffrey Sachs have proposed contests so companies can compete to develop effective vaccines. The Clinton administration included in its 2001 budget proposal a tax credit to U.S. pharmaceutical companies that developed vaccines for diseases prevalent in the developing world. However, the vast needs in this area suggest that multilateral financing will be needed to compensate private firms for such initiatives.

Integration National borders, as they are currently conceived, make nations artificially more distant and only accentuate the costs already imposed by geographical conditions. Borders limit the movement of goods, capital, and labor and thus limit access to markets. Some regions—most notably Western Europe—have already begun eliminating internal borders. But for the last 50 years we have witnessed the creation of more and more nations in the developing world, with their own new borders, making these countries effectively more distant than their physical geography implies. Can poor nations afford this additional source of remoteness?

If shipping goods across the U.S.-Canadian border adds the equivalent of thousands of miles in transportation costs, then the commercial logistics of trading between countries with weak political institutions and a history of cross-border animosity will prove to be infinitely more expensive problems for importers and exporters. And borders do not merely complicate the movement of goods and the coordination of cross-country infrastructure; capital also has trouble crossing borders. Since investment contracts are often enforced at the national level, sovereignty can shelter borrowers who are able but unwilling to repay. This situation introduces "sovereign risk" into financial markets, limiting capital movements and rendering them increasingly fickle.

Borders also prevent people in poorer areas from moving to more prosperous regions. For example, the decline in agricultural employment in the United States prompted significant regional migration, and when Europe went through a similar process at the end of the 19th and beginning of the 20th century, it had an escape valve in the form of a wide-open immigration policy in the United States. Today's geographically trapped peoples seldom enjoy such opportunities. Not that they don't search for them: About one third of the landlocked Burkinabes and one fifth of Bolivians work in neighboring nations. Not only does immigration offer poor people a chance to have a better life but it also allows them to send money to their families at home. For nations such as El Salvador, the Dominican Republic, and Egypt, worker remittances from abroad often exceed the value of those countries' annual manufacturing exports.

Finally, borders limit the possibilities for risk-sharing in the face of natural disasters. In the United States, the Federal Emergency Management Agency is funded by federal taxes; therefore, when disaster strikes a particular state or region, the rest of the nation can help mitigate the damage. Small countries have a smaller geographical space than large countries in which to share risks. When earthquakes destroyed Managua, Nicaragua, in 1972, and when a hurricane devastated Honduras in 1998, the national tax base was destroyed, making it impossible to marshal national resources to deal with the lost infrastructure. Countries that are small and vulnerable to hurricanes, floods, and earthquakes may become nonviable after a major disaster wipes out their productive capacity. Poor nations usually bear the brunt of such emergencies: Ninety-six percent of all deaths from natural disasters occur in developing countries.

The current conceptions of borders compound the problems attributable to geography. The world has been quite willing to create new nation-states under the banner of self-determination. But unless borders can be made less problematic for economic integration, they may condemn geographically distant countries to an independent oblivion.

GEO-GLOBALIZATION

If distance and geography did not matter for economic development, then we would witness much greater convergence of income levels and standards of living across regions and countries. Instead, we are witnessing divergence, because geography prevents poor nations from fully participating in the global division of labor. If current trends persist, countries that face high transportation costs and a high dependence on tropical agriculture will be left far behind, mired in poverty and income inequality. Will the rest of the world find this outcome morally acceptable? Will it find it efficient? Or will the fallout from these destitute regions be seen as endangering the quality of life for the rest of us? In a sense, we have already asked and answered these questions; the existence of myriad development institutions around the globe attests to the world's desire to meet the challenges of economic development. But all our answers have fallen short. The gap between rich and poor has only widened.

Many people blame economic globalization for poverty and injustice in the developing world. Yet it is the absence of globalization—or an insufficient dose of it—that is truly to blame for these inequities. The solution to geography's poverty trap is for developing countries to become more globalized. We need transnational arrangements to make borders less of an impediment to moving people, goods, and capital. We need agreements that can facilitate the development of international transportation infrastructure. And we need global mechanisms to harness the R&D capabilities of the world in health and agricultural technology. In short, we need more globalized governance.

Want to Know More?

Recent major works have emphasized the crucial role of geography in human history. In particular, see David S. Landes's *The Wealth and Poverty of Nations: Why Some Are So Rich and Some So Poor* (New York: W.W. Norton & Company, 1999); William H. McNeill's *Plagues and Peoples* (New York: Anchor Books, 1998); Jared Diamond's *Guns, Germs, and Steel: The Fates of Human Societies* (New York: W.W. Norton & Company, 1997); and Lawrence E. Harrison and Samuel Huntington, eds. *Culture Matters: How Values Shape Human Progress* (New York: Basic Books, 2000), particularly Jeffrey Sachs's chapter titled **"Notes on a New Sociology of Economic Development."**

John Luke Gallup, Jeffrey D. Sachs, and Andrew D. Mellinger explore the link between geographical factors and socioeconomic progress in **"Geography and Economic Development"** (Cambridge: National Bureau of Economic Research Working Paper W6849, December 1998). The Inter-American Development Bank's *Development Beyond Economics: Economic and Social Progress in Latin America 2000 Report* (Washington: Inter-American

Development Bank, 2000) also examines this relationship, particularly in Chapter 3, titled **"Geography and Development."** Raymond Arsenault describes the impact of the air conditioner on socioeconomic conditions in the southern United States in **"The Cooling of the South"** (*Wilson Quarterly*, Summer 1984).

For an assessment of the connection between geography and income distribution, see **"Nature, Development, and Distribution in Latin America: Evidence on the Role of Geography, Climate, and Natural Resources"** (Washington: Inter-American Development Bank Working Paper No. 378, August 1998) by Michael Gavin and Ricardo Hausmann. Stanley Engerman and Kenneth Sokoloff assess the impact of natural resources on institutional development in **"Factor Endowments, Institutions, and Differential Paths of Growth Among New World Economies: A View from Economic Historians of the United States"** in Stephen Haber, ed. *How Latin America Fell Behind: Essays on the Economic Histories of Brazil and Mexico, 1800–1914* (Stanford: Stanford University Press, 1997). Paul Krugman surveys the interplay between geography and contemporary economic thought in *Development, Geography, and Economic Theory* (Cambridge: MIT Press, 1995).

* For links to relevant Web sites, as well as a comprehensive index of related FOREIGN POLICY articles, access **www.foreignpolicy.com.**

Ricardo Hausmann is professor of the practice of economic development at the John F. Kennedy School of Government at Harvard University and former chief economist of the Inter-American Development Bank.

UNIT 3

Weapons of Mass Destruction

Unit Selections

9. **Nuclear Nightmares**, Bill Keller
10. **Return of the Nuclear Debate**, Leon Fuerth
11. **In North Korea and Pakistan, Deep Roots of Nuclear Barter**, David E. Sanger
12. **Towards an Internet Civil Defence Against Bioterrorism**, Ronald E. LaPorte et al.

Key Points to Consider

- What type of weapons or methods do you believe future terrorists are most likely to use in attempting to launch attacks in the United States?

- Explain how the current efforts to strengthen existing arms-control agreements or design new arms-control agreements are likely to help slow down the proliferation of weapons of mass destruction.

- What measures can nation-states or international organizations take to prevent future deals such as the past nuclear barter deal between North Korea and Pakistan from occurring?

- What additional measures would you propose to attempt to prevent terrorists or other nation-states from obtaining or developing weapons of mass destruction?

- Do you think the proposal for an Internet-based civil defense monitoring program would be more effective as a national or international program? Explain.

 **Links: www.dushkin.com/online/**
These sites are annotated in the World Wide Web pages.

The Bulletin of the Atomic Scientists
http://www.bullatomsci.org

Federation of American Scientists
http://www.fas.org

ISN International Relations and Security Network
http://www.isn.ethz.ch

The RMA Debate: Terrorism and Counter-terrorism
http://www.comw.org/rma/fulltext/terrorism.html

Terrorism Research Center
http://www.terrorism.com

The chilling prospect that the terrorists who attacked America on September 11, 2001, might have been carrying biological or chemical weapons remains a strong issue. The deaths of five Americans and the possible exposure of many others to anthrax spores from letters delivered through the U.S. Postal Service and the discovery of al Qaeda training manuals that contained information about chemical, biological, and nuclear (CBN) weapons dramatically underscore how far proliferation of weapons of mass destruction has progressed. In "Nuclear Nightmares" Bill Keller describes the reasons why most experts are in agreement that a future terrorist attack on the United States will involve the use of nuclear materials. When and how such an attack will occur remains a hotly debated issue. How to protect citizens from weapons of mass destruction—chemical, biological, and nuclear—and prevent additional proliferation is a central security concern for policymakers worldwide.

Scholars and practitioners, however, also continue to disagree about whether nation-states, terrorists, or lone deviants are most likely to use weapons of mass destruction against civilians in the future. Consequently, there is even less agreement now than in past years about how best to deter or counter weapons of mass destruction and whether the spread of CBN weapons encourages or acts as a disincentive to interstate war.

During the cold war, four major nuclear arms control regimes were developed: the Moscow-Washington negotiations that led to the Strategic Arms Limitation Treaty (SALT) and the Anti-Ballistic Missile (ABM) Treaty of 1972; the Nonproliferation Treaty (NPT); and the grandfather of them all, the Atmospheric Test Ban Treaty negotiated during the Eisenhower administration. Collectively, these agreements were core components of the "nuclear arms control regime." This regime suffered a major setback in December 2001 when President Bush announced that the United States planned to withdraw unilaterally from the 1972 Anti-Ballistic Missile (ABM) Treaty in order to pursue the development of an anti-ballistic missile system. Bush justified the U.S. policy change on the grounds that a new ABM system was required to deter future missile attacks by rogue states and possibly even by terrorist groups. The decision triggered intense criticism at home and abroad. Many worried that the world was on the eve of a new arms race.

Even before this decision, many observers were already concerned about the unilateral thrust of U.S. strategic policies after the United States failed to support multinational efforts to bolster the enforcement mechanism of the 1972 Biological Weapons Convention and the failure of the Clinton administration to win support from the U.S. Senate for the Comprehensive Test Ban Treaty (CTBT) that would have banned underground nuclear testing worldwide. The Bush administration decided not to support multinational efforts to bolster the enforcement mechanism of the 1972 Biological Weapons Convention. Continued vocal protests from America's friends and foes to these changes in U.S. policies were muted by the U.S. war on terrorism in 2001. However, as Leon Fuerth, a former senior official in the Clinton administration, predicts in his article, "Return of the Nuclear Debate," these protests are likely to return in the future. Fuerth explains how the comprehensive defense review, initiated by the Bush administration during its first year in office, is leading to a much broader revision of U.S. strategic vision than most critics of the ABM Treaty withdrawal decision realize. The Bush defense review reconsidered existing policies on nuclear modernization, arms control, ballistic missile defense, space dominance, and the "transformation" of conventional forces. Fuerth is critical of the administration's unilateral approach on the grounds that strategic stability cannot be imposed; he believes it must be set in place by mutual consent.

As faith in arms control has weakened, several nation-states have reconsidered their national strategies. Today at least 25 countries either have, or are in the process of developing, weapons of mass destruction. Two dozen are researching, developing, or stockpiling chemical weapons. It is difficult to see how the genie of nuclear weapons technology will be kept in the bottle. How real the threat of a regional war that engulfs the world may be was dramatically illustrated at the end of 2001 when two declared nuclear powers, India and Pakistan, massed troops on their shared borders, recalled diplomats, and issued thinly veiled threats about what might happen in the event of another armed conflict.

Some of the complexities involved in contemporary counter-proliferation efforts were illustrated by the policy compromises that the United States had to make in order to pursue the war on terrorism. During the military campaign in Afghanistan several senior officials in the Bush administration supported extending the military campaign to Iraq in order to topple Saddam Hussein's regime. Secretary of State Colin Powell, along with senior military advisers, counseled restraint in considering future military actions against a regime that is widely believed to have significant covert biochemical and possibly even nuclear capabilities. Plans for an invasion of Iraq continued throughout 2002 despite widespread opposition to a military invasion of Iraq by all the heads of the U.S. uniform services. Senior officials in the Bush administration, who continued to support an invasion of Iraq, claimed that they had proof that Iraq had continued to pursue weapons of mass destruction covertly since expelling UN weapons inspectors in 1998.

As pressure mounted on the U.S. administration to pursue nonmilitary options first, the United States secured a resolution from the UN Security Council that called on Iraq to readmit UN weapons inspectors. In the face of a united world, Saddam Hussein relented and permitted UN weapons inspectors to enter Iraq at the end of 2002. The U.S. government agreed to wait for the first report by this new team of inspectors regarding the status of alleged covert weapons of mass destruction before launching military actions.

During the closing months of 2002, the North Korean regime escalated an ongoing conflict by reopening the nuclear reactor at Yongbyon. This act was preceded by earlier U.S. public disclosures that North Korea was pursuing covert nuclear weapons research in violation of a 1994 agreement and the United States was terminating financial aid to North Korea intended to purchase alternative energy supplies. The decision to reopen the Yongbyon nuclear power plant set off alarm bells because the installation is capable of extracting plutonium from reactor waste that can be used in nuclear weapons. Next, the North Korean leader, Kim Jong Il, requested that UN weapons inspectors leave the country. North Korea then formally withdrew from the 1994 multinational agreement and unilaterally withdrew from the Nuclear Nonproliferation Treaty. To diffuse the situation, the Bush administration backed away from a longstanding declaration that it would not tolerate a North Korean nuclear arsenal, and announced, instead, that the United States would be willing to engage North Korea in informal talks. A few days later, the U.S. government said that bombing North Korea was no longer an immediate option, since the country had acquired nuclear weapons.

How North Korea acquired the fissile material necessary to build two bombs illustrates one of the core security dilemmas of the current era. In "In North Korea and Pakistan, Deep Roots of Nuclear Barter," David Sanger describes a perfect marriage of interest. Pakistan provided North Korea with many of the designs for gas centrifuges and machinery needed to make highly enriched uranium. In exchange, North Korea gave Pakistan the plans to build a missile capable of hitting India. While experts disagree about whether North Korea already has nuclear weapons, they tend to agree that within a few years, the North Korean regime will have a nuclear arsenal that is capable of putting U.S. citizens and troops at risk in South Korea, Japan, and much of Asia while Pakistan will soon be able to threaten to attack all parts of India with a missile-launched nuclear warhead.

Many security analysts continue to stress the potential threats created by the seepage of weapons, materials, and expertise from the former Soviet Union in recent years as one of the most serious proliferation issues in the world today. Additional proliferation concerns were raised at the end of 2001 about the effectiveness of controls in the United States over biochemical agents produced during former biochemical weapons programs. These concerns emerged after FBI–commissioned tests identified the strain used in the anthrax letter attacks as being a strain that was developed in the past by the U.S. military. Although the perpetrators of the anthrax letter attacks remain at large, one question is on the minds of many people today: How difficult would it be for terrorists to get hold of chemical, biological, and even nuclear materials?

While do-it-yourself mass destruction remains a challenge, this obstacle may be resolved in the future as terrorists engage in more "joint ventures" and as additional states become willing to sell their missile exports to terrorist groups. Another real concern is the possibility that terrorists will use radiation and biological or chemical materials as weapons designed to instill fear rather than to cause mass casualties. The first documented report of a completed act of nuclear terrorism involved Chechen rebels who obtained a canister of cesium in 1995. More recent reports that al Qaeda operatives were seeking to build and use radiation "dirty" bombs led the Bush administration to take steps to increase security around America's 103 nuclear power plants. Recent recognition that terrorists could attempt to use livestock and food crops as biological warfare weapons has also been fueling efforts to rethink how best to protect the nation's food chain. Intense efforts are also under way to increase U.S. security, especially around the nation's seaports and borders. Beefing up border security is one of the most important missions of the newly formed U.S. Department of Homeland Security. Other countries are moving to implement similar measures in an effort to protect citizens from future terrorist attacks.

Throughout the 1990s, many analysts wondered why terrorists had not used biological weapons. Earlier incidents involving terrorist attempts to use biological weapons suggested that the requirement of having sophisticated high-tech delivery methods was an obstacle to terrorists interested in causing mass casualties. However, the anthrax letter attacks that led to the deaths of 5 people and placed more than 3,000 Americans at risk for possible anthrax infection may have signaled that this barrier will no longer prevent bioterrorist attacks.

Even though no one really knows whether another biological attack is in the offing, most individuals agree that it is prudent to prepare for this terrifying eventuality. The Bush administration decision to administer smallpox vaccine to essential government workers is one recent effort to prepare for future contingencies. Ronald LaPorte and his associates warn that there is little evidence that the large resources currently being put into bioterrorism preparedness will work. Instead, citizens in countries throughout the world must face the disturbing fact that it is very difficult to predict and guard against a bioterrorist attack. Instead of building an inflexible Maginot Line of defense as the United States is now attempting to do, LaPorte and his associates advocate relying upon an ever-alert, flexible electronic-matrix of civil defense.

Nuclear Nightmares

Experts on terrorism and proliferation agree on one thing:
Sooner or later, an attack will happen here. When and how is what robs them of sleep.

By Bill Keller

The panic that would result from contaminating the Magic Kingdom with a modest amount of cesium would probably shut the place down for good and constitute a staggering strike at Americans' sense of innocence.

Not If But When Everybody who spends much time thinking about nuclear terrorism can give you a scenario, something diabolical and, theoretically, doable. Michael A. Levi, a researcher at the Federation of American Scientists, imagines a homemade nuclear explosive device detonated inside a truck passing through one of the tunnels into Manhattan. The blast would crater portions of the New York skyline, barbecue thousands of people instantly, condemn thousands more to a horrible death from radiation sickness and—by virtue of being underground—would vaporize many tons of concrete and dirt and river water into an enduring cloud of lethal fallout. Vladimir Shikalov, a Russian

nuclear physicist who helped clean up after the 1986 Chernobyl accident, envisioned for me an attack involving highly radioactive cesium-137 loaded into some kind of homemade spraying device, and a target that sounded particularly unsettling when proposed across a Moscow kitchen table—Disneyland. In this case, the human toll would be much less ghastly, but the panic that would result from contaminating the Magic Kingdom with a modest amount of cesium—Shikalov held up his teacup to illustrate how much—would probably shut the place down for good and constitute a staggering strike at Americans' sense of innocence. Shikalov, a nuclear enthusiast who thinks most people are ridiculously squeamish about radiation, added that personally he would still be happy to visit Disneyland after the terrorists struck, although he would pack his own food and drink and destroy his clothing afterward.

Another Russian, Dmitry Borisov, a former official of his country's atomic energy ministry, conjured a suicidal pilot. (Suicidal pilots, for obvious reasons, figure frequently in these fantasies.) In Borisov's scenario, the hijacker dive-bombs an Aeroflot jetliner into the Kurchatov Institute, an atomic research center

in a gentrifying neighborhood of Moscow, which I had just visited the day before our conversation. The facility contains 26 nuclear reactors of various sizes and a huge accumulation of radioactive material. The effect would probably be measured more in property values than in body bags, but some people say the same about Chernobyl. Maybe it is a way to tame a fearsome subject by Hollywoodizing it, or maybe it is a way to drive home the dreadful stakes in the arid-sounding business of nonproliferation, but in several weeks of talking to specialists here and in Russia about the threats an amateur evildoer might pose to the homeland, I found an unnerving abundance of such morbid creativity. I heard a physicist wonder whether a suicide bomber with a pacemaker would constitute an effective radiation weapon. (I'm a little ashamed to say I checked that one, and the answer is no, since pacemakers powered by plutonium have not been implanted for the past 20 years.) I have had people theorize about whether hijackers who took over a nuclear research laboratory could improvise an actual nuclear explosion on the spot. (Expert opinions differ, but it's very unlikely.) I've been instructed how to disperse plutonium into the ventilation system of an office building.

The realistic threats settle into two broad categories. The less likely but far more devastating is an actual nuclear explosion, a great hole blown in the heart of New York or Washington, followed by a toxic fog of radiation. This could be produced by a black-market nuclear warhead procured from an existing arsenal. Russia is the favorite hypothetical source, although Pakistan, which has a program built on shady middlemen and covert operations, should not be overlooked. Or the explosive could be a homemade device, lower in yield than a factory nuke but still creating great carnage.

The second category is a radiological attack, contaminating a public place with radioactive material by packing it with conventional explosives in a "dirty bomb" by dispersing it into the air or water or by sabotaging a nuclear facility. By comparison with the task of creating nuclear fission, some of these schemes would be almost childishly simple, although the consequences would be less horrifying: a panicky evacuation, a gradual increase in cancer rates, a staggeringly expensive cleanup, possibly the need to demolish whole neighborhoods. Al Qaeda has claimed to have access to dirty bombs, which is unverified but entirely plausible, given that the makings are easily gettable.

Nothing is really new about these perils. The means to inflict nuclear harm on America have been available to rogues for a long time. Serious studies of the threat of nuclear terror date back to the 1970's. American programs to keep Russian nuclear ingredients from falling into murderous hands—one of the subjects high on the agenda in President Bush's meetings in Moscow this weekend—were hatched soon after the Soviet Union disintegrated a decade ago. When terrorists get around to trying their first nuclear assault, as you can be sure they will, there will be plenty of people entitled to say I told you so.

All Sept. 11 did was turn a theoretical possibility into a felt danger. All it did was supply a credible cast of characters who hate us so much they would thrill to the prospect of actually doing it—and, most important in rethinking the probabilities, would be happy to die in the effort. All it did was give our nightmares legs.

Tom Ridge cupped his hands prayerfully and pressed his fingertips to his lips. "Nuclear," he said simply.

And of the many nightmares animated by the attacks, this is the one with pride of place in our experience and literature—and, we know from his own lips, in Osama bin Laden's aspirations. In February, Tom Ridge, the Bush administration's homeland security chief, visited The Times for a conversation, and at the end someone asked, given all the things he had to worry about—hijacked airliners, anthrax in the mail, smallpox, germs in crop-dusters—what did he worry about most? He cupped his hands prayerfully and pressed his fingertips to his lips. "Nuclear," he said simply.

My assignment here was to stare at that fear and inventory the possibilities. How afraid should we be, and what of, exactly? I'll tell you at the outset, this was not one of those exercises in which weighing the fears and assigning them probabilities laid them to rest. I'm not evacuating Manhattan, but neither am I sleeping quite as soundly. As I was writing this early one Saturday in April, the floor began to rumble and my desk lamp wobbled precariously. Although I grew up on the San Andreas Fault, the fact that New York was experiencing an earthquake was only my second thought.

The best reason for thinking it won't happen is that it hasn't hap-

pened yet, and that is terrible logic. The problem is not so much that we are not doing enough to prevent a terrorist from turning our atomic knowledge against us (although we are not). The problem is that there may be no such thing as "enough."

25,000 Warheads, and It Only Takes One My few actual encounters with the Russian nuclear arsenal are all associated with Thomas Cochran. Cochran, a physicist with a Tennessee lilt and a sense of showmanship, is the director of nuclear issues for the Natural Resources Defense Council, which promotes environmental protection and arms control. In 1989, when glasnost was in flower, Cochran persuaded the Soviet Union to open some of its most secret nuclear venues to a roadshow of American scientists and congressmen and invited along a couple of reporters. We visited a Soviet missile cruiser bobbing in the Black Sea and drank vodka with physicists and engineers in the secret city where the Soviets first produced plutonium for weapons.

Not long ago Cochran took me cruising through the Russian nuclear stockpile again, this time digitally. The days of glasnost theatrics are past, and this is now the only way an outsider can get close to the places where Russians store and deploy their nuclear weapons. On his office computer in Washington, Cochran has installed a detailed United States military map of Russia and superimposed upon it high-resolution satellite photographs. We spent part of a morning mouse-clicking from missile-launch site to submarine base, zooming in like voyeurs and contemplating the possibility that a terrorist could figure out how to steal a nuclear warhead from one of these places.

"Here are the bunkers," Cochran said, enlarging an area the size of a football stadium holding a half-dozen elongated igloos. We were hovering over a site called Zhukovka, in western Russia. We were pleased to see it did not look ripe for a hijacking.

"You see the bunkers are fenced, and then the whole thing is fenced again," Cochran said. "Just outside you can see barracks and a rifle range for the guards. These would be troops of the 12th Main Directorate. Somebody's not going to walk off the street and get a Russian weapon out of this particular storage area."

In the popular culture, nuclear terror begins with the theft of a nuclear weapon. Why build one when so many are lying around for the taking? And stealing tends to make better drama than engineering. Thus the stolen nuke has been a staple in the literature at least since 1961, when Ian Fleming published "Thunderball," in which the malevolent Spectre (the Special Executive for Counterintelligence, Terrorism, Revenge and Extortion, a strictly mercenary and more technologically sophisticated precursor to al Qaeda) pilfers a pair of atom bombs from a crashed NATO aircraft. In the movie version of Tom Clancy's thriller "The Sum of All Fears," due in theaters this week, neo-Nazis get their hands on a mislaid Israeli nuke, and viewers will get to see Baltimore blasted to oblivion.

Eight countries are known to have nuclear weapons—the United States, Russia, China, Great Britain, France, India, Pakistan and Israel. David Albright, a nuclear-weapons expert and president of the Institute for Science and International Security, points out that Pakistan's program in particular was built almost entirely through black markets and industrial espionage, aimed at circumventing Western export controls. Defeating the discipline of nuclear nonproliferation is ingrained in the culture. Disaffected individuals in Pakistan (which, remember, was intimate with the Taliban) would have no trouble finding the illicit channels or the rationalization for diverting materials, expertise—even, conceivably, a warhead.

But the mall of horrors is Russia, because it currently maintains something like 15,000 of the world's (very roughly) 25,000 nuclear warheads, ranging in destructive power from about 500 kilotons, which could kill a million people, down to the one-kiloton land mines that would be enough to make much of Manhattan uninhabitable. Russia is a country with sloppy accounting, a disgruntled military, an audacious black market and indigenous terrorists.

It's easier to take the fuel and build an entire weapon from scratch than it is to make one of these things go off.

There is anecdotal reason to worry. Gen. Igor Valynkin, commander of the 12th Main Directorate of the Russian Ministry of Defense, the Russian military sector in charge of all nuclear weapons outside the Navy, said recently that twice in the past year terrorist groups were caught casing Russian weapons-storage facilities. But it's hard to know how seriously to take this. When I made the rounds of nuclear experts in Russia earlier this year, many were skeptical of these near-miss anecdotes, saying the security forces tend to exaggerate such incidents to dramatize their own prowess (the culprits are always caught) and enhance their budgets. On the whole, Russian and American military experts sound not very alarmed about the vulnerability of Russia's nuclear warheads. They say Russia takes these weapons quite seriously, accounts for them rigorously and guards them carefully. There is no confirmed case of a warhead being lost. Strategic warheads, including the 4,000 or so that President Bush and President Vladimir Putin have agreed to retire from service, tend to be stored in hard-to-reach places, fenced and heavily guarded, and their whereabouts are not advertised. The people who guard them are better paid and more closely vetted than most Russian soldiers.

Eugene E. Habiger, the four-star general who was in charge of American strategic weapons until 1998 and then ran nuclear antiterror programs for the Energy Department, visited several Russian weapons facilities in 1996 and 1997. He may be the only American who has actually entered a Russian bunker and inspected a warhead *in situ*. Habiger said he found the overall level of security comparable to American sites, although the Russians depend more on people than on technology to protect their nukes.

The image of armed terrorist commandos storming a nuclear bunker is cinematic, but it's far more plausible to think of an inside job. No observer of the unraveling Russian military has much trouble imagining that a group of military officers, disenchanted by the humiliation of serving a spent superpower, embittered by the wretched conditions in which they spend much of their military lives or merely greedy, might find a way to divert a warhead to a terrorist for the right price. (The Chechen warlord Shamil Basayev, infamous for such ruthless exploits as taking an entire hospital hostage, once hinted that he had an opportunity to buy a nuclear warhead from the stockpile.) The anecdotal evidence of desperation in the military is plentiful and disquieting. Every year the Russian press provides stories like that of the 19-year-old sailor who went on a rampage aboard an Akula-class nuclear submarine, killing eight people and threatening to blow up the boat and its nuclear reactor; or the five soldiers at Russia's nuclear-weapons test site who killed a guard, took a hostage and tried to hijack an aircraft, or the officers who reportedly stole five assault helicopters, with their weapons pods, and tried to sell them to North Korea.

The Clinton administration found the danger of disgruntled nuclear caretakers worrisome enough that it considered building better housing for some officers in the nuclear rocket corps. Congress, noting that the United States does not build housing for its own officers, rejected the idea out of hand.

If a terrorist did get his hands on a nuclear warhead, he would still face the problem of setting it off. American warheads are rigged with multiple PAL's ("permissive action links")—codes and self-disabling devices designed to frustrate an unauthorized person from triggering the explosion. General Habiger says that when he examined Russian strategic weapons he found the level of protection comparable to our own. "You'd have to literally break the weapon apart to get into the gut," he told me. "I would submit that a more likely scenario is that there'd be an attempt to get hold of a warhead and not explode the warhead but extract the plutonium or highly enriched uranium." In other words, it's easier to take the fuel and build an entire weapon from scratch than it is to make one of these things go off.

Then again, Habiger is not an expert in physics or weapons design. Then again, the Russians would seem to have no obvious reason for misleading him about something that important. Then again, how many times have computer hackers hacked their way into encrypted computers we were assured were impregnable? Then again, how many computer hackers does al Qaeda have? This subject drives you in circles.

The most troublesome gap in the generally reassuring assessment of Russian weapons security is those tactical nuclear warheads—smaller, short-range weapons like torpedoes, depth charges, artillery shells, mines. Although their smaller size and greater number makes them ideal candidates for theft, they have gotten far less attention simply because, unlike all of our long-range weapons, they happen not to be the subject of any formal treaty. The first President Bush reached an informal understanding with President Gorbachev and then with President Yeltsin that both sides would gather and destroy thousands of tactical nukes. But the agreement included no inventories of the stockpiles, no outside monitoring, no verification of any kind. It was one of those trust-me deals that, in the hindsight of Sept. 11, amount to an enormous black hole in our security.

Did I say earlier there are about 15,000 Russian warheads? That number includes, alongside the scrupulously counted strategic warheads in bombers, missiles and submarines, the commonly used estimate of 8,000 tactical warheads. But that figure is at best an educated guess. Other educated guesses of the tactical nukes in Russia go as low as 4,000 and as high as 30,000. We just don't know. We don't even know if the Russians know, since they are famous for doing things off the books. "They'll tell you they've never lost a weapon," said Kenneth Luongo, director of a private antiproliferation group called the Russian-American Nuclear Security Advisory Council. "The fact is, they don't know. And when you're talking about warhead counting, you don't want to miss even one."

And where are they? Some are stored in reinforced concrete bunkers like the one at Zhukovka. Others are deployed. (When the submarine Kursk sank with its 118 crewmen in August 2000, the Americans' immediate fear was for its nuclear armaments. The standard load out for a submarine of that class includes a couple of nuclear torpedoes and possibly some nuclear depth charges.) Still others are supposed to be in the process of being dismantled under terms of various formal and informal arms-control agreements. Some are in transit. In short, we don't really know.

The other worrying thing about tactical nukes is that their anti-use devices are believed to be less sophisticated, because the weapons were designed to be employed in the battlefield. Some of the older systems are thought to have no permissive action links at all, so that setting one off would be about as complicated as hot-wiring a car.

Efforts to learn more about the state of tactical stockpiles have been frustrated by reluctance on both sides to let visitors in. Viktor Mikhailov, who ran the Russian Ministry of Atomic Energy until 1998 with a famous scorn for America's nonproliferation concerns, still insists that the United States programs to protect Russian nuclear weapons and material mask a secret agenda of intelligence-gathering. Americans, in turn, sometimes balk at reciprocal access, on the grounds that we are the ones paying the bills for all these safety upgrades, said the former Senator Sam Nunn, co-author of the main American program for securing Russian nukes, called Nunn-Lugar.

People in the field talk of a nuclear 'conex' bomb, using the name of those shack-size steel containers—2,000 of which enter America every hour on trains, trucks and ships. Fewer than 2 percent are cracked open for inspection.

"We have to decide if we want the Russians to be transparent—I'd call it cradle-to-grave transparency with nuclear material and inventories and so forth," Nunn told me. "Then we have to open up more ourselves. This is a big psychological breakthrough we're talking about here, both for them and for us."

The Garage Bomb One of the more interesting facts about the atom bomb dropped on Hiroshima is that it had never been tested. All of those spectral images of nuclear coronas brightening the desert of New Mexico— those were to perfect the more complicated plutonium device that was dropped on Nagasaki. "Little Boy," the Hiroshima bomb, was a rudimentary gunlike device that shot one projectile of highly enriched uranium into another, creating a critical mass that exploded. The mechanics were so simple that few doubted it would work, so the first experiment was in the sky over Japan.

The closest thing to a consensus I heard among those who study nuclear terror was this: building a nuclear bomb is easier than you think, probably easier than stealing one. In the rejuvenated effort to prevent a terrorist from striking a nuclear blow, this is where most of the attention and money are focused.

A nuclear explosion of any kind "is not a sort of high-probability thing," said a White House official who follows the subject closely. "But getting your hands on enough fissile material to build an improvised nuclear device, to my mind, is the least improbable of them all, and particularly if that material is highly enriched uranium in metallic form. Then I'm really worried. That's the one."

To build a nuclear explosive you need material capable of explosive nuclear fission, you need expertise, you need some equipment, and you need a way to deliver it.

Delivering it to the target is, by most reckoning, the simplest part. People in the field generally scoff at the mythologized suitcase bomb; instead they talk of a "conex bomb," using the name of those shack-size steel containers that bring most cargo into the United States. Two thousand containers enter America every hour, on trucks and trains and especially on ships sailing into more than 300 American ports. Fewer than 2 percent are cracked open for inspection, and the great majority never pass through an X-ray machine. Containers delivered to upriver ports like St. Louis or Chicago pass many miles of potential targets before they even reach customs.

"How do you protect against that?" mused Habiger, the former chief of our nuclear arsenal. "You can't. That's scary. That's very, very scary. You set one of those off in Philadelphia, in New York City, San Francisco, Los Angeles, and you're going to kill tens of thousands of

people, if not more." Habiger's view is "It's not a matter of *if*; it's a matter of *when*"—which may explain why he now lives in San Antonio.

The Homeland Security office has installed a plan to refocus inspections, making sure the 2 percent of containers that get inspected are those without a clear, verified itinerary. Detectors will be put into place at ports and other checkpoints. This is good, but it hardly represents an ironclad defense. The detection devices are a long way from being reliable. (Inconveniently, the most feared bomb component, uranium, is one of the hardest radioactive substances to detect because it does not emit a lot of radiation prior to fission.) The best way to stop nuclear terror, therefore, is to keep the weapons out of terrorist hands in the first place.

Fabricating a nuclear weapon is not something a lone madman—even a lone genius—is likely to pull off in his hobby room.

The basic know-how of atom-bomb-building is half a century old, and adequate recipes have cropped up in physics term papers and high school science projects. The simplest design entails taking a lump of highly enriched uranium, about the size of a cantaloupe, and firing it down a big gun barrel into a second lump. Theodore Taylor, the nuclear physicist who designed both the smallest and the largest American nuclear-fission warheads before becoming a remorseful opponent of all things nuclear, told me he recently looked up "atomic bomb" in the World Book Encyclopedia in the upstate New York nursing home where he now lives, and he found enough basic information to get a careful reader started. "It's accessible all over the place," he said. "I don't

mean just the basic principles. The sizes, specifications, things that work."

Most of the people who talk about the ease of assembling a nuclear weapon, of course, have never actually built one. The most authoritative assessment I found was a paper, "Can Terrorists Build Nuclear Weapons?" written in 1986 by five experienced nuke-makers from the Los Alamos weapons laboratory. I was relieved to learn that fabricating a nuclear weapon is not something a lone madman—even a lone genius—is likely to pull off in his hobby room. The paper explained that it would require a team with knowledge of "the physical, chemical and metallurgical properties of the various materials to be used, as well as characteristics affecting their fabrication; neutronic properties; radiation effects, both nuclear and biological; technology concerning high explosives and/or chemical propellants; some hydrodynamics; electrical circuitry; and others." Many of these skills are more difficult to acquire than, say, the ability to aim a jumbo jet.

The schemers would also need specialized equipment to form the uranium, which is usually in powdered form, into metal, to cast it and machine it to fit the device. That effort would entail months of preparation, increasing the risk of detection, and it would require elaborate safeguards to prevent a mishap that, as the paper dryly put it, would "bring the operation to a close."

Still, the experts concluded, the answer to the question posed in the title, while qualified, was "Yes, they can."

David Albright, who worked as a United Nations weapons inspector in Iraq, says Saddam Hussein's unsuccessful crash program to build a nuclear weapon in 1990 illustrates how a single bad decision can mean a huge setback. Iraq had extracted highly enriched uranium from research-reactor fuel and had, maybe, barely enough for a bomb. But the manager in charge of casting the metal was so afraid the stuff would spill or get contaminated that he decided to melt it in tiny batches. As a result, so much of the uranium was wasted that he ended up with too little for a bomb.

"You need good managers and organizational people to put the elements together," Albright said. "If you do a straight-line extrapolation, terrorists will all get nuclear weapons. But they make mistakes."

On the other hand, many experts underestimate the prospect of a do-it-yourself bomb because they are thinking too professionally. All of our experience with these weapons is that the people who make them (states, in other words) want them to be safe, reliable, predictable and efficient. Weapons for the American arsenal are designed to survive a trip around the globe in a missile, to be accident-proof, to produce a precisely specified blast.

But there are many corners you can cut if you are content with a big, ugly, inefficient device that would make a spectacular impression. If your bomb doesn't need to fit in a suitcase (and why should it?) or to endure the stress of a missile launch; if you don't care whether the explosive power realizes its full potential; if you're willing to accept some risk that the thing might go off at the wrong time or might not go off at all, then the job of building it is immeasurably simplified.

"As you get smarter, you realize you can get by with less," Albright said. "You can do it in facilities that look like barns, garages, with simple machine tools. You can do it with 10 to 15 people, not all Ph.D.'s, but some engineers, technicians. Our judgment is that a gun-type device is well within the capability of a terrorist organization."

All the technological challenges are greatly simplified if terrorists are in league with a country—a place with an infrastructure. A state is much better suited to hire expertise (like dispirited scientists from decommissioned nuclear installations in the old Soviet Union) or to send its own scientists for M.I.T. degrees.

Thus Tom Cochran said his greatest fear is what you might call a bespoke nuke—terrorists stealing a quantity of weapons-grade uranium and taking it to Iraq or Iran or Libya, letting the scientists and engineers there fashion it into an elementary weapon and then taking it away for a delivery that would have no return address.

That leaves one big obstacle to the terrorist nuke-maker: the fissile material itself.

To be reasonably sure of a nuclear explosion, allowing for some material being lost in the manufacturing process, you need roughly 50 kilograms—110 pounds—of highly enriched uranium. (For a weapon, more than 90 percent of the material should consist of the very unstable uranium-235 isotope.) Tom Cochran, the master of visual aids, has 15 pounds of depleted uranium that he keeps in a Coke can; an eight-pack would be plenty to build a bomb.

Only 41 percent of Russia's weapon-usable material has been secured... So the barn door is still pretty seriously ajar. We don't know whether any horses have gotten out.

The world is awash in the stuff. Frank von Hippel, a Princeton physicist and arms-control advocate, has calculated that between 1,300 and 2,100 metric tons of weapons-grade uranium exists—at the low end, enough for 26,000 rough-hewed bombs. The largest stockpile is in Russia, which Senator Joseph Biden calls "the candy store of candy stores."

Until a decade ago, Russian officials say, no one worried much about the safety of this material. Viktor Mikhailov, who ran the atomic energy ministry and now presides over an affiliated research institute, concedes there were glaring lapses.

"The safety of nuclear materials was always on our minds, but the focus was on intruders," he said. "The system had never taken account of the possibility that these carefully screened people in the nuclear sphere could themselves represent a danger. The system was not designed to prevent a danger from within."

Then came the collapse of the Soviet Union and, in the early 90's, a few frightening cases of nuclear materials popping up on the black market.

If you add up all the reported attempts to sell highly enriched uranium or plutonium, even including those that have the scent of security-agency hype and those where the material was of uncertain quality, the total amount of material still falls short of what a bomb-maker would need to construct a single explosive.

But Yuri G. Volodin, the chief of safeguards at Gosatomnadzor, the Russian nuclear regulatory agency, told me his inspectors still discover one or two instances of attempted theft a year, along with dozens of violations of the regulations for storing and securing nuclear material. And as he readily concedes: "These are the detected cases. We can't talk about the cases we don't know." Alexander Pikayev, a former aide to the Defense Committee of the Russian Duma, said: "The vast majority of installations now have fences. But you know Russians. If you walk along the perimeter, you can see a hole in the fence, because the employees want to come and go freely."

The bulk of American investment in nuclear safety goes to lock the stuff up at the source. That is clearly the right priority. Other programs are devoted to blending down the highly enriched uranium to a diluted product unsuitable for weapons but good as reactor fuel. The Nuclear Threat Initiative, financed by Ted Turner and led by Nunn, is studying ways to double the rate of this diluting process.

Still, after 10 years of American subsidies, only 41 percent of Russia's weapon-usable material has been secured, according to the United States Department of Energy. Russian officials said they can't even be sure how much exists, in part because the managers of nuclear facilities, like everyone else in the Soviet industrial complex, learned to cook their books. So the barn door is still pretty seriously ajar. We don't know whether any horses have gotten out.

And it is not the only barn. William C. Potter, director of the Center for Nonproliferation Studies at the Monterey Institute of International Studies and an expert in nuclear security in the former Soviet states, said the American focus on Russia has neglected other locations that could be tempting targets for a terrorist seeking bomb-making material. There is, for example, a bomb's worth of weapons-grade uranium at a site in Belarus, a country with an erratic president and an anti-American orientation. There is enough weapons-grade uranium for a bomb or two in Kharkiv, in Ukraine. Outside of Belgrade, in a research reactor at Vinca, sits sufficient material for a bomb—and there it sat while NATO was bombarding the area.

"We need to avoid the notion that because the most material is in Russia, that's where we should direct all of our effort," Potter said. "It's like assuming the bank robber will target Fort Knox because that's where the most gold is. The bank robber goes where the gold is most accessible."

Weapons of Mass Disruption The first and, so far, only consummated act of nuclear terrorism took place in Moscow in 1995, and it was scarcely memorable. Chechen rebels obtained a canister of cesium, possibly from a hospital they had commandeered a few months before. They hid it in a Moscow park famed for its weekend flea market and called the press. No one was hurt. Authorities

treated the incident discreetly, and a surge of panic quickly passed.

The story came up in virtually every conversation I had in Russia about nuclear terror, usually to illustrate that even without splitting atoms and making mushroom clouds a terrorist could use radioactivity—and the fear of it—as a potent weapon.

The idea that you could make a fantastic weapon out of radioactive material without actually producing a nuclear bang has been around since the infancy of nuclear weaponry. During World War II, American scientists in the Manhattan Project worried that the Germans would rain radioactive material on our troops storming the beaches on D-Day. Robert S. Norris, the biographer of the Manhattan Project director, Gen. Leslie R. Groves, told me that the United States took this threat seriously enough to outfit some of the D-Day soldiers with Geiger counters.

No country today includes radiological weapons in its armories. But radiation's limitations as a military tool—its tendency to drift afield with unplanned consequences, its long-term rather than short-term lethality—would not necessarily count against it in the mind of a terrorist. If your aim is to instill fear, radiation is anthrax-plus. And unlike the fabrication of a nuclear explosive, this is terror within the means of a soloist.

If your aim is to instill fear, radiation is anthrax-plus. And unlike the fabrication of a nuclear explosive, this is terror within the means of a soloist.

That is why, if you polled the universe of people paid to worry about weapons of mass destruction (W.M.D., in the jargon), you would find a general agreement that this is

probably the first thing we'll see. "If there is a W.M.D. attack in the next year, it's likely to be a radiological attack," said Rose Gottemoeller, who handled Russian nuclear safety in the Clinton administration and now follows the subject for the Carnegie Endowment. The radioactive heart of a dirty bomb could be spent fuel from a nuclear reactor or isotopes separated out in the process of refining nuclear fuel. These materials are many times more abundant and much, much less protected than the high-grade stuff suitable for bombs. Since Sept. 11, Russian officials have begun lobbying hard to expand the program of American aid to include protection of these lower-grade materials, and the Bush administration has earmarked a few million dollars to study the problem. But the fact is that radioactive material suitable for terrorist attacks is so widely available that there is little hope of controlling it all.

The guts of a dirty bomb could be cobalt-60, which is readily available in hospitals for use in radiation therapy and in food processing to kill the bacteria in fruits and vegetables. It could be cesium-137, commonly used in medical gauges and radiotherapy machines. It could be americium, an isotope that behaves a lot like plutonium and is used in smoke detectors and in oil prospecting. It could be plutonium, which exists in many research laboratories in America. If you trust the security of those American labs, pause and reflect that the investigation into the great anthrax scare seems to be focused on disaffected American scientists.

Back in 1974, Theodore Taylor and Mason Willrich, in a book on the dangers of nuclear theft, examined things a terrorist might do if he got his hands on 100 grams of plutonium—a thimble-size amount. They calculated that a killer who dissolved it, made an aerosol and introduced it into the ventilation system of an office building could deliver a lethal dose to the entire floor area of a large skyscraper. But plutonium dispersed outdoors in the open air, they estimated, would be far less effective. It would blow away in a gentle wind.

The Federation of American Scientists recently mapped out for a Congressional hearing the consequences of various homemade dirty bombs detonated in New York or Washington. For example, a bomb made with a single footlong pencil of cobalt from a food irradiation plant and just 10 pounds of TNT and detonated at Union Square in a light wind would send a plume of radiation drifting across three states. Much of Manhattan would be as contaminated as the permanently closed area around the Chernobyl nuclear plant. Anyone living in Manhattan would have at least a 1-in-100 chance of dying from cancer caused by the radiation. An area reaching deep into the Hudson Valley would, under current Environmental Protection Agency standards, have to be decontaminated or destroyed.

Frank von Hippel, the Princeton physicist, has reviewed the data, and he pointed out that this is a bit less alarming than it sounds. "Your probability of dying of cancer in your lifetime is already about 20 percent," he said. "This would increase it to 20.1 percent. Would you abandon a city for that? I doubt it."

Indeed, some large portion of our fear of radiation is irrational. And yet the fact that it's all in your mind is little consolation if it's also in the minds of a large, panicky population. If the actual effect of a radiation bomb is that people clog the bridges out of town, swarm the hospitals and refuse to return to live and work in a contaminated place, then the impact is a good deal more than psychological. To this day, there is bitter debate about the actual health toll from the Chernobyl nuclear accident. There are researchers who claim that the people who evacuated are actually in worse health over all from the trauma of relocation, than those who stayed put and marinated in the residual radiation. But the fact is, large swaths of developed land around the Chernobyl site still lie abandoned,

much of it bulldozed down to the subsoil. The Hart Senate Office Building was closed for three months by what was, in hindsight, our society's inclination to err on the side of alarm.

There are measures the government can take to diminish the dangers of a radiological weapon, and many of them are getting more serious consideration. The Bush administration has taken a lively new interest in radiation-detection devices that might catch dirty-bomb materials in transit. A White House official told me the administration's judgment is that protecting the raw materials of radiological terror is worth doing, but not at the expense of more catastrophic threats.

"It's all over," he said. "It's not a winning proposition to say you can just lock all that up. And then, a bomb is pretty darn easy to make. You don't have to be a rocket scientist to figure about fertilizer and diesel fuel." A big fertilizer bomb of the type Timothy McVeigh used to kill 168 people in Oklahoma City, spiced with a dose of cobalt or cesium, would not tax the skills of a determined terrorist.

"It's likely to happen, I think, in our lifetime," the official said. "And it'll be like Oklahoma City plus the Hart Office Building. Which is real bad, but it ain't the World Trade Center."

The Peril of Power Plants Every eight years or so the security guards at each of the country's 103 nuclear power stations and at national weapons labs can expect to be attacked by federal agents armed with laser-tag rifles. These mock terror exercises are played according to elaborate rules, called the "design basis threat," that in the view of skeptics favor the defense. The attack teams can include no more than three commandos. The largest vehicle they are permitted is an S.U.V. They are allowed to have an accomplice inside the plant, but only one. They are not allowed to improvise. (The mock assailants at one Department of Energy

lab were ruled out of order because they commandeered a wheelbarrow to cart off a load of dummy plutonium.) The mock attacks are actually announced in advance. Even playing by these rules, the attackers manage with some regularity to penetrate to the heart of a nuclear plant and damage the core. Representative Edward J. Markey, a Massachusetts Democrat and something of a scourge of the nuclear power industry, has recently identified a number of shortcomings in the safeguards, including, apparently, lax standards for clearing workers hired at power plants.

One of the most glaring lapses, which nuclear regulators concede and have promised to fix, is that the design basis threat does not contemplate the possibility of a hijacker commandeering an airplane and diving it into a reactor. In fact, the protections currently in place don't consider the possibility that the terrorist might be willing, even eager, to die in the act. The government assumes the culprits would be caught while trying to get away.

A nuclear power plant is essentially a great inferno of decaying radioactive material, kept under control by coolant. Turning this device into a terrorist weapon would require cutting off the coolant so the atomic furnace rages out of control and, equally important, getting the radioactive matter to disperse by an explosion or fire. (At Three Mile Island, the coolant was cut off and the reactor core melted down, generating vast quantities of radiation. But the thick walls of the containment building kept the contaminant from being released, so no one died.)

One way to accomplish both goals might be to fly a large jetliner into the fortified building that holds the reactor. Some experts say a jet engine would stand a good chance of bursting the containment vessel, and the sheer force of the crash might disable the cooling system—rupturing the pipes and cutting off electricity that pumps the water through the core. Before nearby residents had begun to evacuate, you could have a meltdown that would spew a volcano of radioactive isotopes into the air, causing fatal radiation sickness for those exposed to high doses and raising lifetime cancer rates for miles around.

This sort of attack is not as easy, by a long shot, as hitting the World Trade Center. The reactor is a small, low-lying target, often nestled near the conspicuous cooling towers, which could be destroyed without great harm. The reactor is encased in reinforced concrete several feet thick, probably enough, the industry contends, to withstand a crash. The pilot would have to be quite a marksman, and somewhat lucky. A high wind would disperse the fumes before they did great damage.

Invading a plant to produce a meltdown, even given the record of those mock attacks, would be more complicated, because law enforcement from many miles around would be on the place quickly, and because breaching the containment vessel is harder from within. Either invaders or a kamikaze attacker could instead target the more poorly protected cooling ponds, where used plutonium sits, encased in great rods of zirconium alloy. This kind of sabotage would take longer to generate radiation and would be far less lethal.

Discussion of this kind of potential radiological terrorism is colored by passionate disagreements over nuclear power itself. Thus the nuclear industry and its rather tame regulators sometimes sound dismissive about the vulnerability of the plants (although less so since Sept.11), while those who regard nuclear power as inherently evil tend to overstate the risks. It is hard to sort fact from fear-mongering.

Nuclear regulators and the industry grumpily concede that Sept. 11 requires a new estimate of their defenses, and under prodding from Congress they are redrafting the so-called design basis threat, the one plants are required to defend against. A few members of Congress have proposed installing ground-to-air missiles at nuclear plants, which most experts think is a recipe for a disastrous mishap.

"Probably the only way to protect against someone flying an aircraft into a nuclear power plant," said Steve Fetter of the University of Maryland, "is to keep hijackers out of cockpits."

Being Afraid For those who were absorbed by the subject of nuclear terror before it became fashionable, the months since the terror attacks have been, paradoxically, a time of vindication. President Bush, whose first budget cut $100 million from the programs to protect Russian weapons and material (never a popular program among conservative Republicans), has become a convert. The administration has made nuclear terror a priority, and it is getting plenty of goading to keep it one. You can argue with their priorities and their budgets, but it's hard to accuse anyone of indifference. And resistance—from scientists who don't want security measures to impede their access to nuclear research materials, from generals and counterintelligence officials uneasy about having their bunkers inspected, from nuclear regulators who worry about the cost of nuclear power, from conservatives who don't want to subsidize the Russians to do much of anything—has become harder to sustain. Intelligence gathering on nuclear material has been abysmal, but it is now being upgraded; it is a hot topic at meetings between American and foreign intelligence services, and we can expect more numerous and more sophisticated sting operations aimed at disrupting the black market for nuclear materials. Putin, too, has taken notice. Just before leaving to meet Bush in Crawford, Tex., in November, he summoned the head of the atomic energy ministry to the Kremlin on a Saturday to discuss nuclear security. The subject is now on the regular agenda when Bush and Putin talk.

These efforts can reduce the danger but they cannot neutralize the fear, particularly after we have been so vividly reminded of the hostility some of the world feels for us, and of our vulnerability.

Fear is personal. My own—in part, because it's the one I grew up with, the one that made me shiver through the Cuban missile crisis and "On the Beach"—is the horrible magic of nuclear fission. A dirty bomb or an assault on a nuclear power station, ghastly as that would be, feels to me within the range of what we have survived. As the White House official I spoke with said, it's basically Oklahoma City plus the Hart Office Building. A nuclear explosion is in a different realm of fears and would test the country in ways we can scarcely imagine.

A mushroom cloud of irradiated debris would blossom more than two miles into the air. Then highly lethal fallout would begin drifting back to earth, riding the winds into the Bronx or Queens or New Jersey.

As I neared the end of this assignment, I asked Matthew McKinzie, a staff scientist at the Natural Resources Defense Council, to run a computer model of a one-kiloton nuclear explosion in Times Square, half a block from my office, on a nice spring workday. By the standards of serious nuclear weaponry, one kiloton is a junk bomb, hardly worthy of respect, a fifteenth the power of the bomb over Hiroshima.

A couple of days later he e-mailed me the results, which I combined with estimates of office workers and tourist traffic in the area. The blast and searing heat would gut buildings for a block in every direction, incinerating pedestrians and crushing people at their desks. Let's say 20,000 dead in a matter of seconds. Beyond this, to a distance of more than a quarter mile, anyone directly exposed to the fireball would die a gruesome death from radiation sickness within a day—anyone, that is, who survived the third-degree burns. This larger circle would be populated by about a quarter million people on a workday. Half a mile from the explosion, up at Rockefeller Center and down at Macy's, unshielded onlookers would expect a slower death from radiation. A mushroom cloud of irradiated debris would blossom more than two miles into the air, and then, 40 minutes later, highly lethal fallout would begin drifting back to earth, showering injured survivors and dooming rescue workers. The poison would ride for 5 or 10 miles on the prevailing winds, deep into the Bronx or Queens or New Jersey.

A terrorist who pulls off even such a small-bore nuclear explosion will take us to a whole different territory of dread from Sept. 11. It is the event that preoccupies those who think about this for a living, a category I seem to have joined.

"I think they're going to try," said the physicist David Albright. "I'm an optimist at heart. I think we can catch them in time. If one goes off, I think we will survive. But we won't be the same. It will affect us in a fundamental way. And not for the better."

Bill Keller is a Times columnist and a senior writer for the magazine.

From the *New York Times* Magazine, May 26, 2002, pp. 22, 24-29, 51, 54-55, 57. © 2002 by Bill Keller. Distributed by The New York Times Special Features. Reprinted by permission.

Return of the Nuclear Debate

Leon Fuerth

In time, President George W. Bush's administration may release much more detail concerning its intentions for nuclear modernization, arms control, ballistic missile defense, space dominance, and the "transformation" of conventional forces.

These initiatives, though presented separately, actually relate to each other and can be best understood with reference to an organized concept of the United States' security needs moving forward into the twenty-first century. But the administration itself has announced no such overall concept. And yet, even if the administration really sees no connections between its choices about nuclear weapons and arms control, defenses and offense, and nuclear and conventional forces—even if the administration actually insists there are no connections—they do interact. It is a good thing for people to be conscious of how these interactions might work out.

The Administration's Case

The following, then, is an interpretation of what the administration appears to have in mind.

- Arms control was at best a highly imperfect effort to regulate the military rivalry between the United States and the Soviet Union. In many ways, such agreements were little more than contracts allowing both sides to pursue next steps in the arms race that each had elected as desirable, arms control or not. At worst, arms control agreements may actually have channeled competition into even more dangerous technologies. Moreover, the price for these agreements was high in terms of unending squabbles about fine points, compliance, and verification—along with dismantling and inspection procedures that literally cost billions of dollars to operate.
- But by far the highest cost of arms control was that it locked the United States into the doctrine of mutually assured destruction (MAD). The objective of limiting offensive nuclear weapons was to find a way to make each side feel certain that, under even the most difficult conditions, it would always be able to destroy an aggressor. The purpose of the Anti-Ballistic Missile (ABM) Treaty was to reinforce this "security" by

making it impossible for either side to develop a defense of its national territory. However, the ABM Treaty also perpetuated MAD, which was fine so long as MAD was the best arrangement both sides could envision to deter the other.

- Things are now radically different. The Cold War is over; Russia is too weak to be a threat; and, besides, relations with it are basically good. Therefore, the levels and kinds of nuclear weapons available to the United States and to Russia are no longer reasons for mutual fear. In view of this, the United States and Russia should stop trying to regulate their strategic nuclear weapons by way of explicit, formal agreements and instead rely on "understandings" informally worked out and not legally binding. Both sides would be free to shape their nuclear forces as they please; political and economic realism will assure that this translates into spontaneous deep reductions all around.
- On the other hand, the United States faces an emerging new threat from weapons of mass destruction (WMD) in the hands of countries such as North Korea or Iran. To deal with this threat, the United States needs to develop and deploy ABM weapons. The Clinton administration was working on a limited defense—the so-called national missile defense (NMD)—but its heart was never really in the program, and the design was anemic. The United States should shorten the timetable to initial deployment and aim to make the system much more powerful from the outset. The ABM Treaty prevents the United States from doing this and is therefore an obstacle to self-defense. The treaty should be abandoned if Russia will not agree to liberally rewriting it.
- Russia was ready to fight against even the last administration's very limited NMD on the grounds that it constituted a first step toward a system ultimately big enough to threaten the credibility of their nuclear deterrent. In ratifying the second Strategic Arms Reduction Talks (START II) agreement, the Russian Duma attached a condition according to which the provisions of the treaty would not be executed if the United States did not commit itself to the continuance of the ABM Treaty. The United States should not be deterred by Russian opposition. Perhaps, in time, the U.S. government can persuade Russia to accept its word that U.S. defenses are not in-

tended to be a threat. But if not, the United States should be prepared to pull out of the ABM Treaty. Doing so will produce no untoward consequences for the security of the United States. Russia cannot afford very impressive countermeasures. Russia has threatened that, if the United States pulls out of the ABM Treaty, Russia will pull out of all existing arms control agreements, but the United States can discount that as bluster. But if this warning turns out to be real, the United States is better off without such agreements in any event and, therefore, can dispense with them.

Highly questionable assumptions hold the administration's chain of reasoning together.

- The United States' allies are extremely nervous about all this and are very reluctant to give the United States their cooperation. But the United States can eventually persuade them to give up their reservations and to come along with it. Washington will have to promise them protection from ballistic missiles under a U.S. or a joint defensive shield. Honoring that commitment will require that the United States explore technologies and systems that cannot in any way be reconciled with the basic logic of the ABM Treaty, but the allies will value that promise more than they value the existing system of arms control. Besides, in the last analysis, what choice do they have?

- The United States does not much care what China thinks about this or what the Chinese government's responses might be. China has a relatively small number of intercontinental ballistic missiles (ICBMs), and the United States expects them to modernize that force with or without a U.S. decision to build missile defenses. Building defenses is unlikely, therefore, to result in any increase of the Chinese nuclear threat in excess of that which we already anticipate.

- On the other hand, building ballistic missile defenses would better prepare the United States for a confrontation with China over Taiwan. The United States should not only accept that a confrontation with China is in the cards, but should act on that assumption by profoundly reshaping—"transforming"—U.S. conventional military forces.

- By its nature, the military systems needed for a full-scale transformation of U.S. conventional forces will intensify U.S. dependence on space-based intelligence and battle management systems. Space must therefore be viewed as a zone of military operations where the United States must attain dominance. The United States may even need to develop space-based offensive weapons for use against ground targets.

- In particular, the United States should attempt to develop technologies for space-based defense against ballistic missiles and, if successful, should deploy them. The fact that space-based defenses really could threaten Russia's nuclear retaliatory forces ought not to prevent U.S. action. Russia will trust the United States. Besides, given the tremendous prob-

lems besetting the Russian government, what is its recourse? Once the Russian government acquiesces to earlier phases of development and deployment of U.S. defenses, the precedents will be set anyway.

Critique: The Yardstick of Strategic Stability

It is fairly easy to pick out some of the highly questionable assumptions that hold this chain of reasoning together. First, there is a readiness to gamble hugely on unproven technologies and even on technologies not yet invented. Second, there is eagerness, even zeal, to be rid of the ABM Treaty long before the workability of an elaborate defense system can be estimated, much less demonstrated. Third, there is a complacent view of U.S. trustworthiness, to the point where one assumes that all the rest of the world, friends and enemies alike, have merely to hear what U.S. intentions are, and they will be reconciled. Fourth, there is a facile dismissal of the ability of other countries to take actions against the United States that this country would find truly damaging. Fifth, there is a naïve model of history that sees the United States and Russia as having attained durable friendship, and a very dangerous corollary that fatalistically accepts China as the next great threat. Sixth, there is radical impatience with the difficulties of negotiating agreements. And seventh, there is an unproven and utterly reckless belief that the United States does not need to regard nuclear weapons, ballistic missile defenses, and space warfare as mutually interactive. The administration may believe that a framework encompassing strategic and defensive weapons must be accepted to mollify Russia, but it is very likely to view that framework as a means to relax rather than to clarify linkages between offense and defense.

Strategic stability cannot be imposed; it must be set in place by mutual consent.

It is this last point that may be the most serious deficiency of all because it denies the relevance of any effort to gauge the net impact of a complex set of proposed changes against some standard value of what is good for the United States. There is such a standard. It is called strategic stability. One ought not to pretend that strategic stability is a scientific unit of measure. It does involve modeling and calculations at some level, but then moves on to a heavy mixture of history, political science, and even intuition. Despite these imprecisions, however, strategic stability offers a clear statement of what the goals of the United States should be: that its nuclear weapons and its doctrines about how and when to use them should decrease, rather than increase, the likelihood of being forced into an otherwise undesired conflict. A fundamental corollary to this definition of strategic stability is that it does not exist for any one party unless it exists for all. Hence, it cannot be imposed; it must be set in place by mutual consent.

The concept of strategic stability evolved over a lengthy period of time. Until the advent of modern arms control in the late 1960s, there was no means in place to constrain rivalry between the United States and the Soviet Union over nuclear weapons and delivery systems. Each side imputed the worst intentions to the other and reacted to every new weapons development—rumored or real—as a further proof of hostility. Consequently, each side sought to offset any perceived advantage accruing to the other. It was soon clear that, because neither side could ever allow the other to enjoy what it would regard as marked overall supremacy, the arms race might continue indefinitely in a kind of rolling stalemate.

The first Strategic Arms Limitation Talks (SALT I) treaty, signed in 1972, attempted to end this cycle, but mainly rechanneled it. Limits on the number of launchers for ICBMs perversely encouraged both sides to speed the deployment of multiple independently targetable reentry vehicles (MIRVs) on those launchers remaining. Each side watched the other deploying more and more warheads on fewer missiles. The net result was to increase a sense of vulnerability on each side. As the number of MIRVs deployed per launcher increased, and as improvements in accuracy were made, each side feared the other was acquiring the means to carry out a first strike against the other's ICBMs.

Strategic arithmetic is the same as always; and so is human nature.

Prospective defenses against ballistic missiles added to this growing sense of insecurity, because if either side were able to effectively protect its national territory, it might be even more tempted to think of launching a first strike. To be more concrete, the idea was that one side might think that it would be able to destroy a substantial portion of the other's retaliatory force in a first strike, and then use its ABM shield to deal with the response. The ABM Treaty was negotiated to settle this fear, by denying each party the means to build the necessary shield, but it also locked both sides into a condition of permanent mutual vulnerability.

The START II agreement, signed in 1993, was a deliberate effort to break this cycle by shaping the offensive forces of both sides in such a way as to negate even the hypothetical advantages of a first strike. Both sides agreed to begin de-MIRVing their ballistic missiles, while reducing them in number. The result would be a smaller overall force but, perhaps even more importantly, a force that both sides believed would be intrinsically more stable. That is to say, there would be a balance between the nuclear forces of both sides such that neither would have reason to believe that the other could objectively benefit from launching a first strike. Defenses were to remain bounded by the ABM Treaty, which now acquired a new value: as a means to facilitate deeper reductions by reducing the fear that radically improved and expanded defenses might create a first-strike capability at a much lower overall offensive weapons level.

START II responded to three fundamental axioms about strategic stability: first, that reductions alone may not necessarily produce a better, more crisis-resistant situation; second, that the relationship between offensive forces needs to be consciously designed to produce this effect; and third, that reductions may continue to progressively lower levels, providing that (1) the resultant balance is always arranged with a view toward stability and that (2) defenses are not allowed to become a wild card.

If all this sounds arcane, it was. The possibility that either Washington or Moscow would actually think it wise to launch a first strike may have been very small. But the fear that each side felt about this was real because it was based on the actual capabilities of weapons and on a reading of human psychology. START II recognized that this fear had to be addressed through a mutually acceptable, ergo negotiated, arrangement.

Obviously, the world has changed tremendously since the days when START II was negotiated, but strategic arithmetic is the same as always. And so is human nature. Strategic arithmetic tells us that deep reductions can lead to results that damage strategic stability, if they are not worked out on a mutual basis. Unilateral reductions are a role of the dice, by comparison. Chances that things will work out for the worse are at least 50/50. And if they do work out for the worse, basic human instincts will take over: the fear of the other's intentions, based on the fear of the other's capabilities, will again acquire the power to drive events. If, at the same time, the ABM Treaty falls away so that the defensive side of the stability equation becomes completely unpredictable, paranoia about first strikes will be reestablished.

The administration argues that, because the Cold War is over, the need for arms control has disappeared. In doing so, the administration uses profound changes in the real world to rationalize old policies dressed up as new. The administration asserts complete freedom of action for the United States, including the right to dispense with the Comprehensive Test Ban Treaty (CTBT), the ABM Treaty, and any other arms control agreements that might perish along with them.

In the real world, however, attention must be paid to the interests and considerations of others if the United States is not to end up isolated and unable to marshal support for issues that matter greatly to U.S. principles and even U.S. security. Arms control remains necessary precisely in order to build durable structures for order and stability in the twenty-first century. The United States needs it to regulate and balance out strategic nuclear forces; it needs it to open a path toward progressively deeper reductions in those forces; it needs it to have the transparency and legal structure needed to dispel uncertainty and stop paranoia from putting down new roots. "Good fences make good neighbors" remains a simple truth.

The administration's course of action can literally reignite the arms race.

Consequences: A Renewed Arms Race

Following Bush's meeting with Russian president Vladimir Putin at the G-8 meeting in Genoa, the administration is now able to test its theories with Russia. The stakes are extremely high. In the worst case, the administration's course of action can literally reignite the arms race. One should think carefully about what that would mean.

Even though the Russian government is strapped for cash, the fact is that ballistic missiles and nuclear warheads are relatively cheap (land-based systems, that is). Russia has throw-weight to spare on their existing ICBM force and could load up to maximum, something that it did not do even in the Cold War. The Russian nuclear weapons establishment still has the capacity to manufacture at arms-race levels. The U.S. weapons establishment, by the way, is drastically smaller than in the past and is in no position to compete, nor would it be for at least a decade.

So Russia could rapidly build up its inventory of online strategic warheads without fielding new missiles. That would be the obvious, brute-force response to a U.S. move that potentially threatened the credibility of Russia's nuclear deterrent in its eyes. Credibility is at least as much a matter of subjective impression as it is of force-planners' calculations. It does not matter what the United States thinks of Russia's deterrent, or vice versa; it only matters what each country thinks of itself—and in existential calculations of this sort, too much is never enough. The prompt Russian reaction could thus easily be to build up their forces.

Such a buildup would also obliterate what progress the two countries have made toward strategic stability because increasing the numbers of warheads loaded on ICBMs would progressively add to the value of the Russian land-based launchers as targets for the United States, or so Russia would reason. Now, any American may think it would be an act of insanity for the United States to contemplate a first strike, but the whole point is that our opinion does not matter: it is the opinion of the Russian side that is dispositive.

There would also be immense opportunity costs. In the atmosphere of a renewed arms race, it is unlikely that the United States would be able to make progress toward what should be its follow-on goals at this moment: START III, with lower final levels than those agreed at Helsinki; reduction of the Russian nuclear arms complex to post–Cold War size; transparency arrangements for nuclear weapons in storage; accountability for inventories of bomb-grade fissionable materials; continuity of the joint program for U.S. purchases of Russian bomb-grade uranium blended down into fuel-grade for U.S. power reactors; execution of agreements just completed for the safe disposition of plutonium; and negotiation of a treaty for the cessation of further production of bomb-grade fissionable material. In theory, one supposes, programs premised on remarkable intimate cooperation might continue even in the midst of a new arms race, but that flies in the face of any sensible reading of political and psychological reality.

To this list of costs, one should also add the possibility that Russia will act on its threat to abrogate all other arms control agreements, including perhaps the Intermediate-Range Nuclear Force (INF) Treaty, which eliminated SS-20s, ground-launched cruise missiles, and Pershing IIs aimed at the heart of Europe. Never mind, of course, that the United States will promise to protect Europe under a U.S. ballistic missile shield. Who will pay for that additional safety? The North Atlantic Treaty Organization (NATO) allies have already reduced military spending so much that they are running hollowed-out armed forces. If the NATO allies were prepared to spend billions more on defense, one would hope that the money would go first to remedy their tremendous weaknesses in conventional forces. Or is the United States supposed to foot the bill? And if so, then at what cost to the U.S. military posture?

It is folly to imagine that the history of Russia as a great power has been sealed.

Of course, it might be that Russia is simply too short of money to take any of these steps, but that does not mean reconciliation with U.S. behavior. On the contrary, it would inspire a slow-burning desire to even the score. At some point, sooner or later, Russia would have its chance. It is folly to imagine that the history of Russia as a great power has been written and sealed.

As for China, its resources may limit it only to modernization in forms it was already pursuing. In that case, China may deploy road-mobile ICBMs that are harder to target, and push forward until it has the technology to MIRV these, to maximize the chance of overwhelming a U.S. defensive shield. China is, however, a country whose gross domestic product (GDP) grows at about 8 percent a year and will not lack for means for much longer. Thus, one should not ignore the possibility of a major expansion of Chinese ballistic missile forces. Meanwhile, the United States will have built into the Chinese political system a deepening conviction that the United States is an implacable enemy. The United States will therefore be building momentum toward confrontation that could unleash the nuclear war it was fortunate enough to avoid with the Soviet Union.

A final word about our allies. In the end, faced with an atmosphere of inevitability, and the choice of resisting the United States to the point of severely damaging alliances, U.S. friends may swallow their objections and acquiesce. If this happens, however, it will be yet another galling example for the allies of their dependence on the United States and of a style of U.S. leadership they consider both arrogant and reckless. If a new arms race does materialize, the consequences for relations between the United States and its allies will be disastrous.

A Constructive Alternative

All this having been said, there are some valid elements in the administration's position. It is possible to stipulate that (1) the proliferation of weapons of mass destruction and the means to deliver them to the United States represents a potential, important oncoming threat; (2) if these threats continue to develop, it would be appropriate for the United States to have developed and, as necessary, deployed a proportionate, limited defense; and (3) deeper reductions of nuclear weapons than those provided for in the START II agreement are acceptable and desirable.

The United States needs Congress to undertake a national debate.

But there are critical qualifications. Those qualifications should be the organizing issues of the national debate the United States needs the Congress to undertake. This debate would be most meaningful in Congress. From the congressional perspective, possible elements of a constructive alternative are:

- Given a choice between defensive technologies that require a reasonable modification of the ABM Treaty and more aggressive technologies that require the treaty's abrogation, the United States should prefer the former. Such a choice exists. It is the NMD system, whose development commenced under the last administration.
- The United States should explicitly avoid developing technologies that can be scaled up to threaten to neutralize the entire Russian nuclear deterrent. Space-based interception of ballistic missiles is such a technology and should not be on the U.S. agenda.
- The United States should explore technologies for attacking ballistic missiles during boost phase, using land- or sea-based interceptors. However, these technologies should be viewed as hedges in case of the appearance of second-generation ballistic missile delivery systems in the hands of rogue states (e.g., systems MIRVed or with penetration aids). Moreover, the United States should offer to develop such systems jointly with Russia.
- The United States should resist an effort to rush to deploy an improvised boost-phase defense. Haste is unnecessary and, notwithstanding what the administration is likely to say in support of such defenses, they are in fact technologically immature and present very serious issues of crisis stability.
- Vigorous work should continue on theater ballistic missile defenses, up to and including deployment. However, the U.S. Congress should refuse to fund work on theater defenses that would violate the memorandum of understanding on succession to the ABM Treaty signed with Russia in 1997, which established an upper limit for the performance of theater ballistic missile interceptors.
- Any proposal for deeper reductions of nuclear weapons than those agreed at Helsinki (2,000–2,500) must be preceded by a thorough review of the implications for U.S. security and must be accompanied by testimony from the Joint Chiefs of Staff (JCS) as to military sufficiency under existing guidance. The deeper the proposed cut, the more imperative it is to be sure that what the military says truly reflects their private

judgment, as opposed to the political line one can expect to hear from the Pentagon's civilian leadership.

- Unilateral reductions certified by the JCS as militarily sound could be considered, particularly if the case can be made that savings are important for other military applications. Such recommendations should be preceded by persuasive arguments concerning stability, including an assessment of how different end-states for the United States and Russia might influence perceptions of stability. Finally, Congress should not fund such reductions unless the administration either has worked out thorough arrangements for mutual verification with Russia, in binding form, or has presented a convincing case for why the United States should have no concern regarding the disposition of hundreds of Russian strategic nuclear warheads. There is no need to rush into unilateral reductions. Patience should rule.
- The dollar costs of a ballistic missile defense must be convincingly presented, along with the means for financing them. Trade-offs between developing defenses and paying for the "transformation" of our conventional military forces must be presented clearly and discussed. The U.S. government should not agree to a missile defense project with open-ended costs, to be paid for either at the expense of military readiness or modernization/transformation.
- We should not follow an artificially accelerated timetable, or one that deliberately moves to destroy the ABM Treaty simply to explore whether one technology or another is likely to work.
- If, after thoroughly testing technologies that are candidates for deployment, after getting the best cost estimates possible, and if the efforts of rogue states to proliferate continue, the United States should be prepared to deploy. At that juncture, if Russia—having been thoroughly consulted beforehand—still objects and has been unwilling to work out an amendment to the ABM Treaty, the United States would be obliged to let national interest govern its actions, notwithstanding Russian views. The U.S. government should use the intervening time to make similar efforts to talk through these issues with China. As a nonsignatory to the ABM Treaty, China has no say in its fate, but every effort should be made to enter into a serious strategic dialogue with China.
- Despite the administration's announced intention not to seek ratification of the CTBT, Congress should refuse to fund new programs for nuclear weapons development or testing. Congress should, however, be prepared to spend any reasonable amount on stockpile management in order to assure the viability of the U.S. reserve. It should also make clear that legal restraints on funding for testing will expire if another of the CTBT's original nuclear state signatories were to start testing.
- Whether or not an NMD system is deployed, Congress should both demand and support a comprehensive program to combat proliferation and to protect the U.S. homeland against the possible infiltration of weapons of mass destruction, which could turn out to be a vulnerability as great, if not greater than, exposure to any future threat of ballistic missile attack by a rogue state.

• Given the potential expense of an NMD system of even the most modest proportions, the overall impact of the availability of funds for modernizing conventional military forces needs to be assessed. The United States will continue to need conventional military forces with balanced capabilities for dealing with a wide range of contingencies.

These, then, could be the outlines of a responsible political stance in the face of steamroller tactics from the administration, and they could also, ultimately, be the basis for an emerging compromise. Those who take the lead in this debate may be characterized as lacking vision. On the contrary, they will be looking out for all of us.

Leon Fuerth is the Shapiro Visiting Professor at the Elliott School of International Affairs at the George Washington University. He served as national security advisor to former vice president Al Gore during 1993–2001.

From *The Washington Quarterly,* Autumn 2001, Vol 24, No. 4, pp. 97-108. © 2001 by the Center for Strategic and International Studies (CSIS) and the Massachusetts Institute of Technology. Reprinted by permission.

In North Korea and Pakistan, Deep Roots of Nuclear Barter

By David E. Sanger

SEOUL, South Korea, Nov. 21

Last July, American intelligence agencies tracked a Pakistani cargo aircraft as it landed at a North Korean airfield and took on a secret payload: ballistic missile parts, the chief export of North Korea's military.

The shipment was brazen enough, in full view of American spy satellites. But intelligence officials who described the incident say even the mode of transport seemed a subtle slap at Washington: the Pakistani plane was an American-built C-130.

It was part of the military force that President Pervez Musharraf had told President Bush last year would be devoted to hunting down the terrorists of Al Qaeda, one reason the administration was hailing its new cooperation with a country that only a year before it had labeled a rogue state.

But several times since that new alliance was cemented, American intelligence agencies watched silently as Pakistan's air fleet conducted a deadly barter with North Korea. In transactions intelligence agencies are still unraveling, the North provided General Musharraf with missile parts he needs to build a nuclear arsenal capable of reaching every strategic site in India.

In a perfect marriage of interests, Pakistan provided the North with many of the designs for gas centrifuges and much of the machinery it needs to make highly enriched uranium for the country's latest nuclear weapons project, one intended to put at risk South Korea, Japan and 100,000 American troops in Northeast Asia.

The Central Intelligence Agency told members of Congress this week that North Korea's uranium enrichment program, which it discovered only this summer, will produce enough material to produce weapons in two to three years. Previously it has estimated that North Korea probably extracted enough plutonium from a nuclear reactor to build one or two weapons, until that program was halted in 1994 in a confrontation with the United States.

Yet the C.I.A. report—at least the unclassified version—made no mention of how one of the world's poorest and most isolated nations put together its new, complex uranium project.

In interviews over the past three weeks, officials and experts in Washington, Pakistan and here in the capital of South Korea described a relationship between North Korea and Pakistan that now appears much deeper and more dangerous than the United States and its Asian allies first suspected.

The accounts raise disturbing questions about the nature of the uneasy American alliance with General Musharraf's government. The officials and experts described how, even after Mr. Musharraf sided with the United States in ousting the Taliban and hunting down Qaeda leaders, Pakistan's secretive A. Q. Khan Nuclear Research Laboratories continued its murky relationship with the North Korean military. It was a partnership linking an insecure Islamic nation and a failing Communist one, each in need of the other's expertise.

Pakistan was desperate to counter India's superior military force, but encountered years of American-imposed sanctions, so it turned to North Korea. For its part, North Korea, increasingly cut off from Russia and China, tried to replicate Pakistan's success in developing nuclear weapons based on uranium, one of the few commodities that North Korea has in plentiful supply.

Yet while the United States has put tremendous diplomatic pressure on North Korea in the past two months to abandon the project, and has cut off oil supplies to the country, it has never publicly discussed the role of Pakistan or other nations in supplying that effort.

American and South Korean officials, when speaking anonymously, say the reason is obvious: the Bush administration has determined that Pakistan's cooperation in the search for Al Qaeda is so critical—especially with new evidence suggesting that Osama bin Laden is still alive, perhaps on Pakistani soil.

So far, the White House has ignored federal statutes that require President Bush to impose stiff economic penalties on any country involved in nuclear proliferation or, alternatively, to issue a public waiver of those penalties in the interest of national security. Mr. Bush last year removed penalties that were imposed on Pakistan after it set off a series of nuclear tests in 1998.

White House officials would not comment on the record for this article, saying that discussing Pakistan's role could compromise classified intelligence. Instead, they noted that General Musharraf, after first denying Pakistani involvement in North Korea's nuclear effort, has assured Secretary of State Colin L. Powell that no such trade will occur in the future.

"He said, 'Four hundred percent assurance that there is no such interchange taking place now,'" Secretary Powell said in a briefing late last month. Pressed about Pakistan's contributions to the nuclear program that North Korea admitted to last month, Secretary Powell smiled tightly and said, "We didn't talk about the past."

A State Department spokesman, Philip Reeker, said, "We are aware of the allegations" about Pakistan, though he would not comment on the substance. "This administration will abide by the law," he said.

Intelligence officials say they have seen no evidence of exchanges since Washington protested the July missile shipment. Even in that incident, they cannot determine if the C-130 that picked up missile parts in North Korea brought nuclear-related goods to North Korea.

But American and Asian officials are far from certain that Pakistan has cut off the relationship, or even whether General Musharraf is in control of the transactions.

Yet in the words of one American official who has reviewed the intelligence, North Korea's drive in the past year to begin full-scale enrichment of uranium uses technology that "has 'Made in Pakistan' stamped all over it." They doubt that North Korea will end its effort even if Pakistan cuts off its supplies.

"In Kim Jong Il's view, what's the difference between North Korea and Iraq?" asked one senior American official with long experience dealing with North Korea. "Saddam doesn't have one, and look what's happening to him."

A Meeting of Minds in 1993

Pakistan's military ties to North Korea go back to the 1970's. But they took a decisive turn in 1993, just as the United States was forcing the North to open up its huge nuclear reactor facilities at Yongbyon. Yongbyon was clearly a factory for producing bomb-grade plutonium from spent nuclear fuel.

When North Korea refused to allow in inspectors headed by Hans Blix, the man now leading the inspections in Iraq, President Bill Clinton went to the United Nations to press penalties and the Pentagon drew up contingency plans for a strike against the plant in case North Korea removed the fuel rods to begin making bomb-grade plutonium.

In the midst of that face-off, Benazir Bhutto, then the prime minister of Pakistan, arrived in Pyongyang, the North Korean capital. It was the end of December, freezing cold, and yet the North Korean government arranged for tens of thousands of the city's well-trained citizens to greet her on the streets. At a state dinner, Ms. Bhutto complained about the American penalties imposed on her country and North Korea.

"Pakistan is committed to nuclear nonproliferation," she said, according to a transcript issued at the time. However, she added, states still have "their right to acquire and develop nuclear technology for peaceful purposes, geared to their economic and social developments."

Ms. Bhutto's delegation left with plans for North Korea's Nodong missile, according to former and current Pakistani officials.

The Pakistani military had long coveted the plans, and by April 1998, it successfully tested a version of the Nodong, renamed the Ghauri. Its flight range of about 1,000 miles put much of India within reach of Pakistan's nuclear warheads.

A former senior Pakistani official recalled in an interview that the Bhutto government planned to pay North Korea "from the invisible account" for covert programs. But events intervened.

Months after Ms. Bhutto's visit, the Clinton administration and North Korea reached a deal that froze all nuclear activity at Yongbyon, where international inspectors still live year-round.

In return, the United States and its allies promised North Korea a steady flow of fuel oil and the eventual delivery of two proliferation-resistant nuclear reactors to produce electric power. That was important in a country so lacking in power that, from satellite images taken at night, it appears like a black hole compared to the blazing lights of South Korea.

But within three years, Kim Jong Il grew disenchanted with the accord and feared that the nuclear power plants would never be delivered. He never allowed the International Atomic Energy Agency to begin the wide-ranging inspections required before the critical parts of the plants could be delivered.

By 1997 or 1998, American intelligence has now concluded, he was searching for an alternative way to build a bomb, without detection. He found part of the answer in Pakistan, which along with Iran, Libya, Yemen, Syria and Egypt was now a regular customer for North Korean missile parts, American military officials said.

A. Q. Khan, the father of Pakistan's nuclear bomb, who had years ago stolen the engineering plans for gas centrifuges from the Netherlands, visited North Korea several times. The visits were always cloaked in secrecy.

But several things are now clear. Pakistan was running out of hard currency to pay the North Koreans, who were in worse shape. North Korea feared that without a nuclear weapon it would eventually be absorbed by the economic might of the South, or squeezed by the military might of the United States.

In 1997 or 1998, Kim Jong Il and his generals decided to begin a development project for a bomb based on highly enriched uranium, a slow and difficult process, but relatively easy to hide.

They did so even while sporadically pursuing a better relationship with Washington. In the last days of the Clinton administration, the North negotiated with Secretary of State Madeleine K. Albright for a deal to restrict North Korean missile exports in return for a removal of economic penalties, a de-listing from the State Department's account of countries that sponsor terrorism and talks about diplomatic recognition. The deal was never reached.

President Bill Clinton even considered an end-of-term trip to North Korea, but was talked out of it by aides who feared that the North was not ready to make real concessions. The nuclear revelations of the past few weeks suggest those aides saved Mr. Clinton from embarrassment.

"Lamentably, North Korea never really changed," said one senior Western official here with long experience in the topic. "They came to the conclusion that the nuclear card was their one ace in the hole, and they couldn't give it up."

Caves and Clues

American intelligence agencies, meanwhile, suspected that North Korea was restarting a secret program. In 1998, satellites were focused on a huge underground site where the C.I.A. believed Kim Jong Il was trying to build a second plutonium-reprocessing center. But they were looking in the wrong place: after American officials negotiated access to the suspect site, they found only a series of man-made caves with no nuclear-related equipment, and no apparent purpose.

"World's largest underground parking lot," one American intelligence official joked at the time.

Rumors of a secret enriched-uranium project persisted, however. The C.I.A. and the Oak Ridge National Laboratory in Tennessee evaluated the evidence but reached no firm conclusion.

But there were hints. One Western diplomat who visited North Korea in May 1998, just as world attention focused on Pakistan, which had responded to India's underground nuclear tests by setting off six of its own, recalled witnessing an odd celebration. "I was in the Foreign Ministry," the official recalled last week. "About 10 minutes into our meeting, the North Korean diplomat we were seeing broke into a big smile and pointed with pride to these tests. They were all elated.

"Here was a model of a poor state getting away with developing a nuclear weapon."

When the Clinton administration raised the rumors of a Pakistan-North Korea link with Prime Minister Nawaz Sharif, who succeeded Ms. Bhutto, he denied them. It was

only after General Musharraf overthrew Mr. Sharif's government, and after Mr. Bush took office, that South Korean intelligence agencies picked up strong evidence that North Korea was buying components for an enriched-uranium program.

The agencies passed the evidence along to Washington, according to South Korean and American officials. It looked suspiciously similar to the gas centrifuge technology used in Pakistan. "My guess is that Pakistan was the only available partner," said Lee Hong Koo, a former South Korean prime minister and unification minister.

A. H. Nayya, a physics professor at Quaid-e-Azam University in Islamabad, who has no role in the country's nuclear program, agreed: "The clearest possibility is that the Pakistanis gave them the blueprint. 'Here it is. You make it on your own.'"

> North Korea's effort a decade ago to extract plutonium from a reactor at Yongbyon ended when the country froze the activities in return for energy help. American officials do not know the site of a later program to develop a weapon using enriched uranium.

Under American pressure, Dr. Khan was removed from the operational side of the Pakistani nuclear program. He was made an "adviser to the president" on nuclear technology.

Here in Seoul, nuclear experts working for the government of President Kim Dae Jung say they were subtly discouraged from publicly writing or speculating about the North's secret programs because the Korean government feared that it would derail President Kim's legacy: the "sunshine policy" of engagement with North Korea and encouraging investment there.

By this summer, however, the C.I.A. concluded that the North had moved from research to production. The intelligence agency took the evidence to Condoleezza Rice, the president's national security adviser, who asked for a review by all American intelligence agencies.

Such a request is usually a prescription for conflicting interpretations. Instead, the agencies came back with a unanimous opinion: the North Korean program was well under way, and had to be stopped.

Telling the North, 'You're Busted'

After sending senior officials to Japan and South Korea in August to present the new evidence, Mr. Bush decided to confront the North Koreans. On Oct. 4, James A. Kelly, the assistant secretary of state for East Asian and Pacific affairs, was in North Korea and told his counterparts that the United States had detailed information about the enriched-uranium program.

"We wanted to make it clear to them that they were busted," a senior administration official said.

The North Koreans initially denied the accusation, but the next day, after what they told the American visitors was an all-night discussion, they admitted that they were pursuing the secret weapons program, several officials said.

"We need nuclear weapons," Kang Sok Joo, the North Korean senior foreign policy official, said, arguing that the program was a result of the Bush administration's hostility.

Mr. Kelly responded that the program began at least four years ago, when Mr. Bush was governor of Texas. The Americans left after one North Korean official declared that dialogue on the subject was worthless and said, "We will meet sword with sword."

Since then, the North Koreans have been more circumspect. They have talked publicly about having the right to a nuclear weapon, even though they have signed the Nuclear Nonproliferation Treaty and an agreement with South Korea to keep the Korean Peninsula free of nuclear weapons.

The Bush administration has been uncharacteristically restrained. President Bush led the push for an oil cutoff, but also issued a statement on Nov. 15 saying that the United States had no intention of invading North Korea. His aides hoped that the statement would give Kim Jong Il the kind of security guarantee he had long demanded— and a face-saving way to end the nuclear program.

Mr. Bush's aides say the way to deal with North Korea, in contrast to their approach to Iraq, is to exploit its economic vulnerabilities and offer carrots, essentially the strategy the Clinton administration used. Many here in Seoul believe it may work this time.

"The North Koreans are a lot more dependent on us, and on the West, than they were in the 1994 nuclear crisis," said Han Sung Joo, who served as South Korea's foreign minister then.

But the reality, officials acknowledged, is that Mr. Bush has little choice but to pursue a diplomatic solution with North Korea.

Kim Jong Il has 11,000 artillery tubes dug in around the demilitarized zone, all aimed at Seoul. In the opening hours of a war, tens of thousands of people could die, military officials here say.

"Here's the strategy," one American official said. "Tell the North Koreans, quite publicly, that they can't get away with it. And say the same thing to Pakistan, but privately, quietly."

From the *New York Times,* November 24, 2002, pp. A1, A16. © 2002 by The New York Times Company. Reprinted by permission.

Article 12

Towards an internet civil defence against bioterrorism

Approaches towards the public-health prevention of bioterrorism are too little, and too late. New information-based approaches could yield better homeland protection. An internet civil defence is presented where millions of eyes could help to identify suspected cases of bioterrorism, with the internet used to report, confirm, and prevent outbreaks.

Lancet Infectious Diseases 2001; **1:** 125–127

Ronald E LaPorte, Francois Sauer, Steve Dearwater, Akira Sekikawa, Eun Ryoung Sa, Deborah Aaron, and Eugene Shubnikov

Despite the large resources put into bioterrorism preparedness,[1] there is little evidence that they work. The West Nile virus outbreak showed how vulnerable our homeland is to attacks. Confusion, error, miscommunication, and misdiagnosis reigned supreme in this near-calamity.[2] By the time the cause was discovered the epidemic was well past its tipping point. Bioterrorists would use more deadly agents, and would spread them over much greater areas, creating extraordinary havoc. As shown by this incident, public health is ill prepared to handle attacks by bioterrorists.

More than US$250 million is to be spent this year to prepare existing US public health systems for bioterrorism. The UK is also spending large amounts of money to thwart bioterrorism. However, even with the increase in funding, the West Nile virus episode and others show how poorly prepared our public health institutions are for an attack. We argue that improvement will not happen by pouring more money into established paradigms of public health, instead new and more effective information technologies need to be tested and introduced. We need to move forward towards a new approach, an internet civil defence to prevent bioterrorism.

We must face the disturbing fact that it is very difficult to predict and guard against a bioterrorist attack because there are too many targets, too many means to penetrate the targets, the occurrences are too rare, and the bioterrorists are too crafty. Pouring more money into a decaying system is like shoring up the Maginot line as the cunning enemy easily slips around our defences. We need to develop a broader, "flatter," and nimbler organisation to combat bioterrorism, with thousands more eyes. The existing hierarchical public health system is not sufficiently agile to compete with bioterrorists.

A tradition of civil defence

The military defence of the US and UK in warfare has always partly been the responsibility of their citizens. We would argue that this is also the case for the war against bioterrorism, where citizens should be the first and most important line of defence. We have the opportunity now to build an internet-based outbreak-prevention network, by connecting citizens to each other on the internet to form a civil defence.

Civil defence and home guard during World War II consisted in part of individuals who were trained to prevent battle damage within their country. Civil defence represented the identification of leaders in the community who would oversee activities during the time of air raids. These were the trusted agents who would guide people to shelters, or help to dig people out from ruined buildings. We knew when the alarm went off above the firehouse that the leaders would be a trusted source of information. During the Cold War of the 1950s and 1960s, civil defence became more coordinated; there were specific guidelines as to where to take people for protection, civilians were trained to watch over the neighbourhoods to prevent looting, and mitigation of panic was an essential component.

Towards the end of the Cold War, civil defence and home guard in our countries started to wane. A primary reason for this was that the arsenals on both sides became so destructive that a war would almost certainly yield mutual annihilation, and any attempt at civil defence would be futile.

However, with the fall of communism the prospect of complete annihilation was lessened. Focal disasters, whether they are manmade (terrorism, nuclear reactors) or infectious (bioterrorism, emerging disease), will continue to plague our citizens, and will be rare and difficult to predict.

As civil defence waned, a new form of home guard grew against an internal enemy, crime. The neighbourhood watch notion (eg, http://www.nwatch.org.uk) developed as a result of increasing incidence and fear of crime, and involves people in communities uniting to watch over each other to deter crime. Is this not what we want for bioterrorism? Why not establish the same principle, whereby neighbours watch out for each other, but do this on the internet with a global health-network neighbourhood watch to act as a deterrent against bioterrorism, and to mitigate damage in the event of an attack. The beauty of such a system is that with the internet, there is a death to distance. Thus bioterrorist experts from Russia could be brought in over the internet within minutes after the identification of an attack. The internet community leaders could provide updated information to others, to alleviate fear, reduce panic, and to guide people to safety.

The electronic matrix

Instead of building an inflexible Maginot line defence as we are now, perhaps we should consider an ever alert, flexible electronic-matrix civil defence as our first line

of defence. This public-health system would collect locally information of possible bioterrorist activity, allowing suspicious behaviour to be investigated and planned bioterrorist activity to be stopped. Rather than the present 2000 public health experts worldwide on the outlook for bioterrorism, there would be 20 million brains on internet bioterrorism watch. We thus have millions of citizen watchdogs.

There are many advantages of developing a bioterrorism watch. First, we have many more "eyeballs" on the lookout for bioterrorist activity. Second, the implementation of an internet bioterrorism watch would help to reduce panic (the "War of the Worlds" effect) if a major bioterrorism event happened. What better to alleviate fear than a visit from your brother, or neighbour, or friend who says that everything is all right, because this is what had been heard from the civil-defence-network portal? We believe information from trusted agents; we do not from people we do not know. In this way, the internet civil-defence team acts as a secure information pipeline that can filter accurate information through a matrix of people we trust. Lastly, in the days and weeks following the outbreak, this network will act to bring accurate information to the people who need it the most.

The internet is being used to monitor bioterrorism and emerging diseases. There are many relevant websites; two of the most reliable are ProMed (http://www.fas.org/promed) and WHO's outbreak verification list (http://www.cdc.gov/ncidod/eid/vol6no2/grein.htm). These websites differ from the one we propose because they are either list servers or portals and can be viewed as one-to-many transmissions of bioterrorism information—eg, reports of an attack sent to a list of 10 000 people or posted for millions to see. The proposed internet civil defence is internet community-based with basically a one-to-one transmission of information, and trusted agents spreading the message to each other. Based on the idea of six degrees of separation, we can reach large numbers of people worldwide through interpersonal relationships. List servers or websites are more likely to be fraudulent than a message from your mother that can be verified with a phone call.

The idea of a neighbourhood watch can be more specific—eg, a school watch. Children are well suited to monitoring for bioterrorism, because they are some of the most susceptible targets, they transmit agents readily through the school and fam-

ily, and they are among the easiest groups to monitor. In many ways they are like the "canary in the coal mine," the earliest indication of an attack. However, there is virtually no monitoring of children in either the US or UK for bioterrorism, but there is for "crime terrorism". The launch of a school watch on the internet could lead to a local-to-global watchdog system being used to guard our children.

The first step for an internet watch clearly needs to be the identification and verification of a possible bioterrorist attack. There will be false positives with this approach, but it is better to be more sensitive than specific when dealing with weapons of mass destruction. The false positive rate can be reduced, however, with certified training programmes based on the experiences of civil-defence systems from past decades.

How would we grow an internet civil-defence network? This would be quite simple, and has been done to a lesser extent with the global health network (http://www.pitt.edu/~super1/), which consists of 4600 academics who are experts in bioterrorism-prevention and the internet. The goal is to identify a core of people who have web access, and who want to protect their communities against bioterrorism. A second example would be to use existing networks of people such as Medscape (http://www.medscape.com), which has over 2 million subscribers in health, of whom 75 000 are in infectious disease. Certification systems similar to those used during World War II could evolve.

There is perhaps evidence that a civil-defence model will work more effectively than the current approaches. On the internet people are networked together looking for viruses—computer viruses—so that at first sight notification is spread worldwide. Many pairs of eyes then try to find the person who developed and unleashed the virus. This system works very well. Another example is the case of ten escaped convicts from Texas who were found as the result of people in the heart of the USA talking together through the internet.

Disinformation, where there is a variety of fraud, hacking, and malevolent cyberwar or infowar activities, presents a considerable drawback in the prevention of bioterrorism, and can happen irrespective of the system that is implemented. However, with an internet civil defence the risk can be lessened by very simple measures. The first is that we warn individuals that the only good information is that from a trusted agent from a select group of perhaps ten friends and family members. Sophisticated bioterrorists could spread misinformation using your father's name but an internet-based civil-defence system, with routine telephone call verification to a relative, would greatly reduce the probability of hacking and causing disinformation to a large percentage of the 20 million in the system.

Ideally, we would reach at least one person in each local community, as with the neighbourhood watch. This person would then set up the model for the inter-

net neighbourhood watch by doing the exact same procedure used in traditional neighbourhood watch, but over the internet. This would create a local area civil-defence watch that, when combined with the other census tracts, and other countries, quickly branches upward to a wider area global watch.

Conclusions

Present public health approaches to the prevention of bioterrorism have too few participants, cannot cover enough contingencies to ensure that bioterrorism is thwarted, and are too costly. Little is done to address the panic that would follow an attack. Bringing the eyes, minds, and ears of citizens of the US, the UK, and the rest of the world to fight bioterrorism offers a unique opportunity to thwart and mitigate outbreaks. The civil defence/home guard model has proven to be effective in our countries. It is likely that the development of an internet civil-defence model would be very effective as first line of defence against bioterrorism. This is not to say that the internet civil-defence programme should replace current approaches, but it should be considered as a new weapon system to enhance our current homeland defence.

References

1. Khan AS, Morse S, Lillibridge S. Public-health preparedness for biological terrorism in the USA. *Lancet* 2000; **356:** 1179–82.

2. Steinhauer J, Miller J. In New York outbreak, glimpse of gaps in biological defenses. *The New York Times* 1999 Oct 11; Sect. Metropolitan desk.

UNIT 4
North America

Unit Selections

Key Points to Consider

- Explain why you agree or disagree with the shift in U.S. policy evidenced in the new National Security Strategy released in 2002.

- What additional measures could the United States and Canada take to increase security along their 5,500-mile-long shared border?

- Will Canada continue to scale back its troop commitments to future international peacekeeping missions?

 **Links: www.dushkin.com/online/**
These sites are annotated in the World Wide Web pages.

The Henry L. Stimson Center   Peace Operations and Europe
http://www.stimson.org/fopo/?SN=FP20020610372
The North American Institute
http://www.northamericaninstitute.org

After a contentious period, George W. Bush was declared the winner of the 2000 U.S. presidential election. Initially the new president focused more on domestic priorities, including passage of a tax cut and the formulation of measures to cope with a developing recession. Foreign policy was a distant-second policy priority. Senior Bush officials began implementing several changes in U.S. multilateral commitments that generated sharp criticism at home and abroad.

The terrorist attacks on the World Trade Center in New York and the Pentagon in Washington D.C., dramatically changed the foreign policy landscape. Americans of all political persuasions came together quickly to support the President's "war on terrorism." For many Americans, September 11, 2001, marked the end of the post–cold war era and the start of a new "war" against global terrorism. As the war on terrorism became the central organizing principle of America's foreign and defense policies, the Bush administration developed a new National Security Strategy in September 2002. John Lewis Gaddis in "A Grand Strategy of Transformation" describes the new policy as the most sweeping shift in U.S. grand strategy since the beginning of the cold war. According to Gaddis, its success will depend on the willingness of the rest of the world to welcome U.S. power with open arms.

Early indications are not promising for those who want to avoid the widespread misperceptions and national stereotypes that characterized the cold war era. A public opinion poll conducted by the Pew Foundation at the end of 2001 found a large gap between how Americans view themselves and how others in the world view Americans. In random samples of political, media, and business elite on five continents, a large majority of foreigners believed that the United States is mostly acting unilaterally in the fight against terrorism and believed that U.S. foreign policies in the world were responsible for the September 11 attacks. In sharp contrast, 70 percent of American opinion-makers in the same survey said that the United States was acting jointly with its friends and is taking into account the interest of its partners in the war on terror. Even more Americans continue to have difficulties understanding why foreigners wanted to kill so many innocent American civilians.

Despite divisions among their publics, western European governments moved quickly to support the war on terrorism with proclamations of solidarity in the aftermath of September 11 despite growing differences between the United States and Europe on a diverse range of issues including global warming, biotechnology, peacekeeping, and national missile defense.

The war on terrorism also transformed the status of Russia in the minds of many senior officials in the Bush administration. Republican thinking about Russia was changed rapidly by the September 11 terrorist attack. Bush officials began to see Russia as a potentially key player in critical areas—from fighting the war on terrorism to providing a backup oil reserve.

Alternative sources of Middle East oil emerged as an important issue for U.S. policymakers as doubts surfaced about the durability of the close relationship between the United States and the Saudi Arabian government. An important change in this relationship is that the United States can no longer separate the Arab-Israeli conflict from bilateral relations with Saudi Arabia or U.S. strategy in the Persian Gulf.

It is too early to tell how long the U.S. Congress and American public will support the war on terrorism and be willing to pay for increased defense spending. New national defense priorities promise to be expensive since America's war on terrorism involves the entire array of foreign policy instruments including diplomacy, intelligence, and military strikes. Homeland defense commitments promise to be among the most expensive because this priority requires federal aid to hundreds of "first line defender" agencies at the state and local levels and financial support to help rebuild a broken public health system.

Immanuel Wallerstein in "The Eagle Has Crash Landed," takes a longer-term perspective on the effects of growing U.S. commitments abroad and concludes that America is a declining world power whose slide began during the Vietnam era. Wallerstein argues that the pace of decline is more rapid today because of military "overreach" that has seriously hurt the country's economy. For Wallerstein, the key question is whether the United States can devise a way to descend gracefully.

Immediately after the terrorist attacks in 2001 there was strong bipartisan support in Congress and among the American public for far-reaching anti-terrorist legislation that gave the executive branch new powers. Surprisingly few voices were raised within the United States to protest the introduction of military tribunals or expanded federal powers that carry the potential of eroding civil liberties and privacy rights of citizens in addition to those of foreigners. Instead, the strongest objections came from abroad. At the end of 2001, most European governments refused to extradite suspected terrorists to the United States unless the U.S. government guaranteed that the suspects would not be tried in military tribunals or given death sentences if convicted.

Canada immediately committed military forces to the war on terrorism after the U.S.–led air campaign against al Qaeda began. The day after U.S. and British forces launched their military assault on Afghanistan, Canada announced plans to send ships, planes, and about 2,000 service personnel to Afghanistan. The U.S. and Canadian governments undertook new security measures along the world's longest undefended border. The short-term response was to increase barriers along the 5,500-mile long frontier. However, as Michael Grunwald describes in "Economic Crossroads on the Line," the response led to long waits and costly delays at the border, especially for Canadians. As a longer-term response, both countries agreed to explore ways to rely on technology to create a "smarter border."

Tensions between the two states flared again as U.S. officials detained Canadian citizens, without charge, as suspected terrorists. In a second incident, two U.S. air force pilots accidentally bombed and killed a small group of Canadian soldiers in Afghanistan. The incidents triggered sharp diplomatic protests and increased anti-American feeling in public opinion polls in Canada.

A gap between the leaders of the two countries was also evident in the rhetoric of senior Canadian officials. At the close of 2002, Canadian prime minister Jean Chretien observed that perceived Western arrogance had played a part in the September 11 attacks and warned the United States and other wealthy nations against "humiliating" poorer countries.

By the end of 2002, financial constraints on increasing the Canadian expenditures for defense raised serious questions about whether Canada would be able to continue an activist foreign policy and a long-standing commitment to contribute troops to UN international peacekeeping missions. Since 2000, the number of troops committed to peacekeeping operations were reduced as part of cost-cutting measures. At the same time, Canada purchased new military hardware and implemented new tax cuts. Neither policy change was enough to stop a continuing degradation of Canadian force readiness or an exodus of experienced personnel from the intelligence service. After studying the current state of the national forces, a Canadian senate committee recently recommended that the military immediately withdraw its forces from overseas duty for 2 years and spend billions more to stop the armed forces from collapsing.

A GRAND STRATEGY OF TRANSFORMATION

President George W. Bush's national security strategy could represent the most sweeping shift in U. S. grand strategy since the beginning of the Cold War. But its success depends on the willingness of the rest of the world to welcome U.S. power with open arms.

By John Lewis Gaddis

It's an interesting reflection on our democratic age that nations are now expected to publish their grand strategies before pursuing them. This practice would have surprised Metternich, Bismarck, and Lord Salisbury, though not Pericles. Concerned about not revealing too much, most great strategists in the past have preferred to concentrate on implementation, leaving explanation to historians. The first modern departure from this tradition came in 1947 when George F. Kennan revealed the rationale for containment in *Foreign Affairs* under the inadequately opaque pseudonym "Mr. X," but Kennan regretted the consequences and did not repeat the experiment. Not until the Nixon administration did official statements of national security strategy became routine. Despite his reputation for secrecy, Henry Kissinger's "State of the World" reports were remarkably candid and comprehensive—so much so that they were widely regarded at the time as a clever form of disinformation. They did, though, revive the Periclean precedent that in a democracy even grand strategy is a matter for public discussion.

That precedent became law with the Goldwater-Nichols Department of Defense Reorganization Act of 1986, which required the president to report regularly to Congress and the American people on national security strategy (NSS) [see sidebar]. The results since have been disappointing. The Reagan, Bush, and Clinton administrations all issued NSS reports, but these tended to be restatements of existing positions, cobbled together by committees, blandly worded, and quickly forgotten. None sparked significant public debate.

George W. Bush's report on "The National Security Strategy of the United States of America," released on September 17, 2002, has stirred controversy, though, and surely will continue to do so. For it's not only the first strategy statement of a new administration; it's also the first since the surprise attacks of September 11, 2001. Such attacks are fortunately rare in American history—the only analogies are the British burning of the White House and Capitol in 1814 and the Japanese attack on Pearl Harbor in 1941—but they have one thing in common: they prepare the way for new grand strategies by showing that old ones have failed. The Bush NSS, therefore, merits a careful reading as a guide to what's to come.

WHAT THE NSS SAYS

Beginnings, in such documents, tell you a lot. The Bush NSS, echoing the president's speech at West Point on June 1, 2002, sets three tasks: "We will defend the peace by fighting terrorists and tyrants. We will preserve the peace by building good relations among the great powers. We will extend the peace by encouraging free and open societies on every continent." It's worth comparing these goals with the three the Clinton administration put forth in its final NSS, released in December 1999: "To enhance America's security. To bolster America's economic prosperity. To promote democracy and human rights abroad."

The differences are revealing. The Bush objectives speak of defending, preserving, and extending peace; the Clinton statement seems simply to assume peace. Bush calls for cooperation among great powers; Clinton never uses that term. Bush specifies the encouragement of free and open societies on every continent; Clinton contents himself with "promoting" democracy and human rights "abroad." Even in these first few lines, then, the Bush NSS comes across as more forceful, more carefully crafted, and—unexpectedly—more multilateral than its immediate predecessor. It's a tip-off that there're interesting things going on here.

The first major innovation is Bush's equation of terrorists with tyrants as sources of danger, an obvious outgrowth of September 11. American strategy in the past, he notes, has concentrated on defense against tyrants. Those adversaries required "great armies and great industrial capabilities"—resources only states could provide—to threaten U.S. interests. But now, "shadowy networks of individuals can bring great chaos and suffering to our shores for less than it costs to purchase a single tank." The strategies that won the Cold War—containment and deterrence—won't work against such dangers, because those strategies assumed the existence of identifiable regimes led by identifiable leaders operating by identifiable means from identifiable territories. How, though, do you contain a shadow? How do you deter someone who's prepared to commit suicide?

There've always been anarchists, assassins, and saboteurs operating without obvious sponsors, and many of them have risked their lives in doing so. Their actions have rarely shaken the stability of states or societies, however, because the number of victims they've targeted and the amount of physical damage they've caused have been relatively small. September 11 showed that terrorists can now inflict levels of destruction that only states wielding military power used to be able to accomplish. Weapons of mass destruction were the last resort for those possessing them during the Cold War, the NSS points out. "Today, our enemies see weapons of mass destruction as weapons of choice." That elevates terrorists to the level of tyrants in Bush's thinking, and that's why he insists that preemption must be added to—though not necessarily in all situations replace—the tasks of containment and deterrence: "We cannot let our enemies strike first.

The NSS is careful to specify a legal basis for preemption: international law recognizes "that nations need not suffer an attack before they can lawfully take action to defend themselves against forces that present an imminent danger of attack." There's also a preference for preempting multilaterally: "The United States will constantly strive to enlist the support of the international community." But "we will not hesitate to act alone, if necessary, to exercise our right of self-defense by acting preemptively against such terrorists, to prevent them from doing harm against our people and our country."

Preemption in turn requires hegemony. Although Bush speaks, in his letter of transmittal, of creating "a balance of power that favors human freedom" while forsaking "unilateral advantage," the body of the NSS makes it clear that "our forces will be strong enough to dissuade potential adversaries from pursuing a military build-up in hopes of surpassing, or equaling, the power of the United States." The West Point speech put it more bluntly: "America has, and intends to keep, military strengths beyond challenge." The president has at last approved, therefore, Paul Wolfowitz's controversial recommendation to this effect, made in a 1992 "Defense Planning Guidance" draft subsequently leaked to the press and then disavowed by the first Bush administration. It's no accident that Wolfowitz, as deputy secretary of defense, has been at the center of the new Bush administration's strategic planning.

How, though, will the rest of the world respond to American hegemony? That gets us to another innovation in the Bush strategy, which is its emphasis on cooperation among the great powers. There's a striking contrast here with Clinton's focus on justice for small powers. The argument also seems at odds, at first glance, with maintaining military strength beyond challenge, for don't the weak always unite to oppose the strong? In theory, yes, but in practice and in history, not necessarily. Here the Bush team seems to have absorbed some pretty sophisti-cated political science, for one of the issues that discipline has been wrestling with recently is why there's still no anti-American coalition despite the overwhelming dominance of the United States since the end of the Cold War.

Bush suggested two explanations in his West Point speech, both of which most political scientists—not all—would find plausible. The first is that other great powers *prefer* management of the international system by a single hegemon as long as it's a relatively benign one. When there's only one superpower, there's no point for anyone else to try to compete with it in military capabilities. International conflict shifts to trade rivalries and other relatively minor quarrels, none of them worth fighting about. Compared with what great powers have done to one another in the past, this state of affairs is no bad thing.

There's a coherence in the Bush strategy that the Clinton national security team—notable for its simultaneous cultivation and humiliation of Russia—never achieved.

U.S. hegemony is also acceptable because it's linked with certain values that all states and cultures—if not all terrorists and tyrants—share. As the NSS puts it: "No people on earth yearn to be oppressed, aspire to servitude, or eagerly await the midnight knock of the secret police." It's this association of power with universal principles, Bush argues, that will cause other great powers to go along with whatever the United States has to do to preempt terrorists and tyrants, even if it does so alone. For, as was the case through most of the Cold War, there's something worse out there than American hegemony.

The final innovation in the Bush strategy deals with the longer term issue of removing the causes of terrorism and tyranny. Here, again, the president's thinking parallels an emerging consensus within the academic community. For it's becoming clear now that poverty wasn't what caused a group of middle-class and reasonably well-educated Middle Easterners to fly three airplanes into buildings and another into the ground. It was, rather, resentments growing out of the absence of representative institutions in their own societies, so that the only outlet for political dissidence was religious fanaticism.

Hence, Bush insists, the ultimate goal of U.S. strategy must be to spread democracy everywhere. The United States must finish the job that Woodrow Wilson started. The world, quite literally, must be made safe for democracy, even those parts of it, like the Middle East, that have so far resisted that tendency. Terrorism—and by implication the authoritarianism that breeds it—must become as obsolete as slavery, piracy, or genocide: "behavior that no respectable government can condone or support and that all must oppose."

The Bush NSS, therefore, differs in several ways from its recent predecessors. First, it's proactive. It rejects the

Clinton administration's assumption that since the movement toward democracy and market economics had become irreversible in the post—Cold War era, all the United States had to do was "engage" with the rest of the world to "enlarge" those processes. Second, its parts for the most part interconnect. There's a coherence in the Bush strategy that the Clinton national security team—notable for its simultaneous cultivation and humiliation of Russia—never achieved. Third, Bush's analysis of how hegemony works and what causes terrorism is in tune with serious academic thinking, despite the fact that many academics haven't noticed this yet. Fourth, the Bush administration, unlike several of its predecessors, sees no contradiction between power and principles. It is, in this sense, thoroughly Wilsonian. Finally, the new strategy is candid. This administration speaks plainly, at times eloquently, with no attempt to be polite or diplomatic or "nuanced." What you hear and what you read is pretty much what you can expect to get.

WHAT THE NSS DOESN'T SAY

There are, however, some things that you won't hear or read, probably by design. The Bush NSS has, if not a hidden agenda, then at least one the administration isn't advertising. It has to do with why the administration regards tyrants, in the post-September 11 world, as at least as dangerous as terrorists.

Bush tried to explain the connection in his January 2002 State of the Union address when he warned of an "axis of evil" made up of Iraq, Iran, and North Korea. The phrase confused more than it clarified, though, since Saddam Hussein, the Iranian mullahs, and Kim Jong II are hardly the only tyrants around, nor are their ties to one another evident. Nor was it clear why containment and deterrence would not work against these tyrants, since they're all more into survival than suicide. Their lifestyles tend more toward palaces than caves.

Both the West Point speech and the NSS are silent on the "axis of evil." The phrase, it now appears, reflected overzealous speechwriting rather than careful thought. It was an ill-advised effort to make the president sound, simultaneously, like Franklin D. Roosevelt and Ronald Reagan, and it's now been given a quiet burial. This administration corrects its errors, even if it doesn't admit them.

That, though, raises a more important question: Why, having buried the "axis of evil," is Bush still so keen on burying Saddam Hussein? Especially since the effort to do so might provoke him into using the weapons of last resort that he's so far not used? It patronizes the administration to seek explanations in filial obligation. Despite his comment that this is "a guy that tried to kill my dad," George W. Bush is no Hamlet, agonizing over how to meet a tormented parental ghost's demands for revenge. Shakespeare might still help, though, if you shift the analogy to Henry V. That monarch understood the psychological value of victory—of defeating an adversary sufficiently thoroughly that you shatter the confidence of others, so that they'll roll over themselves before you have to roll over them.

For Henry, the demonstration was Agincourt, the famous victory over the French in 1415. The Bush administration got a taste of Agincourt with its victory over the Taliban at the end of 2001, to which the Afghans responded by gleefully shaving their beards, shedding their burkas, and cheering the infidels—even to the point of lending them horses from which they laser-marked bomb targets. Suddenly, it seemed, American values were transportable, even to the remotest and most alien parts of the earth. The vision that opened up was not one of the clash among civilizations we'd been led to expect, but rather, as the NSS puts it, a clash "inside a civilization, a battle for the future of the Muslim world."

How, though, to maintain the momentum, given that the Taliban is no more and that al Qaeda isn't likely to present itself as a conspicuous target? This, I think, is where Saddam Hussein comes in: Iraq is the most feasible place where we can strike the next blow. If we can topple this tyrant, if we can repeat the Afghan Agincourt on the banks of the Euphrates, then we can accomplish a great deal. We can complete the task the Gulf War left unfinished. We can destroy whatever weapons of mass destruction Saddam Hussein may have accumulated since. We can end whatever support he's providing for terrorists elsewhere, notably those who act against Israel. We can liberate the Iraqi people. We can ensure an ample supply of inexpensive oil. We can set in motion a process that could undermine and ultimately remove reactionary regimes elsewhere in the Middle East, thereby eliminating the principal breeding ground for terrorism. And, as President Bush did say publicly in a powerful speech to the United Nations on September 12, 2002, we can save that organization from the irrelevance into which it will otherwise descend if its resolutions continue to be contemptuously disregarded.

If I'm right about this, then it's a truly *grand* strategy. What appears at first glance to be a lack of clarity about who's deterrable and who's not turns out, upon closer examination, to be a plan for transforming the entire Muslim Middle East: for bringing it, once and for all, into the modern world. There's been nothing like this in boldness, sweep, and vision since Americans took it upon themselves, more than half a century ago, to democratize Germany and Japan, thus setting in motion processes that stopped short of only a few places on earth, one of which was the Muslim Middle East.

CAN IT WORK?

The honest answer is that no one knows. We've had examples in the past of carefully crafted strategies failing: most conspicuously, the Nixon-Kissinger attempt, during the early 1970s, to bring the Soviet Union within the international system of satisfied states. We've had examples of carelessly improvised strategies succeeding: The Clinton administration accomplished this feat in Kosovo in 1999.

Power's Paper Trail

The National Security Strategy (NSS) report was born out of the Goldwater-Nichols Act of 1986, the fourth major post-World War II reorganization of the U.S. Defense Department. It is one of more than 2,000 reports that federal departments, agencies, commissions, and bureaus must submit to Congress each year.

In requiring a wide-ranging, yet detailed annual report, the 99th Congress that passed Goldwater-Nichols hoped to remedy what it considered a major shortfall of Cold War era executive branches—the inability to formulate and communicate concrete mid- and long-term national security strategy. "Few in the Congress at that time doubted that there existed a grand strategy," Don Snider, a political scientist at the U.S. Military Academy, has noted. "What they doubted, or disagreed with, was its *focus* in terms of values, interests, and objectives; its *coherence* in terms of relating means to ends; its *integration*

in terms of the elements of power; and its *time* horizon." Requiring the report was also a way for Congress to ensure greater civilian control over the military and its planning, a major political theme of the Goldwater-Nichols Act.

Yet the quality of the NSS depends on how willing presidential administrations are to be frank and forthcoming. Over the last 16 years, most have interpreted Congress's mandate loosely. Former President Bill Clinton's first NSS reportedly went through 21 drafts before it was finally submitted a year and a half late. (George W. Bush also missed the deadline for his first report by more than a year. It was due June 15, 2001.) And relevance problems have been common. Several reports in the 1990s were rendered all but obsolete by rapid changes in the global geopolitical picture. They were, in political-speak, overtaken by events. Others, like George Bush Sr.'s 1993 report, turned out to be little more than cheerleading

sessions for an administration's foreign policy accomplishments. "Even when submitted, the National Security Strategy report has generally been late," a frustrated Sen. Strom Thurmond barked to colleagues in 1994. "In addition," Thurmond added, "the... report has seldom met the expectations of those of us who participated in passing the Goldwater-Nichols Act."

That the report has fallen short of the expectations held by Goldwater-Nichols supporters like Thurmond shouldn't come as a surprise, says University of Virginia political scientist Larry Sabato. The result was predictable. Every presidential administration views Congress as a competitor, rather than a partner, "even when the same party controls both," says Sabato. As a result, "information about a critical subject like national security is power—pure, raw power—and the stakes are high. Rarely is power generously shared or given away in Washington."

—*FP*

The greatest theorist of strategy, Carl von Clausewitz, repeatedly emphasized the role of chance, which can at times defeat the best of designs and at other times hand victory to the worst of them. For this reason, he insisted, theory can never really predict what's going to happen.

Does this mean, though, that there's *nothing* we can say? That all we can do is cross our fingers, hope for the best, and wait for the historians to tell us why whatever happened was bound to happen? I don't think so, for reasons that relate, rather mundanely, to transportation. Before airplanes take off—and, these days, before trains leave their terminals—the mechanics responsible for them look for cracks, whether in the wings, the tail, the landing gear, or on the Acela the yaw dampers. These reveal the stresses produced while moving the vehicle from where it is to where it needs to go. If undetected, they can lead to disaster. That's why inspections—checking for cracks—are routine in the transportation business. I wonder if they ought not to be in the strategy business as well. The potential stresses I see in the Bush grand strategy— the possible sources of cracks—are as follows:

Multitasking Critics as unaccustomed to agreeing with one another as Brent Scowcroft and Al Gore have warned against diversion from the war on terrorism if the United States takes on Saddam Hussein. The principle involved here—deal with one enemy at a time—is a sound one. But plenty of successful strategies have violated it. An obvious example is Roosevelt's decision to fight simultaneous wars against Germany and Japan between 1941 and 1945. Another is Kennan's strategy of containment, which worked by deterring the Soviet Union while reviving democracy and capitalism in Western Europe and Japan. The explanation, in both instances, was that these were wars on different fronts against the same enemy: authoritarianism and the conditions that produced it.

The Bush administration sees its war against terrorists and tyrants in much the same way. The problem is not that Saddam Hussein is actively supporting al Qaeda, however much the Bush team would like to prove that. It's rather that authoritarian regimes throughout the Middle East support terrorism indirectly by continuing to produce generations of underemployed, unrepresented, and therefore radicalizable young people from whom Osama bin Laden and others like him draw their recruits.

The Bush administration has depleted the reservoir of support from allies it ought to have in place before embarking on such a high-risk strategy.

Bush has, to be sure, enlisted authoritarian allies in his war against terrorism—for the moment. So did Roosevelt when he welcomed the Soviet Union's help in the war against Nazi Germany and imperial Japan. But the Bush strategy has long-term as well as immediate implications, and these do not assume *indefinite* reliance on regimes like those that currently run Saudi Arabia, Egypt, and Pakistan. Reliance on Yasir Arafat has already ended.

The welcome These plans depend critically, however, on our being welcomed in Baghdad if we invade, as we were in Kabul. If we aren't, the whole strategy collapses, because it's premised on the belief that ordinary Iraqis will *prefer* an American occupation over the current conditions in which they live. There's no evidence that the Bush administration is planning the kind of military commitments the United States made in either of the two world wars, or even in Korea and Vietnam. This strategy relies on getting cheered, not shot at.

Who's to say, for certain, that this will or won't happen? A year ago, Afghanistan seemed the least likely place in which invaders could expect cheers, and yet they got them. It would be foolish to conclude from this experience, though, that it will occur everywhere. John F. Kennedy learned that lesson when, recalling successful interventions in Iran and Guatemala, he authorized the failed Bay of Pigs landings in Cuba. The trouble with Agincourts—even those that happen in Afghanistan—is the arrogance they can encourage, along with the illusion that victory itself is enough and that no follow-up is required. It's worth remembering that, despite Henry V, the French never became English.

Maintaining the moral high ground It's difficult to quantify the importance of this, but why should we need to? Just war theory has been around since St. Augustine. Our own Declaration of Independence invoked a decent respect for the opinions of humankind. Richard Overy's fine history of World War II devotes an entire chapter to the Allies' triumph in what he calls "the moral contest." Kennedy rejected a surprise attack against Soviet missiles in Cuba because he feared losing the moral advantage: Pearl Harbor analogies were enough to sink plans for preemption in a much more dangerous crisis than Americans face now. The Bush NSS acknowledges the multiplier effects of multilateralism: "no nation can build a safer, better world alone." These can hardly be gained though

unilateral action *unless that action itself commands multilateral support*.

The Bush team assumes we'll have the moral high ground, and hence multilateral support, if we're cheered and not shot at when we go into Baghdad and other similar places. No doubt they're right about that. They're seeking U.N. authorization for such a move and may well get it. Certainly, they'll have the consent of the U.S. Congress. For there lies behind their strategy an incontestable moral claim: that in some situations preemption is preferable to doing nothing. Who would not have preempted Hitler or Milosevic or Mohammed Atta, if given the chance?

Will Iraq seem such a situation, though, if we're not cheered in Baghdad? Can we count on multilateral support if things go badly? Here the Bush administration has not been thinking ahead. It's been dividing its own moral multipliers though its tendency to behave, on an array of multilateral issues ranging from the Kyoto Protocol to the Comprehensive Test Ban Treaty to the International Criminal Court, like a sullen, pouting, oblivious, and overmuscled teenager. As a result, it's depleted the reservoir of support from allies it ought to have in place before embarking on such a high-risk strategy.

There are, to be sure, valid objections to these and other initiatives the administration doesn't like. But it's made too few efforts to use *diplomacy*—by which I mean *tact*—to express these complaints. Nor has it tried to change a domestic political culture that too often relishes having the United States stand defiantly alone. The Truman administration understood that the success of containment abroad required countering isolationism at home. The Bush administration hasn't yet made that connection between domestic politics and grand strategy. That's its biggest failure of leadership so far.

The Bush strategy depends ultimately on *not* standing defiantly alone—just the opposite, indeed, for it claims to be pursuing values that, as the NSS puts it, are "true for every person, in every society." So this crack especially needs fixing before this vehicle departs for its intended destination. A nation that sets itself up as an example to the world in most things will not achieve that purpose by telling the rest of the world, in some things, to shove it.

WHAT IT MEANS

Despite these problems, the Bush strategy is right on target with respect to the new circumstances confronting the United States and its allies in the wake of September 11. It was sufficient, throughout the Cold War, to contain without seeking to reform authoritarian regimes: we left it to the Soviet Union to reform itself. The most important conclusion of the Bush NSS is that this Cold War assumption no longer holds. The intersection of radicalism with technology the world witnessed on that terrible morning means that the persistence of authoritarianism anywhere can breed resentments that can provoke terrorism that

Want to Know More?

See the White House's Web site for the **Bush National Security Strategy (NSS) report** of September 17, 2002, the **"axis of evil"** speech of January 29, the **West Point speech** of June 1, and the **U.N. speech** of September 12. The **Clinton administration's last NSS**, issued in December 1999, is available on the Web site of the National Archives and Records Administration.

More on the thinking and record of the Clinton administration can be found in Douglas Brinkley's **"Democratic Enlargement: The Clinton Doctrine"** (FOREIGN POLICY, Spring 1997) and the *FP* editors' **"Think Again: Clinton's Foreign Policy"** (FOREIGN POLICY, November/December 2000).

For Pericles's precedent in publicly discussing grand strategy, see Thucydides's *History of the Peloponnesian War*, translated by Rex Warner (Baltimore: Penguin, 1972). Carl von Clausewitz's *On War*, edited and translated by Michael Howard and Peter Paret (Princeton: Princeton University Press, 1976), is the standard edition of this classic. See John Lewis Gaddis's *Strategies of Containment: A Critical Appraisal of Postwar American National Security Policy* (New York: Oxford University Press, 1982) for a discussion of George F. Kennan's and John F. Kennedy's strategies.

For the career and influence of Paul Wolfowitz, see Bill Keller's **"The Sunshine Warrior"** (*New York Times Magazine*, September 22, 2002). For works by political scientists trying to explain the persistence of U.S. hegemony, see William C. Wohlforth's **"The Stability of a Unipolar World"** (*International Security*, Vol. 24, No. 1, 1999) and G. John Ikenberry's *After Victory: Institutions, Strategic Restraint, and the Rebuilding of Order After Major Wars* (Princeton: Princeton University Press, 2001).

The emerging consensus on the causes of terrorism in the Middle East can be traced in Fouad Ajami's *Dream Palace of the Arabs: A Generation's Odyssey* (New York: Vintage, 1999), Bernard Lewis's *What Went Wrong? Western Impact and Middle Eastern Response* (New York: Oxford University Press, 2002), Gilles Kepel's *Jihad: The Trail of Political Islam* (Cambridge: Harvard University Press, 2002), and the **"Arab Human Development Report: Creating Opportunities for Future Generations"** (New York: United Nations, 2002), available on the U.N. Web site.

The best discussion of Agincourt, apart from Shakespeare, is in Chapter 2 of John Keegan's *The Face of Battle: A Study of Agincourt, Waterloo, and the Somme* (New York: Penguin, 1983). Chapter 9 of Richard Overy's *Why the Allies Won* (London: Pimlico, 1995) discusses the "moral contest." For the influence of Pearl Harbor on Kennedy's rejection of preemption during the Cuban missile crisis, see Ernest R. May and Philip D. Zelikow, eds., *The Kennedy Tapes: Inside the White House During the Cuban Missile Crisis* (Cambridge: Harvard University Press, 1997).

For critiques of the Bush administration's strategy; see Brent Scowcroft's **"Don't Attack Saddam"** (*Wall Street Journal*, August 15, 2002) and Dean E. Murphy's **"Gore Calls Bush's Policy a Failure on Several Fronts"** (*New York Times*, September 24, 2002).

For links to relevant Web sites, access to the *FP* Archive, and a comprehensive index of related FOREIGN POLICY articles, go to **www.foreignpolicy.com**.

can do us grievous harm. There is a compellingly realistic reason now to complete the idealistic task Woodrow Wilson began more than eight decades ago: the world must be made safe for democracy, because otherwise democracy will not be safe in the world.

The Bush NSS report could be, therefore, the most important reformulation of U.S. grand strategy in over half a century. The risks are great—though probably no more than those confronting the architects of containment as the Cold War began. The pitfalls are plentiful—there are cracks to attend to before this vehicle departs for its intended destination. There's certainly no guarantee of suc-

cess—but as Clausewitz would have pointed out, there never is in anything that's worth doing.

We'll probably never know for sure what bin Laden and his gang hoped to achieve with the horrors they perpetrated on September 11, 2001. One thing seems clear though: it can hardly have been to produce this document, and the new grand strategy of transformation that is contained within it.

John Lewis Gaddis is the Robert A. Lovett professor of military and naval history at Yale University, and author, most recently, of The Landscape of History: How Historians Map the Past *(New York: Oxford University Press, 2002).*

From *Foreign Policy*, November/December 2002, pp. 50-57. © 2002 by the Carnegie Endowment for International Peace.

THE EAGLE HAS CRASH LANDED

Pax Americana is over. Challenges from Vietnam and the Balkans to the Middle East and September 11 have revealed the limits of American supremacy. Will the United States learn to fade quietly, or will U.S. conservatives resist and thereby transform a gradual decline into a rapid and dangerous fall?

By Immanuel Wallerstein

The United States in decline? Few people today would believe this assertion. The only ones who do are the U.S. hawks, who argue vociferously for policies to reverse the decline. This belief that the end of U.S. hegemony has already begun does not follow from the vulnerability that became apparent to all on September 11, 2001. In fact, the United States has been fading as a global power since the 1970s, and the U.S. response to the terrorist attacks has merely accelerated this decline. To understand why the so-called Pax Americana is on the wane requires examining the geopolitics of the 20th century, particularly of the century's final three decades. This exercise uncovers a simple and inescapable conclusion: The economic, political, and military factors that contributed to U.S. hegemony are the same factors that will inexorably produce the coming U.S. decline.

INTRO TO HEGEMONY

The rise of the United States to global hegemony was a long process that began in earnest with the world recession of 1873. At that time, the United States and Germany began to acquire an increasing share of global markets, mainly at the expense of the steadily receding British economy. Both nations had recently acquired a stable political base—the United States by successfully terminating the Civil War and Germany by achieving unification and defeating France in the Franco-Prussian War. From 1873 to 1914, the United States and Germany became the principal producers in certain leading sectors: steel and later automobiles for the United States and industrial chemicals for Germany.

The history books record that World War I broke out in 1914 and ended in 1918 and that World War II lasted from 1939 to 1945. However, it makes more sense to consider the two as a single, continuous "30 years' war" between the United States and Germany, with truces and local conflicts scattered in between. The competition for hegemonic succession took an ideological turn in 1933, when the Nazis came to power in Germany and began their quest to transcend the global system altogether,

seeking not hegemony within the current system but rather a form of global empire. Recall the Nazi slogan *ein tausendjähriges Reich* (a thousand-year empire). In turn, the United States assumed the role of advocate of centrist world liberalism—recall former U.S. President Franklin D. Roosevelt's "four freedoms" (freedom of speech, of worship, from want, and from fear)—and entered into a strategic alliance with the Soviet Union, making possible the defeat of Germany and its allies.

World War II resulted in enormous destruction of infrastructure and populations throughout Eurasia, from the Atlantic to the Pacific oceans, with almost no country left unscathed. The only major industrial power in the world to emerge intact—and even greatly strengthened from an economic perspective—was the United States, which moved swiftly to consolidate its position.

But the aspiring hegemon faced some practical political obstacles. During the war, the Allied powers had agreed on the establishment of the United Nations, composed primarily of countries that had been in the coalition against the Axis powers. The organization's critical feature was the Security Council, the only structure that could authorize the use of force. Since the U.N. Charter gave the right of veto to five powers—including the United States and the Soviet Union—the council was rendered largely toothless in practice. So it was not the founding of the United Nations in April 1945 that determined the geopolitical constraints of the second half of the 20th century but rather the Yalta meeting between Roosevelt, British Prime Minister Winston Churchill, and Soviet leader Joseph Stalin two months earlier.

The formal accords at Yalta were less important than the informal, unspoken agreements, which one can only assess by observing the behavior of the United States and the Soviet Union in the years that followed. When the war ended in Europe on May 8, 1945, Soviet and Western (that is, U.S., British, and French) troops were located in particular places—essentially, along a line in the center of Europe that came to be called the Oder-Neisse Line. Aside from a few minor adjustments, they stayed there. In hindsight, Yalta signified the agreement of both sides that they could stay there and that neither side would use force to push the other out. This tacit accord applied to Asia as well, as evinced by U.S. occupation of Japan and the division of Korea. Politically, therefore, Yalta was an agreement on the status quo in which the Soviet Union controlled about one third of the world and the United States the rest.

Washington also faced more serious military challenges. The Soviet Union had the world's largest land forces, while the U.S. government was under domestic pressure to downsize its army, particularly by ending the draft. The United States therefore decided to assert its military strength not via land forces but through a monopoly of nuclear weapons (plus an air force capable of deploying them). This monopoly soon disappeared: By 1949, the Soviet Union had developed nuclear weapons as well. Ever since, the United States has been reduced to trying to prevent the acquisition of nuclear weapons (and chemical and biological weapons) by additional powers, an effort that, in the 21st century, does not seem terribly successful.

Until 1991, the United States and the Soviet Union coexisted in the "balance of terror" of the Cold War. This status quo was tested seriously only three times: the Berlin blockade of 1948–49, the Korean War in 1950–53, and the Cuban missile crisis of 1962. The result in each case was restoration of the status quo. Moreover, note how each time the Soviet Union faced a political crisis among its satellite regimes—East Germany in 1953, Hungary in 1956, Czechoslovakia in 1968, and Poland in 1981—the United States engaged in little more than propaganda exercises, allowing the Soviet Union to proceed largely as it deemed fit.

Of course, this passivity did not extend to the economic arena. The United States capitalized on the Cold War ambiance to launch massive economic reconstruction efforts, first in Western Europe and then in Japan (as well as in South Korea and Taiwan). The rationale was obvious: What was the point of having such overwhelming productive superiority if the rest of the world could not muster effective demand? Furthermore, economic reconstruction helped create clientelistic obligations on the part of the nations receiving U.S. aid; this sense of obligation fostered willingness to enter into military alliances and, even more important, into political subservience.

Finally, one should not underestimate the ideological and cultural component of U.S. hegemony. The immediate post-1945 period may have been the historical high point for the popularity of communist ideology. We easily forget today the large votes for Communist parties in free elections in countries such as Belgium, France, Italy, Czechoslovakia, and Finland, not to mention the support Communist parties gathered in Asia—in Vietnam, India, and Japan—and throughout Latin America. And that still leaves out areas such as China, Greece, and Iran, where free elections remained absent or constrained but where Communist parties enjoyed widespread appeal. In response, the United States sustained a massive anticommunist ideological offensive. In retrospect, this initiative appears largely successful: Washington brandished its role as the leader of the "free world" at least as effectively as the Soviet Union brandished its position as the leader of the "progressive" and "anti-imperialist" camp.

ONE, TWO, MANY VIETNAMS

The United States' success as a hegemonic power in the postwar period created the conditions of the nation's hegemonic demise. This process is captured in four symbols: the war in Vietnam, the revolutions of 1968, the fall of the Berlin Wall in 1989, and the terrorist attacks of September 2001. Each symbol built upon the prior one, culminating in the situation in which the United States currently finds itself—a lone superpower that lacks true power, a world leader nobody follows and few respect, and a nation drifting dangerously amidst a global chaos it cannot control.

What was the Vietnam War? First and foremost, it was the effort of the Vietnamese people to end colonial rule and establish their own state. The Vietnamese fought the French, the Japanese, and the Americans, and in the end the Vietnamese won—quite an achievement, actually. Geopolitically, however, the war represented a rejection of the Yalta status quo by populations then labeled as Third World. Vietnam became such a powerful symbol because Washington was foolish enough to invest its full military might in the struggle, but the United States still lost. True, the United States didn't deploy nuclear weapons (a decision certain myopic groups on the right have long reproached), but such use would have shattered the Yalta accords and might have produced a nuclear holocaust—an outcome the United States simply could not risk.

But Vietnam was not merely a military defeat or a blight on U.S. prestige. The war dealt a major blow to the United States' ability to remain the world's dominant economic power. The conflict was extremely expensive and more or less used up the U.S. gold reserves that had been so plentiful since 1945. Moreover, the United States incurred these costs just as Western Europe and Japan experienced major economic upswings. These conditions ended U.S. preeminence in the global economy. Since the late 1960s, members of this triad have been nearly economic equals, each doing better than the others for certain periods but none moving far ahead.

When the revolutions of 1968 broke out around the world, support for the Vietnamese became a major rhetorical component. "One, two, many Vietnams" and "Ho, Ho, Ho Chi Minh" were chanted in many a street, not least in the United States. But the 1968ers did not merely condemn U.S. hegemony. They condemned Soviet collusion with the United States, they condemned Yalta, and they used or adapted the language of the Chinese cultural revolutionaries who divided the world into two camps—the two superpowers and the rest of the world.

The denunciation of Soviet collusion led logically to the denunciation of those national forces closely allied with the Soviet Union, which meant in most cases the traditional Communist parties. But the 1968 revolutionaries also lashed out against other components of the Old Left—national liberation movements in the Third World, social-democratic movements in Western Europe, and New Deal Democrats in the United States—accusing them, too, of collusion with what the revolutionaries generically termed "U.S. imperialism."

The attack on Soviet collusion with Washington plus the attack on the Old Left further weakened the legitimacy of the Yalta arrangements on which the United States had fashioned the world order. It also undermined the position of centrist liberalism as the lone, legitimate global ideology. The direct political consequences of the world revolutions of 1968 were minimal, but the geopolitical and intellectual repercussions were enormous and irrevocable. Centrist liberalism tumbled from the throne it had occupied since the European revolutions of 1848 and that had enabled it to co-opt conservatives and radicals alike. These ideologies returned and once again represented a real gamut of choices. Conservatives would again become conservatives, and radicals, radicals. The centrist liberals did not disappear, but they were cut down to size. And in the process, the official U.S. ideological position—antifascist, anticommunist, anticolonialist—seemed thin and unconvincing to a growing portion of the world's populations.

THE POWERLESS SUPERPOWER

The onset of international economic stagnation in the 1970s had two important consequences for U.S. power. First, stagnation resulted in the collapse of "developmentalism"—the notion that every nation could catch up economically if the state took appropriate action—which was the principal ideological claim of the Old Left movements then in power. One after another, these regimes faced internal disorder, declining standards of living, increasing debt dependency on international financial institutions, and eroding credibility. What had seemed in the 1960s to be the successful navigation of Third World decolonization by the United States—minimizing disruption and maximizing the smooth transfer of power to regimes that were developmentalist but scarcely revolutionary—gave way to disintegrating order, simmering discontents, and unchanneled radical temperaments. When the United States tried to intervene, it failed. In 1983, U.S. President Ronald Reagan sent troops to Lebanon to restore order. The troops were in effect forced out. He compensated by invading Grenada, a country without troops. President George H.W. Bush invaded Panama, another country without troops. But after he intervened in Somalia to restore order, the United States was in effect forced out, somewhat ignominiously. Since there was little the U.S. government could actually do to reverse the trend of declining hegemony, it chose simply to ignore this trend—a policy that prevailed from the withdrawal from Vietnam until September 11, 2001.

Meanwhile, true conservatives began to assume control of key states and interstate institutions. The neoliberal offensive of the 1980s was marked by the Thatcher and Reagan regimes and the emergence of the International Monetary Fund (IMF) as a key actor on the world scene. Where once (for more than a century) conservative forces had attempted to portray themselves as wiser liberals, now centrist liberals were compelled to argue that they were more effective conservatives. The conservative programs were clear. Domestically, conservatives tried to enact policies that would reduce the cost of labor, minimize environmental Constraints on producers, and cut back on state welfare benefits. Actual successes were modest, so conservatives then moved vigorously into the international arena. The gatherings of the World Economic Forum in Davos provided a meeting ground for elites and the media. The IMF provided a club for finance ministers and central bankers. And the United States pushed for the creation of the World Trade Organization to enforce free commercial flows across the world's frontiers.

While the United States wasn't watching, the Soviet Union was collapsing. Yes, Ronald Reagan had dubbed the Soviet Union an "evil empire" and had used the rhetorical bombast of calling for the destruction of the Berlin Wall, but the United States didn't really mean it and certainly was not responsible for the Soviet Union's downfall. In truth, the Soviet Union and its East European imperial zone collapsed because of popular disillusionment with the Old Left in combination with Soviet leader Mikhail Gorbachev's efforts to save his regime by liquidating Yalta and instituting internal liberalization (perestroika plus glasnost). Gorbachev succeeded in liquidating Yalta but not in saving the Soviet Union (although he almost did, be it said).

The United States was stunned and puzzled by the sudden collapse, uncertain how to handle the consequences. The collapse of communism in effect signified the collapse of liberalism, removing the only ideological justification behind U.S. hegemony, a justification tacitly supported by liberalism's ostensible ideological opponent. This loss of legitimacy led directly to the Iraqi invasion of Kuwait, which Iraqi leader Saddam Hussein would never have dared had the Yalta arrangements remained in place. In retrospect, U.S. efforts in the Gulf War accomplished a truce at basically the same line of departure. But can a hegemonic power be satisfied with a tie in a war with a middling regional power? Saddam demonstrated that one could pick a fight with the United States and get away with it. Even more than the defeat in Vietnam, Saddam's brash challenge has eaten at the innards of the U.S. right, in particular those known as the hawks, which explains the fervor of their current desire to invade Iraq and destroy its regime.

Between the Gulf War and September 11, 2001, the two major arenas of world conflict were the Balkans and the Middle East. The United States has played a major diplomatic role in both regions. Looking back, how different would the results have been had the United States assumed a completely isolationist position? In the Balkans, an economically successful multinational state (Yugoslavia) broke down, essentially into its component parts. Over 10 years, most of the resulting states have engaged in a process of ethnification, experiencing fairly brutal violence, widespread human rights violations, and outright wars. Outside intervention—in which the United States figured most prominently—brought about a truce and ended the most egregious violence, but this intervention in no way reversed the ethnification, which is now consolidated and somewhat legitimated. Would these conflicts have ended differently without U.S. involvement? The violence might have continued longer, but the basic results would probably not h ave been too different. The picture is even grimmer in the Middle East, where, if anything, U.S. engagement has been deeper and its failures more spectacular. In the Balkans and the Middle East alike, the United States has failed to exert its hegemonic clout effectively, not for want of will or effort but for want of real power.

THE HAWKS UNDONE

Then came September 11—the shock and the reaction. Under fire from U.S. legislators, the Central Intelligence Agency (CIA) now claims it had warned the Bush administration of possible threats. But despite the CIA's focus on al Qaeda and the agency's intelligence expertise, it could not foresee (and therefore, prevent) the execution of the terrorist strikes. Or so would argue CIA Director George Tenet. This testimony can hardly comfort the U.S. government or the American people. Whatever else historians may decide, the attacks of September 11, 2001, posed a major challenge to U.S. power. The persons responsible did not represent a major military power. They were members of a nonstate force, with a high degree of determination, some money, a band of dedicated followers, and a strong base in one weak state. In short, militarily, they were nothing. Yet they succeeded in a bold attack on U.S. soil.

George W Bush came to power very critical of the Clinton administration's handling of world affairs. Bush and his advisors did not admit—but were undoubtedly aware—that Clinton's path had been the path of every U.S. president since Gerald Ford, including that of Ronald Reagan and George H.W. Bush. It had even been the path of the current Bush administration before September 11. One only needs to look at how Bush handled the downing of the U.S. plane off China in April 2001 to see that prudence had been the name of the game.

Following the terrorist attacks, Bush changed course, declaring war on terrorism, assuring the American people that "the outcome is certain" and informing the world that "you are either with us or against us." Long frustrated by even the most conservative U.S. administrations, the hawks finally came to dominate American policy. Their position is clear: The United States wields overwhelming military power, and even though countless foreign leaders consider it unwise for Washington to flex its military muscles, these same leaders cannot and will not do anything if the United States simply imposes its will on the rest. The hawks believe the United States should act as an imperial power for two reasons: First, the United States can get away with it. And second, if Washington doesn't exert its force, the United States will become increasingly marginalized.

Today, this hawkish position has three expressions: the military assault in Afghanistan, the de facto support for the Israeli attempt to liquidate the Palestinian Authority, and the invasion of Iraq, which is reportedly in the military preparation stage. Less than one year after the September 2001 terrorist attacks, it is perhaps too early to assess what such strategies will accomplish. Thus far, these schemes have led to the overthrow of the Taliban in Afghanistan (without the complete dismantling of al Qaeda or the capture of its top leadership); enormous destruction in Palestine (without rendering Palestinian leader Yasir Arafat "irrelevant," as Israeli Prime Minister Ariel Sharon said he is); and heavy opposition from U.S. allies in Europe and the Middle East to plans for an invasion of Iraq.

The hawks' reading of recent events emphasizes that opposition to U.S. actions, while serious, has remained largely verbal. Neither Western Europe nor Russia nor China nor Saudi Arabia has seemed ready to break ties in serious ways with the United States. In other words, hawks believe, Washington has indeed gotten away with it. The hawks assume a similar outcome will occur when the U.S. military actually invades Iraq and after that, when the United States exercises its authority elsewhere in the world, be it in Iran, North Korea, Colombia, or perhaps Indonesia. Ironically, the hawk reading has largely become the reading of the international left, which has been screaming about U.S. policies—mainly because they fear that the chances of U.S. success are high.

But hawk interpretations are wrong and will only contribute to the United States' decline, transforming a gradual descent into a much more rapid and turbulent fall. Specifically, hawk approaches will fail for military, economic, and ideological reasons.

Undoubtedly, the military remains the United States' strongest card; in fact, it is the only card. Today, the United States wields the most formidable military apparatus in the world. And if claims of new, unmatched military technologies are to be believed, the U.S. military edge over the rest of the world is considerably greater today than it was just a decade ago. But does that mean, then, that the United States can invade Iraq, conquer it rapidly, and install a friendly and stable regime? Unlikely. Bear in mind that of the three serious wars the U.S. military has fought since 1945 (Korea, Vietnam, and the Gulf War), one ended in defeat and two in draws—not exactly a glorious record.

Saddam Hussein's army is not that of the Taliban, and his internal military control is far more coherent. A U.S. invasion would necessarily involve a serious land force, one that would have to fight its way to Baghdad and would likely suffer significant casualties. Such a force would also need staging grounds, and Saudi Arabia has made clear that it will not serve in this capacity. Would Kuwait or Turkey help out? Perhaps, if Washington calls in all its chips. Meanwhile, Saddam can be expected to deploy all weapons at his disposal, and it is precisely the U.S. government that keeps fretting over how nasty those weapons might be. The United States may twist the arms of regimes in the region, but popular sentiment clearly views the whole affair as reflecting a deep anti-Arab bias in the United States. Can such a conflict be won? The British General Staff has apparently already informed Prime Minister Tony Blair that it does not believe so.

And there is always the matter of "second fronts." Following the Gulf War, U.S. armed forces sought to prepare for the possibility of two simultaneous regional wars. After a while, the Pentagon quietly abandoned the idea as impractical and costly. But who can be sure that no potential U.S. enemies would strike when the United States appears bogged down in Iraq?

Consider, too, the question of U.S. popular tolerance of nonvictories. Americans hover between a patriotic fervor that lends support to all wartime presidents and a deep isolationist urge. Since 1945, patriotism has hit a wall whenever the death toll has risen. Why should today's reaction differ? And even if the hawks (who are almost all civilians) feel impervious to public opinion, U.S. Army generals, burnt by Vietnam, do not.

And what about the economic front? In the 1980s, countless American analysts became hysterical over the Japanese economic miracle. They calmed down in the 1990s, given Japan's well-publicized financial difficulties. Yet after overstating how quickly Japan was moving forward, U.S. authorities now seem to be complacent, confident that Japan lags far behind. These days, Washington seems more inclined to lecture Japanese policymakers about what they are doing wrong.

Such triumphalism hardly appears warranted. Consider the following April 20, 2002, *New York Times* report: "A Japanese laboratory has built the world's fastest computer, a machine so powerful that it matches the raw processing power of the 20 fastest American computers combined and far outstrips the previous leader, an I.B.M.-built machine. The achievement... is evidence that a technology race that most American engineers thought they were winning handily is far from over." The analysis goes on to note that there are "contrasting scientific and technological priorities" in the two countries. The Japanese machine is built to analyze climatic change, but U.S. machines are designed to simulate weapons. This contrast embodies the oldest story in the history of hegemonic powers. The dominant power concentrates (to its detriment) on the military; the candidate for successor concentrates on the economy. The latter has always paid off, handsomely. It did for the United States. Why should it not pay off for Japan as well, perhaps in alliance with China?

Finally, there is the ideological sphere. Right now, the U.S. economy seems relatively weak, even more so considering the exorbitant military expenses associated with hawk strategies. Moreover, Washington remains politically isolated; virtually no one (save Israel) thinks the hawk position makes sense or is worth encouraging. Other nations are afraid or unwilling to stand up to Washington directly, but even their foot-dragging is hurting the United States.

Yet the U.S. response amounts to little more than arrogant arm-twisting. Arrogance has its own negatives. Calling in chips means leaving fewer chips for next time, and surly acquiescence breeds increasing resentment. Over the last 200 years, the United States acquired a considerable amount of ideological credit. But these days, the United States is running through this credit even faster than it ran through its gold surplus in the 1960s.

The United States faces two possibilities during the next 10 years: It can follow the hawks' path, with negative consequences for all but especially for itself. Or it can realize that the negatives are too great. Simon Tisdall of the *Guardian* recently argued that even disregarding international public opinion, "the U.S. is not able to fight a successful Iraqi war by itself without incurring immense damage, not least in terms of its economic interests and its energy supply. Mr. Bush is reduced to talking tough and looking ineffectual." And if the United States still invades Iraq and is then forced to withdraw it will look even more ineffectual.

President Bush's options appear extremely limited, and there is little doubt that the United States will continue to decline as a decisive force in world affairs over the next decade. The real question is not whether U.S. hegemony is waning but whether the United States can devise a way to descend gracefully, with minimum damage to the world, and to itself.

Want to Know More?

This article draws from the research reported in Terence K. Hopkins and Immanuel Wallerstein's, eds., *The Age of Transition: Trajectory of the World-System, 1945–2025* (London: Zed Books, 1996). In his new book, *The Paradox of American Power: Why the World's Only Superpower Can't Go It Alone* (New York: Oxford University Press, 2002), Joseph S. Nye Jr. argues that the United States can remain on top, provided it emphasizes multilateralism. For a less optimistic view, see Thomas J. McCormick's *America's Half-Century: United States Foreign Policy in the Cold War and After*, 2nd ed. (Baltimore: Johns Hopkins University Press, 1995). David Calleo's latest book, *Rethinking Europe's Future* (Princeton: Princeton University Press, 2001), cogently analyzes the ins and outs of the European Union and its potential impact on U.S. power in the world.

In 1993, the Norwegian Nobel Committee convened a meeting of leading international analysts to discuss the role and influence of superpowers throughout history. Their analyses can be found in Geir Lundestad's, ed., *The Fall of Great Powers: Stability, Peace and Legitimacy* (Oslo: Scandinavian University Press, 1994), which includes essays by William H. McNeill, Istvan Deak, Alec Nove, Wolfgang J. Mommsen, Robert Gilpin, Wang Gungwu, John Lewis Gaddis, and Paul Kennedy, among others. Eric Hobsbawm offers a splendid geopolitical analysis of the 20th century in *The Age of Extremes: A History of the World, 1914–1991* (New York: Pantheon, 1994). Giovanni Arrighi, Beverly J. Silver, and their collaborators take a longer view of hegemonic transitions over the centuries—from Dutch to British, from British to American, from American to some uncertain future hegemon—in *Chaos and Governance in the Modern World System* (Minneapolis: University of Minnesota Press, 1999). Finally, it is always useful to return to Andre Fontaine's classic *History of the Cold War* (New York: Pantheon, 1968).

FOREIGN POLICY's extensive coverage of American hegemony and the U.S. role in the world includes, most recently, **"In Praise of Cultural Imperialism?"** (Summer 1997) by David Rothkopf, **"The Benevolent Empire"** (Summer 1998) by Robert Kagan, **"The Perils of (and for) an Imperial America"** (Summer 1998) by Charles William Maynes, **"Americans and the World: A Survey at the Century's End"** (Spring 1999) by John E. Rielly, **"Vox Americani"** (September/October 2001) by Steven Kull, and **"The Dependent Colossus"** (March/April 2002) by Joseph S. Nye Jr.

For links to relevant Web sites, access to the *FP* Archive, and a comprehensive index of related FOREIGN POLICY articles, go to **www.foreignpolicy.com.**

Immanuel Wallerstein is a senior research scholar at Yale University and author of, most recently, The End of the World As We Know It: Social Science for the Twenty-First Century *(Minneapolis: University of Minnesota Press, 1999).*

Wednesday, December 26, 2001

Economic Crossroads on the Line

Security Fears Have U.S. and Canada Rethinking Life at 49th Parallel

By Michael Grunwald
Washington Post Staff Writer

WINDSOR, Ontario

Long lines of trucks are always inching across the Ambassador Bridge, hauling cargo between the United States and Canada like freight trains of infinite length. This is the busiest border crossing in North America—more goods pass between Detroit and Windsor than between the United States and the entire European Union—but David Jolly, the bridge's manager, now watches the relentless march of commerce with two unpleasant thoughts.

The first is: *Hurry up!* The border between Detroit, synonymous with the Big Three automakers, and Windsor, the "Automotive Capital of Canada," basically bisects a huge assembly line. The flow of traffic and trade has slowly recovered since Sept. 11—a day of 20-mile backups, 14-hour waits and multimillion-dollar factory shutdowns—but volume remains down, security hassles up and delays a bit unpredictable. "See that line of trucks just sitting on the bridge?" Jolly asks. "That's bad. That's money."

His second thought is: *Uh-oh.* That's because trucks that enter the bridge are not checked until they reach the other side. So if terrorists decide to blow up the bridge with a truck bomb, Jolly doesn't see any way to stop them. "Everyone on that bridge would be fresh meat," he says. "It's stupid, but it's reality right now."

Reality may change, though: The events of Sept. 11 have left the U.S.-Canadian border in flux. The short-term response has been to increase barriers along the 5,500-mile frontier. But the dramatic costs of doing so have helped energize long-term commitments from both nations to use technology to create a "smarter border," decreasing barriers yet increasing security, reordering life at the 49th parallel for the forseeable future.

The terrorist attacks quickly inspired new security procedures and binational agreements, while redirecting money and manpower to long-neglected border agencies.

Before Sept. 11, for example, half the border's 126 official crossings were unguarded at night; now the orange cones that used to block U.S. entry roads have been supplemented by people. Customs agents in both countries are asking far more questions and searching far more vehicles and asking to see passports or birth certificates as well as driver's licenses. Programs designed to help frequent border-crossers zip through designated customs lanes were suspended for fear that terrorists might already have the required electronic passes.

The Sept. 11 attacks have left the Ambassador Bridge between Detroit and Windsor—and the rest of the U.S.-Canadian border—in flux.

Now, the U.S. Coast Guard stops all boats that cross the border in the Great Lakes and provides escorts for all oil and gas tankers. The Border Patrol shifted 100 agents here from the Mexican border; the National Guard was assigned here temporarily, then indefinitely. The agency that oversees the Ambassador Bridge is spending an unsustainable $50,000 a week on private security. Traffic is no longer allowed to pile up inside the nearby Detroit-Windsor Tunnel, reducing the potential casualty count of a rush-hour bombing but creating gridlock on nearby city streets.

The result is that the world's longest undefended border is slightly more defended, and more complex to cross. So fewer people are coming across, and businesses that depend on easy cross-border traffic are bleeding money. Now U.S. and Canadian leaders eager to preserve the world's largest trading partnership are shifting focus, dusting off a bevy of reports and accords that were supposed to stop the strangling of trade at the border long ago.

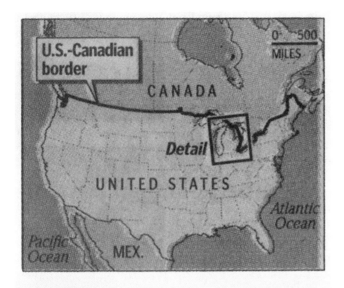

THE WASHINGTON POST

defense minister of Canada who is now president of the Canadian Manufacturers and Exporters. "For years, we've been running this border in the dumbest possible way. Now we need a fundamental rethinking of what this border means."

'Dying on the Vine'

Windsor is just south of Detroit, just across the Detroit River. The Big Three all have factories here, too. But this suburban-style city of 200,000 people will never be confused with Detroit. "It's that chummy, everybody-knows-everybody flavor that we take pride in most," explains Windsor's Web site. "From minor-league hockey teams to charity groups, we're all one big happy family."

In normal times, Windsor hosts 9 million visitors a year, most of them American day-trippers. That's more Americans than visit Montreal or Toronto or Vancouver. But since Sept. 11, in the words of Detroit Regional Chamber of Commerce President Richard Blouse, Windsor has been "dying on the vine." Jim Yanchula, Windsor's downtown revitalization chief, says the city is "hanging on by our fingernails."

Dave Jolly, the Ambassador Bridge manager, says, "Everyone on that bridge would be fresh meat" if terrorists decided to blow it up." It's stupid, but it's reality."

The problem, beyond the general economic slowdown, is that American leisure travelers who used to pop over the border to take advantage of Canada's weaker dollar, lower drinking age or ritzier casinos are staying home. Traffic at the Detroit-Windsor Tunnel is down 40 percent; one-fourth of its workers have been laid off. The Ambassador Bridge, which opened a few weeks after the stock market crash of 1929, has not struggled this badly since it went bankrupt during the Great Depression. At first, Windsor entrepreneurs thought the American bridge and tunnel crowd—especially Arab Americans, who are heavily concentrated around Detroit—just didn't feel like going out. Now it's clear the fear is border-specific.

So Freed's of Windsor, a huge clothing store, is offering customers free alterations while they wait, and a free meal at the restaurant across the street, but its suits have been staying on the racks. Windsor's Raceway has offered 2-for-1 dinner specials, but standardbred racing revenue is down 20 percent and slot machine revenue has fallen 30 percent. At Cheetah's, a downtown Windsor strip club, attendance is down 80 percent; some nights there are more naked women than clothed men. Even the Windsor casino, which covers six city blocks and was one of the world's most profitable gaming operations before Sept.

"Closing this border is not the answer," said U.S. Ambassador Paul Cellucci, a former Massachusetts governor. "We need that border open for business. More open than ever."

For some officials, the ultimate goal is a border so seamless it would barely be noticed. They rattle off staggering statistics from the post-NAFTA era: 200 million vehicle crossings a year, $1.4 billion in cross-border trade a day, one truck over the Ambassador Bridge every six seconds. The consequences of delays and disruptions, they say, can be equally staggering.

"Now everyone realizes that we can't just tweak and tinker," said Perrin Beatty, a former foreign minister and

11, laid off 762 employees, mostly dealers and money-counters.

"There wasn't as much money to count," explained casino spokesman Jim Mundy.

U.S. customs agents search vehicles entering the United States in Detroit via the 80-year-old Ambassador Bridge. Security checks at the crossing, the busiest in North America, sometimes create hours-long traffic backups.

The terrorist attacks also disrupted the daily routines of 30,000 commuters who cross the border every morning, including about 1,600 Canadian nurses who work in Detroit-area hospitals. After Sept. 11, at least two dozen nurses quit, and many others were forced to start their commutes as early as 3 a.m., not an easy time to get a babysitter.

In recent weeks, the overall downturn in traffic volume—along with the swift upturn in border personnel, which means more lanes open at major crossings—has actually reduced delays at the border. Still, border gridlock was an unpredictable problem here before the attacks, and now the fear of it is spreading. Remo Mancini, another Ambassador Bridge official, frets that if officials cannot restore confidence in the border, "vast regions of North America will be deindustrialized."

It was easy to see why one recent morning in the Volvo 18-wheeler that Rick Baker drives for a Canadian firm. Baker was hauling a 7,200-pound load of auto seat parts from Tillsonburg, Ontario, to Plymouth, Mich. That meant a trip across the four-lane Ambassador Bridge, a trip he sometimes makes four times a day.

At 10:19 a.m., Baker had cruised almost all the way to the toll plaza on the Canadian side, with only four trucks ahead of him in line. But at 10:29, there were still four trucks ahead of him, and an orange Coast Guard helicopter was flying ominously overhead.

"Uh-oh," Baker said. "We're stuck."

Stuck is bad for business. During the 1980s, most manufacturers adopted a "just-in-time" delivery system that moves parts directly from trucks to assembly lines. Since Sept. 11, some suppliers have been forced to maintain a "safety stock" in case of a temporary border shutdown, increasing warehousing costs. Truckers have budgeted more time on the road to make their strict "windows," which means steeper costs all around. A headline in Crain's Detroit Business warned: "Just-in-time could become just-in-case."

Baker started moving across the 1.5-mile-long bridge at 10:37; the truck lanes had been shut down for repainting. With 28,000 vehicles crossing the bridge—and occasionally breaking down on it—every day, shutdowns are not unusual these days. Two bomb threats shut down the bridge for over an hour. Another half-dozen times, Jolly closed lanes to move a single truck across the bridge after auto executives called to say they would have to shutter an assembly line—a potential $1 million-an-hour loss—if their shipment did not arrive soon.

In any case, Baker was not done with his morning crossing. He still had to get through customs and to the highway, which required an elaborate series of turns through clogged back roads. "No hats off to the design guys," he grumbled. A customs agent waved him through without asking him or his visitor for ID, but he still didn't reach the highway until 10:56.

For all the hype about security crackdowns, Baker's trailer had been inspected only three times since Sept. 11. But the border was still the time-consuming variable in his daily journeys.

"Thirty-seven minutes and they didn't even check us out," he said. "Don't you feel secure knowing you have such a secure border?"

A Tale of Two Nations

"You seem familiar, yet somehow strange," a man tells his date over martinis in a recent New Yorker cartoon. "Are you by any chance Canadian?"

It would be hard to find two allies closer than the United States and Canada, but the cultural gap between the nations extends beyond socialized medicine, hockey mania, the metric system and "eh"—not to mention trade disputes over potatoes, tomatoes and softwood lumber. It certainly extends to the border, which has always been viewed differently on either side.

For one thing, 90 percent of all Canadians live within 100 miles of the border, so Canada has paid much closer attention to it, while the United States has focused more on the flow of drugs and illegal immigrants from Mexico. The U.S. Border Patrol, for example, had 9,091 agents at the southern border before Sept. 11, and only 334 agents on the northern line, which is more than twice as long.

There was also an attitude gap. Canada sends 87 percent of its exports to the United States, so its emphasis has always been on facilitating trade. The United States sends 25 percent of its exports to Canada, the top trading partner of 38 states, but its economy is not so reliant on cross-border trade. Its emphasis has been on law enforcement, with remote cameras and sensors deployed all along the border. So while Canada has automated and revamped its customs system, for example, the United States has not even integrated the computers of its border agencies.

The gap is especially obvious at the Detroit-Windsor border, where the two countries could not even coordinate on a single electronic pass for frequent travelers. Before Sept. 11, convenience-minded Canada offered its CanPass program free to just about anyone with basic identification. The security-conscious United States fingerprinted and interviewed all applicants for its $129 SENTRI passes and required a slew of extra documents.

Still, when James W. Ziglar, commissioner of the Immigration and Naturalization Service, testified before Congress recently about the northern border, he conceded that it has been "accurately described as porous." The Justice Department's inspector general reported last year that one border sector had identified 65 smuggling corridors but had only 36 sensors to monitor them. The report noted that another sector was "6 to 7 percent effective" in stopping illegal crossings. The border is just too long—and in most areas, too rural—to make sure no one can sneak across. Last year, the Border Patrol caught only 11,000 illegal immigrants up here, vs. more than 1 million down south.

This was not a comforting thought after Sept. 11—not when Canada's intelligence service has warned that nearly every known terrorist organization has a Canadian branch, and when Canada has much more liberal asylum policies than the United States. Americans may remember a Palestinian refugee named Abu Mezer, who was caught three times sneaking into the United States from Canada before he was arrested in 1997 for plotting an attack on New York's subways. Ahmed Ressam, an Algerian who had forged a birth certificate to obtain a Canadian passport, was arrested in 1999 after taking a ferry into Washington state with explosives, which he planned to use to blow up Los Angeles International Airport during millennium celebrations.

In the days after Sept. 11, several U.S. politicians arguing for tighter borders cited rumors that some of the hijackers had crossed from Canada, but the rumors proved to be untrue. And as time has passed, the mood has shifted, with such officials as Treasury Secretary Paul H. O'Neill and Attorney General John D. Ashcroft joining their Canadian counterparts to push for a more cooperative border control system that could thwart terrorists without strangling trade. The two nations have begun taking steps to shift their overall enforcement focus to North American "perimeter security," to stop dangerous foreigners "where the water hits the rocks."

"Nobody's pointing fingers," Cellucci said. "There's just a real sense of urgency. We all recognize that if this border becomes an impediment, that's one more way for the terrorists to win."

Giving Fred a Break

The way to stop the terrorists from winning, Jolly says, is to give Fred a break.

Fred is an American guy, or maybe a Canadian guy. He crosses the border every day to work at, say, General Motors. He's no danger to anyone, but he has to wait in customs lines every morning and evening and answer the same rote questions every morning and evening.

"Look, the guy you've never seen before, knock yourself out, take his car apart, make sure he's not a bad guy,"

Jolly says. "But leave Fred alone. Hassling Fred is not the solution."

One solution is giving electronic passes—with fingerprints or other biometric identifiers to make sure they're not stolen—to frequent travelers and truckers who pass background checks. Tom Ridge, the Bush administration's homeland security director, recently announced that one of the array of suspended electronic-pass programs—a rare binational program limited to the Port Huron-Sarnia crossing—would be reinstated and expanded. Canada is also launching a Customs Self-Assessment program to use transponders and FedEx-style tracking technology to let importers pay duties electronically, with only occasional random inspections at the border to make sure they are reporting cargo honestly. That should help Fred. And if stopping enemies at the border is like finding needles in a haystack, these programs should dramatically shrink the haystack.

Bus driver Kiely Smith, who ferries gamblers between Detroit and Windsor, says border agents have even searched her bus's engine compartment.

Meanwhile, the two countries have agreed to share more intelligence and to beef up seaport and airport security to help keep terrorists out of North America. Canada has begun to tighten some of its asylum policies, and Ridge has floated a plan to streamline the alphabet soup of U.S. border agencies. A Windsor city effort to post bridge and tunnel waiting times at *www.bordernow.com* has helped reduce uncertainty about delays.

But there are still snags. The Bush administration has scaled back congressional proposals to triple U.S. border personnel. And Jolly's *uh-oh* problem, for now, remains a problem: U.S. officials have proposed posting American customs agents in Canada and vice versa so they could detect an explosives-laden truck *before* it reached the Ambassador Bridge or the nearby tunnel, but Canadians protective of their sovereignty have resisted. The National Guard has begun spot-checking a few vehicles as they leave Detroit, but the bridge and tunnel operators say they are still losing sleep over their vulnerability.

Jolly's *hurry up* problem is not really solved, either. Even if increased manpower helps keep more lanes open, even if the electronic passes divert most vehicles into designated lanes, neither Detroit-Windsor crossing is well-equipped to deal with modern traffic streams—and planners expect commercial traffic to triple over the next two decades. The bridge and tunnel, both 80 years old, empty into narrow and jammed city streets. Plans to link them to highways, or add customs booths, or even build a third crossing had gained little traction even before war and recession began squeezing budgets.

"The system was already broken before September 11th, and then it collapsed entirely," Beatty said. "Now we have to resolve the issues that were blocking trade before September 11th."

While they're at it, Elsie Ciaglia mused one recent afternoon, they might want to resolve the issues blocking the teal-green 12:30 p.m. Commuter Express bus to the Windsor casino. Ridership was way down for two months after Sept. 11, but recently the bus was full again—in part because all riders got $10 from the casino plus a $15 meal voucher with their $5 trip. Ciaglia, a retired secretary, was riding it once a week from her local Kmart to the nickel slots, but she was sick of the holdup at the bridge. "I bet this takes 20 minutes," she griped as the bus approached customs at 1:45 p.m.

The 55 elderly passengers all had to disembark and produce documents—the new standard procedure. There was much jostling and griping. "What a mess!" growled Angelo Pederiva, 80, a retired bricklayer with an Italian accent as thick as Chico Marx's. "Do we look like terrorists?" asked Lillian Dobbins, who was out of her house for the first time after spending the two months after Sept. 11 watching Tom Brokaw. "This isn't so bad," replied the driver, Kiely Smith. "One time they even made me open the engine compartment."

At 2:07, the bus was back on the bridge. Ciaglia was right: 22 minutes to check out the old folks. "I should have bet money!" she crowed. "I won't win in Canada, that's for sure."

Article 16

Canada's military 'needs overhaul'

By Lee Carter

BBC correspondent in Toronto

A senate committee in Canada has said it wants the country's military to withdraw all its forces from overseas duty for two years.

In a report just released in Ottawa, the committee says that in addition, the Canadian Government must immediately spend billions of dollars to stop the country's armed forces from collapsing.

> "We talk about the capital equipment crisis, which we refer to as Canada's antique roadshow"
>
> **Senator Colin Kenney**

The report comes amid continuing pressure on Canada from Washington to do the same.

There have been several warnings over the past few years about Canada's underfunded, disintegrating armed forces.

There have also been embarrassing incidents of ageing helicopters crashing into the sea, and cracked submarines taken out of commission.

But this senate committee report is by far the most blunt and scathing.

'Too little training'

Committee chairman Senator Colin Kenney says the senators were appalled by what they found.

"There are too few people doing the work, and there's too little training," he said.

"We talk about the operational crisis, where there's insufficient funding for operations, maintenance and infrastructure. And we talk about the capital equipment crisis, which we refer to as Canada's antique roadshow."

The warning echoes what the United States has been saying for some time.

Worry about the state of Canada's military apparently reaches the highest levels of the Bush administration.

'Weak link'

Interviewed during Monday's remembrance day ceremonies in Ottawa, US ambassador Paul Cellucci repeated his by now familiar refrain—for Canada to spend much more on defence.

At the moment the country spends substantially less than Portugal and yet remains a member of the G8 and Nato.

The US is worried that Canada is a weak link in the security of North America from terrorist threats.

Prime Minister Jean Chretien has consistently shown little interest in the subject.

But he will step down in 2004 and there is evidence that other senior figures in his Liberal government, including the finance minister, have become so embarrassed by the crisis they are prepared to defy him.

UNIT 5

Latin America

Unit Selections

Key Points to Consider

- What are some of the most serious threats to democracy in Latin America today?

- What should the United States, Canada, and European countries do to promote democracy and ease the economic crisis in Argentina?

- Which model do you think is more appropriate for Latin American countries: the development path taken by Mexico or the international role taken by Brazil? Defend your answer.

 **Links: www.dushkin.com/online/**
These sites are annotated in the World Wide Web pages.

Inter-American Dialogue
http://www.iadialog.org

For nearly two centuries, the United States viewed Latin America as being exclusively within the U.S. sphere of influence. But several countries in the region are developing independent international security policies. Over the past two decades, most countries in the region shifted from military dictatorships to democracies. Civilian-led governments adopted neoliberal economic reforms and followed the recommendations of the World Bank and the International Monetary Fund: privatize state-owned corporations, reduce public payrolls and subsidies, and promote free trade and direct foreign investment. Political and economic change occurred during a period of prosperity, but few economic benefits trickled down to most citizens, who remained poor. Increased numbers of citizens in most countries are now demanding a greater share of the wealth created in recent years. Democratic development and economic reform are now firmly established as the overarching norms in the hemisphere but the global economic slowdown, currency and political crises, and continuing insurgencies in several countries are creating challenges. Forrest Colburn in "Fragile Democracies" explains that while Latin America's democracies are not in danger of collapsing, real problems persist. Provocatively, Colburn predicts, "Mexico may be the bellwether rather than an exception in the future."

Few Americans appreciate that North American Free Trade Agreement (NAFTA) countries purchase 40 percent of U.S. exports. The state of the Canadian and Mexican economies is as important for the continued health of the U.S. economy as are economic trends in large states such as Texas. NAFTA has been in existence for nearly a decade. Economists estimate that about 350,000 manufacturing jobs were lost in the United States due to the pact while 2 million better-paying jobs were created. The Bush administration supports creating a larger free-trade zone for all the Americas that would extend from Canada to the tip of South America.

A future Free Trade Area of the Americas would include 34 countries and 800 million people. When and if such a large free-trade zone will come into existence remains uncertain. Increased risks associated with growing economic interdependence throughout the Western Hemisphere may mean that the United States will be more vulnerable to the problems of its neighbors in the future. Recent poll data also indicates strong opposition in Mexico where many farmers believe that their livelihood is threatened by tough competition from U.S. farmers who still receive large subsidies from the U.S. government. Other critics charge that free-trade policies, as long as agricultural subsidies continue, will kill any hope that developing countries would be able to compete with the United States.

The reason why fewer than 30 percent of Mexicans (compared with 48 percent of Americans and 38 percent of Canadians) believe that their country has been a winner as a result of

NAFTA is tied to recent economic trends. In the early 1990s, when free market reforms were adopted, the economic growth rate throughout Central America was approximately 3 percent. This growth plummeted to .8 percent in 1995 after Mexico's economy crashed. The backlash against free trade, government corruption, and the government's $62 billion bank bailout in the late 1980s led to a loss of power by the long-ruling Institutional Revolutionary Party (PRI) in Mexico in 2000. Vicente Fox, a representative from the National Action Party, or PAN, is the first president to be elected from outside the former ruling party in 71 years. Fox must now deal with a Congress that is no longer a rubber stamp because no party holds a majority in either house. The changes represent a fundamentally altered balance of power in Mexico where the president traditionally operated with near omnipotence.

Most analysts hailed the political shift in Mexico as a positive one in its efforts to avoid a social explosion. Continued American aid and the safety valve of emigration to the United States for unemployed Mexicans are other important factors that helped maintain political stability in Mexico. Future American aid, however, is likely to be contingent on how Americans perceive Mexico's role as a source of illegal drugs and immigrants.

Prior to the September 11, 2001, terrorist attacks, President Bush signaled his desire for a "special relationship" with Mexico by visiting Mexico during his first foreign trip. Fox and Bush were making substantial progress toward an agreement that would redefine the status of millions of Mexicans—from illegal immigrants to legal guest workers—who are living in the United States. The historic agreement and many other inter-American issues were shoved to the back burner by the terrorist attacks in 2001. Whether the United States will be able to meet the practical challenges of working with its neighbors is the defining question for inter-American relations in the early part of the new century.

Financial volatility and the global economic downturn have created hardships for large numbers of citizens in Latin American countries along with demands for political changes. Some of these demands have been peaceful; others have not. In Brazil, a leftist opposition leader, Luiz Inacio Lula da Silva, was elected president in 2002 and assumed power in the first peaceful presidential transition in Brazil in over 40 years. Da Silva promised a new style of government and a crusade against hunger, injustice, and want. Many observers are now wondering if da Silva will be willing to help another leftist politician, President Hugo Chavez of Venezuela, who is currently under siege from a nationwide monthlong strike. Chavez, who came to power in a military coup and recently won a presidential election, battled striking workers and middle-class opponents who attempted to bring the country to a standstill in an attempt to force Chavez from office. Despite coup rumors and calls by President Bush to hold early elections, President Chavez refused to step down.

Global economic processes have been instrumental in reshaping the economies and politics of Latin American states. For several decades economic growth in Asia fueled increases in the importance of Maquiladora-produced goods and nontraditional agricultural products in Central America. These products now account for approximately 60 percent of all exports from Guatemala, El Salvador, and Costa Rica. Unfortunately, the small size and openness of Central America's economies meant that Asia's economic slowdown adversely affected the region's economies at the same time that the countries struggled to recover from the storm damage caused by Hurricane Mitch during the fall of 1998. Experts predicted that it might take decades for the region to recover as the losses in Honduras are estimated to be equal to the annual GDP of $4 billion and in Nicaragua to about half of its GDP of $1 billion.

While the larger Latin economies weathered the initial shocks associated with the Asian economic crisis and the reduction in world demand for commodity exports, international confidence in the national currencies declined in part because all countries in the region have balance of payments deficits. After years of borrowing from foreign investors to support a currency system that pegged Argentina's currency to the U.S. dollar, the centrist government of President Fernando de la Rua and its successor government, composed of Peronist politicians, were forced to resign as popular resistance boiled over in violent protests against de la Rua's austerity measures and failures by the new government as well. After 4 years of recession, the default and billions

of dollars in bailout loans from the International Monetary Fund were widely expected by foreign investors. Thus, even though Argentine debt represents 25 percent of all emerging market debt, foreign investors were positioned to weather the losses incurred by Argentina's financial crisis.

In contrast, millions of unemployed and middle-class Argentines and elites in neighboring states must cope with the additional adverse effects generated by the downward spiral of the Argentine economy. In "Argentina May Be Down But I Don't Plan to Get Out," Santiago O'Donnell describes what it is like to live in Argentina today with scavengers who were formerly middle-class citizens, increased numbers of unemployed, rampant inflation, and widespread distrust of banks and the government after the government froze and then devalued assets in banks in 2001. The conditions are the worst that most young Argentineans can remember. The disastrous results of the de la Rua economic program provides a cautionary tale about how market reforms can miss their mark and produce devastating results. Today, millions of citizens throughout Latin America are concerned about the prospect of political unrest and economic collapse. Many of the poorer citizens in these societies appear willing to give up some measure of democracy and accept authoritarian governments that they believe can solve their problems.

Two factors continue to dominate the economic development of Latin America's countries: the region's relatively moderate importance to the world economy, and its continued dependence on international capital markets to fuel local growth. Peter Hakim in "Two Ways to Go Global" describes how Brazil and Mexico have both looked beyond their borders for significant international roles by pursuing very different paths. Mexico has linked its future to the United States and has opened its economy to foreign trade and investment. In contrast, Brazil remains a relatively closed economy and is pursuing an independent leadership role in South America. Which model becomes more dominant throughout the region is likely to be a major factor in Latin America's relations with the United States. The Brazilian strategy has put the country in opposition to the United States on some issues. More confrontations are likely, given the political orientation of the Lula da Silva government.

Until September 11, 2001, illegal drugs were the primary concern of U.S. government officials working in Latin America. Today, new U.S. programs designed to combat terrorist and money-laundering schemes in the region have been added to an already ambitious set of U.S. programs designed to stem the flow of illegal drugs from war-torn countries such as Colombia, and to promote democracy and human rights throughout the Western Hemisphere. Recently, the United States deployed troops to Colombia to help local troops protect a U.S. company-owned pipeline. As U.S. foreign policy priorities shift to support war-on-terrorism operations, many analysts predict that funding for the war on drugs and other programs will be cut.

Fragile Democracies

"Latin America's democracies are not in danger of collapse at this time. But there are many real problems, and not many indications that these problems are being addressed with imagination and determination."

FORREST COLBURN

I asked a young Brazilian in Rio de Janeiro how President Fernando Henrique Cardoso was doing. "Oh, he is doing fine." A pause. "It is the rest of us who are not doing so well." This beguiling response can serve as a leitmotiv for how many Latin Americans view their new democratic regimes. It has just been a few years since authoritarian regimes—most military led—were replaced with popularly elected civilian governments throughout the region. These new regimes have survived. But the fanfare and enthusiasm that marked the transition have given way to a more guarded view of democratic governance in the region, and, indeed, to increasing cynicism and apathy.

The first Latin American country to restore democratic institutions was Ecuador, in 1979. It was followed by its neighbor Peru in 1980, and then Argentina in 1983 and Uruguay in 1984. In Central America, the militaries returned to the barracks in El Salvador in 1984 and in Guatemala in 1986. In 1989 Brazil finally broke free from a military regime that had begun in 1964. Dating the transition in Bolivia and in Honduras is more difficult, but in both countries 1985 seems to have been a pivotal year. Two of the last countries to join the fashionable embracing of democracy, Chile and Nicaragua in 1990, did so after experiencing "socialism." Paraguay closed the cycle with its open and competitive elections in 1993 (some Paraguayans date their transition to the elections of 1989 even though the victor was a general who had a few months earlier overthrown the dictator Alfredo Stoessner). These new democracies joined the three countries where democratic processes had been established earlier: Costa Rica (1948), Colombia (1958), and Venezuela (1958).

As can be expected with such a diverse group of countries, there are many degrees of "success" with democracy. Countries that appear more successful include Chile, Uruguay, Bolivia, Argentina, El Salvador, and Costa Rica. On the other end of the continuum are Peru, Ecuador, Venezuela, Brazil, Honduras, and Nicaragua. Differing outcomes are not easy to explain. There is no correlation between those countries that embraced democracy "early" or "late" and the initial set of outcomes of that transition. Peru, for example, was one of the first countries to return to civilian rule, and its governance remains most problematic. Brazil was one of the last countries to embark on democratic rule, and it, too, is problematic. Similarly, there is no easy correlation between the depth of earlier political conflict and present stability. Bolivia has a history of extreme political instability—characterized by military coups d'état—yet today it seems stable. Venezuela has a history of constitutional rule, but it is floundering. Surprisingly, there are no socioeconomic indicators, such as per capita income, that predict political outcomes.

Certain generalizations, however, can be made for Latin America as a whole.

1. A number of indicators suggest that both new and old democratic regimes are under stress, their legitimacy is being questioned, and their public support is increasingly fragile.

Public opinion polls in individual countries, and for the region as a whole, demonstrate declining public confidence in democracy. A comprehensive survey of attitudes on government and politics in 17 Latin American countries undertaken in 1996 by a private polling organization based in Chile, Barómetro Latinoamericano, showed only a minority of those polled expressing satisfaction with the performance of their country's democracy. The same frustration—or disappointment—was registered in polls conducted in 1997 and 1998. Support for the performance of democratic regimes is highest in Uruguay and Costa Rica, and lowest in such countries as Brazil, Paraguay, Ecuador, and Honduras. The sweeping conclusion from the 1998 polling was that only 35 percent of Latin Americans are satisfied with the performance of their democracy. Public opinion polls also suggest widespread disappointment with the institutions of democracy, in particular with legislatures and political parties.

Also indicative of the fragility of democracy in Latin America is the declining voter turnout in elections. In Brazil, for example, voting is obligatory, yet in the 1998 presidential elections, 36 percent of the 106 million Brazilians eligible to vote stayed home or cast blank or invalid votes. Fernando Henrique Cardoso was reelected president by only 34 percent of the total voting population. In Colombia, abstention has climbed to over 50 percent. In Guatemala's 1996 presidential elections, in the departments of Huehuetenango and Quiché, where the population is overwhelming indigenous, 74 percent of eligible voters stayed home. In the country's 1999 national referendum, the abstention rate was 81 percent.

Another disturbing indicator is the lack of support offered to candidates of traditional political parties. In some countries, such as Peru and Venezuela, important parties have all but collapsed. Alternatively, candidates with no political experience have won elections. Alberto Fujimori in Peru is the best-known case. There also appears to be an increasing tendency—accepted by the public—to concentrate power in the executive, who rules by decrees instead of by laws. Finally, many in the media and the academic community openly express doubts about the desirability of democracy. They talk of the "Pinochet model," a "democracy of low intensity," and of a "facade of democracy."

What is happening? Why the declining support for democracy so soon after the scandals of "dirty wars," "*los desaparecidos*," and incompetent economic management by befuddled generals?

2. Public dissatisfaction is not directed at the economic model of the day, unfettered markets.

Public disenchantment is instead focused on corruption and ineptitude in governing. A regionwide survey of public attitudes by the Spanish magazine *Cambio 16* revealed that for the overwhelming majority the most serious problem in their countries is corruption. In impoverished Bolivia, most Bolivians say that corruption is a greater problem than unemployment. Election results also do not demonstrate public rejection of the liberal economic model. Wealthy businessmen with programs of economic liberalization win elections, such as Juan Carlos Wasmosy in Paraguay. In the 1993 Chilean election, 90 percent of the votes went to candidates who accepted the reigning economic model. Even the 1998 election in Venezuela of Hugo Chávez—a former army colonel and coup plotter—can be interpreted not as a vote against liberalism, but instead a vote against corruption in the country's traditional political parties. The dominant attitude in the region is best captured by a Mexican intellectual who said he no longer cared about governments of the right or the left; he would settle for any honest government.

3. The public has sweeping (and perhaps unfair) criteria for granting political legitimacy.

Intellectuals may separate regime type from everyday public administration, but the public makes no such distinction. Legitimacy is accorded governors not on the basis of how they acquired power, but what they do with it. The abrupt transition from authoritarianism to democracy in the 1980s was not accompanied by a wholesale remaking of the bureaucracies that provide (or do not provide) important services to a needy public. In too many instances, the state remains inefficient, unresponsive, and corrupt. Costa Rica has been a democracy since 1948, but a bumper sticker suggests the challenge facing winners of elections in Latin America: "Every time I hit a pothole, I think of the president." In neighboring Nicaragua, a congressional leader told me that a survey of Nicaraguans revealed that the expectation of members of Congress is that they help secure jobs and visas to the United States. Throughout Latin America, there is a need not just to improve the machinery of government, but also to explain what are reasonable expectations of government—and democratic government in particular. The Uruguayan statesman Julio María Sanguinetti goes further, suggesting that democracy demands a supportive political culture. As he puts it, in the most telling of phraseology, "Consumers need to be citizens too."

4. Political parties are the weakest link in Latin America's democracies.

In contemporary democratic theory, political parties are the essential building blocks. They have responsibility for aggregating public preferences, formulating plans for governing, and fielding candidates. With only limited exceptions, political parties in Latin America are not institutionalized, they are not stable, they do not have roots in society, they are not independent of ambitious leaders, and they are not democratic in their internal organization. In the aftermath of military rule in Ecuador, the country has seemingly made a habit of electing a president from a different political party. Between 14 and 17 parties routinely compete in legislative elections, with at least 10 of them winning seats. The strongest party in Congress at any one time has never represented anything close to a majority of voters, polling between 15 and 30 percent of the vote, and usually about 20 percent.

None of Ecuador's political parties has deep roots in society. Indeed, even politicians have shallow roots in their own political parties. One of the features of Ecuador's politics is what is called *cambio de camiseta* (change of shirt): candidates for positions in Congress run for office under a party banner, but once elected they commonly change parties, or just proclaim themselves independents. At times, more than a fourth of the members of Congress claim no party affiliation. Party affiliation is shed so as to enhance the bargaining power of individual members of Congress. I asked an Ecuadorian congressman, in the privacy of his office, how much a congressional vote was worth. "If it is not a vote of great importance, something like two jobs in the customs authority."

Other countries, notably Brazil, also suffer from political fragmentation. Brazil has 20 political parties, so everything depends on "coalitions." And the stability of coalitions is continually undermined by members of Congress changing parties at will. Even in regimes that are essentially two-party systems, such as Honduras, there are no strong connections between society and party. Electoral competition has not been sufficient to end cronyism, indifference, and corruption. In Honduras, public sentiment toward the country's two political parties is captured by the quip, "They eat from the same plate." In the Dominican Republic there is an even more disturbing maxim about the country's political elite: "They are white and they understand one another."

5. The reform of political parties is stymied by a lack of political involvement on the part of many of the middle class, and the managers and professionals whose skills are sorely needed.

Throughout Latin America, there is a common fatalism about politics, as if all outcomes are foreordained. Political parties are left to *politicos*. As a Guatemalan manager put it, *"Los buenos no se meten"* (Decent people do not get involved). This lack of political participation in the management of political parties, and in politics in general, is at odds with common ideas about the economy, which is held to need constant scrutiny and calibration. A similar attitude toward politics is lacking.

6. Political participation is retarded in part by the lack of ideas, of conflicting paradigms on how best to organize state and society.

As a member of Venezuela's Congress put it, "There is no debate because there are no alternatives." A Peruvian journalist, Raúl Wiener, has argued that Peruvian politics has suffered from a "general crisis of intellectual production." It is telling—and depressing, too—that no one reads anymore. Before, those on the left were the most studious, the most inclined to read political theory. But now they do not read: not Marx or Lenin, and not any substitutes. A former guerrilla in El Salvador, now a member of the Democratic Party, told me: "There is nothing to read; there are no reference points." From her and her colleagues, one learns only from the experience of governing.

The silence of intellectuals and politicians is striking. They both are terribly short of ideals that can mobilize people. In particular, there is little evidence of energy and creativity in searching for ways in which state and society can ameliorate inequality and poverty. Innovation in public administration is largely confined to public finance, where the guiding aspiration is inevitably to rein in government spending (and so avoid inflation-spawning deficits). Private initiatives are few. The poor are seen as an inevitable part of the social landscape.

Even Latin America's last guerrillas, those of Colombia, are seemingly bereft of ideas. Guerrillas operate in nearly half of Colombia and control a sizable part of the country. They are fought by the military and by paramilitary forces. The fighting is savage and has generated a major social crisis. But what is the fighting about? The largest rebel group is the Revolutionary Armed Forces of Colombia (FARC). What, if anything, does it advocate? It has not articulated an ideology or a set of proposals, leaving it open to charges that its fights for the spoils of the trade in narcotics and the gains from such crimes as kidnapping. The FARC has not contributed to political debate in Colombia, let alone elsewhere in Latin America.

7. In the absence of ideas—and passions—there are only interests.

These, of course, have always existed. But perhaps they are more transparent now. Latin America has always been characterized as being both more culturally homogeneous and more radically inegalitarian than other parts of the world. In fact, inequality is real enough, homogeneity less so. The ethnic schism between indigenous people and those referred to as *ladinos* is pronounced. Indigenous populations are better organized and more independent today, especially in countries where they constitute a substantial part of the population. Afro-Brazilians—who represent perhaps 40 percent of the population of Brazil—are also increasingly able to make demands for a redress in their subordinate status.

There are now more pronounced—and growing—divisions, not based on ethnicity, but rather between the capital and the rest of the country (the "interior" as it is still sometimes called in Havana). The economic and cultural disparities between the capital and the remainder of the country introduce an odd disjunct into the politics of representation. In Guatemala, for example, it can be argued that the winner of the 1996 presidential election was the candidate of Guatemala City; the loser was the candidate of the rest of the country. Increasingly, the route to the presidency in Latin America includes serving as mayor of the capital.

In some countries regional disparities are pronounced. The wealthiest state in Mexico, Nueva Leon (located in the north), has a per capita income 10 times that of the poorest Mexican state, Chiapas (located in the south). In a number of countries, including Colombia, Ecuador, Peru, and Bolivia, there is a rivalry between the sierra and the lowlands. (Ecuadorians from the *sierra* call their brethren on the coast "monkeys.") It remains to be seen how ethnic, regional, and urban-rural divisions will play themselves out, but the basic political fault lines are becoming decidedly less ideological.

8. In the fragile democracies of Latin America, the political risk is not, as in decades past, a return to military rule.

The military realizes that the international context has changed and that the pressures on anyone seeking to usurp power would be overwhelming. In any case, the soldiers have no inclination to govern. There are no visible dangers to the nation or their own prerogatives. Like so many others today in Latin America, ranking military officers are interested in making money. Many officers throughout the region have business interests on the side. And the military as an institution often owns and operates businesses, sometimes of an ominous kind, such as the cemetery run by the armed forces in Honduras.

Instead, the greater political danger to democracy in Latin America comes from a growing number of voters who either abstain from voting altogether or vote for antidemocratic candidates who offer "simple solutions" to complex problems. "Simple solutions" are a populist mantra. Populism in Latin America is an elastic concept, but it is most commonly associated with redistributive economic policies, usually financed in an unsustainable manner. Populism has a less well-understood political dimension, however: the "politics of antipolitics." Populists who come to power through democratic elections govern in an authoritarian style justified by continual attacks on traditional political elites and established institutions.

Today, economic populism is not the threat it was in earlier decades. There has been considerable "economic learning" by elites, including those of the private sector and governmental bureaucracies, and the international financial community is always on guard. A political leader bent on pursuing populist economic policies would quickly confront opposition from a daunting coalition of institutionalized actors who can put unbearable pressure on any maverick leader.

Shifts in economic policies bring sharp and sudden shifts in economic fortunes. Resilient democracies can cope, but can fragile democracies?

Much more likely are forms of political populism á la Alberto Fujimori—antiparty politics. Alberto Fujimori, the son of Japanese immigrants, assumed the presidency of Peru in 1990 after winning a competitive election. But he augmented his power in 1992 by what is now known throughout Latin America as a "*fujimorazo*"—dissolving Congress under the pretext that legislators were corrupt and inept. Democracy's "checks and balances" were dismantled with the argument that public faith in Peru's traditional political parties had declined. Fujimori offered "administration," not "politics." Surely all Peruvians recognized that this gambit was antidemocratic. Nonetheless, in Peru—and elsewhere in Latin America—voters can become so frustrated with the traditional parties that they opt for exciting if dangerous leaders. Even elites, fearful of economic populism, may acquiesce in a political leadership that is determined to govern. But this kind of heady and combative government inevitably contributes to the further weakening of democratic institutions, including political parties, legislatures, and the judiciary.

9. The democracies of Latin America are hostage to economic trends and are vulnerable to economic shocks.

There is statistical evidence that the percentage of poor people in Chile has declined because of robust economic growth, up to an enviable 6 to 8 percent per year. In Costa Rica, the successful promotion of exports and the development of tourism have resulted in a low unemployment rate. In many other countries, though, the "new" economic strategy has brought painful dislocations, such as the surge in unemployment in Argentina and the political crisis that has gripped the country since December. And there is scattered but persuasive evidence that economic liberalization and the development of new export commodities are exacerbating inequality. These un-

equal gains—or losses, as the case may be—translate into political tensions. For example, in El Salvador, the National Republican Alliance is threatened with fragmentation because of divisions between the agro-export elite, who founded the party but are not faring so well, and the financial-sector elite, who increasingly dominate the party. Shifts in economic policies bring sharp and sudden shifts in economic fortunes. Resilient democracies can cope, but can fragile democracies?

Another, more nettlesome question: How dependent has Latin America become on a healthy world economy, and, in particular, on capital inflows? For a number of years Latin America has had a very large capital inflow, billions of dollars annually. This investment, along with healthy markets for many of the region's exports, has caused a spurt of economic growth and patched over many problems—including increased inequality. What happens if capital dries up, or if world trade contracts? The economic difficulties in Latin America during the 1980s were exacerbated by indebtedness. Latin American countries continue to have considerable foreign debt, which is constraining and which makes the region especially vulnerable to a contraction in the world economy. There is an expression in financial circles: When the United States sneezes, Europe gets a cold. Perhaps the corollary is, and Latin America gets pneumonia.

10. Mexico may be a bellwether rather than an exception.

In Latin America there is an expression, "*Como México no hay dos*" (There is no place like Mexico). Mexico is sui generis. Before the wave of democracy in Latin America, Mexico did not look so bad: the military was nowhere to be seen, there were elections, and there was a periodic change of presidents. After the "transitions" elsewhere, though, Mexico looked like what the Peruvian writer Mario Vargas Llosa labeled it: "the perfect dictatorship." Elections and a new president every six years no longer masked the fact that the country was essentially a one-party state, the kingdom of the Institutional Revolutionary Party (PRI), founded in 1929. The party mixed paternalistic populism with electoral skullduggery to become the longest-ruling political party in the world.

The PRI provided a facade of legitimacy for cronyism, clientelism, and corruption. The 1982 economic crisis in Mexico stimulated some effort at political reform, as did—again—the 1994 crisis. Although the leadership of the PRI sometimes gave the impression that its attitude was "things have to change so that nothing changes," a variety of circumstances and some statesmanship led to elections in 2000 that were free and competitive. And the PRI lost.

The candidate of the National Action Party (PAN), Vicente Fox, won the election. His campaign appearances included his stomping a plastic dinosaur, representing the PRI, with his cowboy boots. In Fox's December 2000 inauguration address to Congress, he promised there would be, after 71 years of continuity, a "new political future," with a "reform of the state, breaking paradigms." But is a new "paradigm" for governing truly in the offing?

Mexicans voted for an alternative to the PRI only when it was safe to do so, when there was an end to ideological differences that might reverberate throughout state and society. Conflict is being averted, but so is the possibility of profound change. A Mexican president who is not from the PRI is now a threat only to the cronyism of the PRI. Mexico is different, in so many ways, from the rest of Latin America; yet politically Mexico enters the twenty-first century with the same absence of ideological—and thus programmatic—contestation that is lacking elsewhere.

In sum, Latin America's democracies are not in danger of collapse at this time. But there are many real problems—and not many indications—that these problems are being addressed with imagination and determination. It is not clear how tired and cynical voters will react if further deterioration takes place, or if the region is subjected to economic shocks. True, the region's democracies have brought peace and greater protection of basic human rights. Democratic governments have brought a wel-come sense of "normalcy," too. What is notable, though, is not just that the democracies are "rickety," hardly prepared to tackle difficult economic and social problems, but that the populace is not prepared to engage in deep discussions about state and society, about needs and aspirations, and about how these needs and aspirations could be met. The considerable promises of democracy have yet to be fulfilled in Latin America.

FORREST COLBURN, *a* Current History *contributing editor, is a professor of political science at Lehman College, City University of New York. This essay is adapted with permission from his forthcoming book,* Latin America at the End of Politics *(Princeton, N.J.: Princeton University Press, 2002).*

Argentina May Be Down But I Don't Plan to Get Out

By SANTIAGO O'DONNELL

BUENOS AIRES

Even at its worst, this is a beautiful city. It has wide avenues, European-style six-story buildings with wrought-iron balconies, plazas with statues, manicured parks. But these days Buenos Aires stinks. Every neighborhood, no matter how elegant, is littered with reeking garbage, and this is not because the garbage men aren't trying to do their jobs. It's because every night at about 10, more than 100,000 scavengers pour in from the nearby shantytowns. Hurrying to beat the trash trucks that begin their routes at midnight, they rip open and rummage through the bulging plastic bags that have been put out on the sidewalks.

Entire families come, working like military units: the mother at point, sending four or five children in different directions, the father bringing up the rear, pushing a shopping cart holding a baby or two and whatever scraps the kids collect. They rip the bags and scatter the contents because they have to work fast. These days, two or three pairs of hands go through each bag before the garbage truck arrives. Many of the scavengers look as if they once belonged to the middle class: the mother in a flowered dress, the children in Adidas sweats. But their eyes are the eyes of the hopeless poor.

For the middle class, scavenging has become a survival skill on the streets of Buenos Aires.

In Buenos Aires, 10 p.m. to midnight isn't late—it's dinnertime, and for most of the rest of us life is going on at its usual chaotic pace. Traffic is thick, steaks are sizzling, and espressos are steaming. People like me are buying chicken ravioli at the neighborhood rotisserie, or waiting for the bus, or running for a train or a subway.

The army of scavengers descends before our very eyes, and we don't say a word. When the desperate families hurry by, we turn away. "Don't look at me, don't talk to me. I won't expect any help from you, and you won't get mugged," is the unspoken message to those who still have trash to put out from those who need it to survive.

That's how it is, eight months after our government went through five presidents in a week, defaulted on $140 billion in international debt and devalued the savings accounts of 7 million Argentines to less than a third of their worth. That's how we deal with scavenger families, the new symbol of bankrupt, lawless, cynical Argentina: We look away and make them disappear.

Twenty years ago, when I finished high school, Argentina was a nation haunted by the ghosts of others who had vanished. These were the *desaparecidos*, the tens of thousands of citizens who were kidnapped and murdered by right-wing and military death squads between 1973 and 1982. Some were shot. Many were thrown alive from airplanes into the ocean. Few people talked about "the disappeared" during the blood-soaked years of the dictatorship for fear of becoming one of them. Some of the mothers of missing persons staged a silent march of protest every Thursday, but hardly anyone seemed to notice.

I felt I had wasted my youth living under fascists who outlawed street gatherings of more than three people, who thought editing a school newspaper was a "subversive" activity, who banned Donald Duck from newsstands because his relationship with Daisy and his nephews wasn't altogether "proper" or "clear." Then, as a graduation present, the Class of 1982 was awarded the lead role in a senseless, bloody and deeply humiliating war over the Falkland Islands. I was doing my mandatory military service in the Argentine Coast Guard. Luckily I didn't get sent to the front, where elite British forces equipped with state-of-the-art NATO technology overwhelmed draftees like myself armed with World War II-era rifles. I saw friends and neighbors come back in wooden caskets.

When the war ended, I couldn't wait to get out of spooky, hateful Argentina.

In September 1982, I came to the United States, and I stayed for 12 years. I went to the University of California at Berkeley and to Notre Dame and the University of Southern California, then went to work as a news reporter, first for the Los Angeles

Times and later for The Washington Post. I had friends and a 401(k). When I moved to Washington, I rented a nice townhouse on Euclid Street in Adams Morgan and played soccer every Sunday at a public park on Chesapeake Street, where we used trash cans as goal posts. The players included Argentine émigrés from all walks of life: chauffeurs, World Bank economists, gym teachers, even, for a game or two, the ambassador.

After years of weekend basketball in California and softball in Indiana, those soccer games brought back powerful feelings of home. When a family crisis took me back to Buenos Aires, I found myself overwhelmed by the desire to stay. The ghosts of my past had pretty much been laid to rest: The generals were in jail, and the mothers of the disappeared had become national heroes. I wanted to go home, to my streets and my people and my family, to write in my own language about my own country's joys and its problems.

And there *were* problems, though they were masked by a false sense of prosperity. In 1994, the world economy was booming, "emerging markets" were trendy, the Argentine middle class had access to credit for the first time in decades and was off on a shopping spree, buying refrigerators with credit cards, new cars with personal loans and new homes with 10- or 15-year mortgages.

I went back with my eyes open, though. From the States, it had been easy to see that Argentina was still a foreign-debt junkie with bloated budgets and closed factories, a country crippled by rampant corruption that imported everything and produced almost nothing. The government had only two ways to keep afloat—international loans and the sale of state-owned resources. The nation's oil reserves and telephone networks and prime real estate and airline routes were all privatized at fire-sale prices—and once gone, they couldn't be sold again. Meanwhile, the flood of imported products swamped regional economies, blue-collar unemployment soared and shantytowns grew like fungus around the big cities.

But it was home. And we Argentines are sort of used to catastrophes. At least the death squads weren't running loose and the generals weren't planning any new wars. With a glad heart, I moved back, married my high school sweetheart, got a job at a local newspaper, took out a 10-year mortgage and moved into a six-story apartment building with a black marble facade in San Telmo, the oldest neighborhood of Buenos Aires.

In America, people lose track of each other: They're born in Chicago but move to L.A., they move in and out of Washington every three or four years. In Argentina, it's not like that. When I returned, most of the friends I grew up with were still there, meeting for pickup soccer games on the same Astroturf field we had used 20 years before.

We still play there every Sunday. But lately the lives of my soccer buddies have changed. Coyote, the architect, has been unemployed for seven months now. Lucious, the chef, closed his trendy restaurant last November and now works as kitchen supervisor in a pizza chain. Pat, who owns a printing shop, is seriously considering opening a hot dog stand. Matt, the artist, moved to Chile. Nestor, who's in PR, is going to Spain next month to try his luck. Teddy, the corporate lawyer, has worked for a failed insurance company, a failed bank and a failed en-

ergy company and recently spent almost a year in the ranks of the unemployed. Rafi, the car salesman, is hoping to keep his job by making friends with his boss's son.

Everyone agrees: It's never been so bad. The recession is well into its fifth year and showing no signs of slowing down. More than half of the population has sunk below the poverty level; one out of five is unemployed, more than triple the rate of 10 years back. People who had saved money lost it when the government, trying to bail out the country's troubled banks last December, froze all savings accounts for three to seven years and converted any dollar accounts (which was almost all of them) into pesos. The peso had been worth a dollar; it was 70 cents when the accounts were converted; today it's worth less than 30, and with inflation at 60 percent and rising, lifetime savings will be worth peanuts before their owners can touch their money.

A hit that makes you wince: Contestants on Argentina's popular game show called "Human Resources" compete for six months of employment as tour guides, cooks, waiters and window washers.

Inflation is back, like in the 1980s but with an evil new twist—today, salaries don't even try to compete with prices. At the end of the month, each paycheck is worth less than the one before, and you cut something else out of the family budget: garage parking, orange juice for breakfast, cable TV.

You take your paycheck out of the bank the day it goes in, because of course you don't trust the bank. You pay your mortgage in cash. Everybody keeps cash in the house, which has led to the latest crime trend: "express kidnappings." Early in the current crisis, people began to be stopped at gunpoint, taken to the nearest automated teller machine, forced to withdraw their limit and then let go. But now that the banks have no money, the express kidnappers pick people up, call their relatives and order them to pay, say, $200. If things go well, the victim can be free in a few hours.

This in a city that not all that long ago was one of the safest in the world.

Politicians and bankrupt businessmen grow beards and avoid the street because they are terrified of being recognized and showered with rotten eggs, which is the preferred form of heckling these days. Some who dare to go out without enough bodyguards have been publicly beaten.

After the Sunday soccer games, my friends and I talk—about our families, the kidnappings, whatever's in the news. In lowered voices, we talk about who is out of work, who is in trouble. We all worry about our jobs. Many of us think we're kept on only because of a law passed last year that says anybody who's fired during this "social emergency" gets double severance. But that law expires this fall—and besides, the laws in this country don't mean much anymore. It was three weeks after Congress passed a law guaranteeing the safety of bank deposits that those deposits were frozen and devalued.

"What are you still doing here?" my friends ask sometimes. "Don't you ever think of going back to the States?"

It's not a bad question. In a country that once welcomed millions of immigrants (including my Irish forebears a century ago), thousands of Argentines are lining up in front of foreign embassies in search of a way out. Daily flights to Madrid and Rome leave jampacked and return empty. Even the smallest towns in the most remote parts of New Zealand, Canada, Australia and South Africa are coveted destinations. Violence-torn Israel welcomed 20,000 Argentines in the past 12 months.

Teary-eyed emigrants wait in departure lines at Ezeiza Airport here, wrapped in Argentine flags. They're sad, but they're desperate. "I can't live here anymore!" they say. "I want out!"

I know the feeling. But this time I don't want to disappear.

I could get in trouble down here for saying this, but I have to admit I like it when U.S. Treasury Secretary Paul O'Neill hits Argentines with one-liners like, "It's no use sending money down there if it's going to end up in a Swiss bank account." There were certainly things I didn't like about the States, but I miss American-style straight talk.

I get angry at many things here—the fact that good "connections" are more important than merit, the way we've gotten used to cheating and cutting corners in everything from taxes to theater lines, how we allow ourselves to be governed by the same politicians who tell the same old lies, over and over again. It's sad to see what we've managed to do to a vast, resource-rich country.

I worry that I can't leave my 4-year-old son, Jose, the promise of a better future than my parents gave me. There are no quick fixes for this economic disaster, I think; it will be another generation before Argentina can recover. Yet, for all the complicated reasons that human beings cherish their homes, I love this country. I have no regrets about coming back. If I'm hopeful that we may be on the verge of a long, painful rebuilding process, maybe it's just because every night more people are eating trash, and it's getting harder to look away. They won't disappear.

Santiago O'Donnell is an investigative reporter for the Argentine newspaper La Nacion.

Two Ways to Go Global

Peter Hakim

THE DIFFERING PATHS
OF MEXICO AND BRAZIL

FOR THE FIRST TIME EVER, Latin America's two giants, Brazil and Mexico, are both looking beyond their borders for significant international roles. It is striking, however, how differently each is pursuing that goal. Mexico has linked its future to the United States and almost fully opened its economy to foreign trade and investment. Brazil, in contrast, remains a relatively closed economy, pursues an independent leadership role in South America, and is seen by the United States as an opponent on some issues.

Mexico's choices have clearly been influenced by the fact that it sits in the shadow of the world's richest and most powerful nation. Brazil, a continent-sized nation located some 2,400 miles from the United States and surrounded by ten smaller neighbors, has a rather different perspective on the world. Yet until recently, it was Mexico that most zealously shielded its independence from the United States.

Geography, to be sure, has played a major role in the pursuit of these divergent paths. But domestic politics and national ideologies have also been critical in molding the agendas of the two nations. Brazilian political leaders and thinkers, and even ordinary citizens, have long believed that their country should be counted among the world's most important states. Mexicans, meanwhile, historically have been less concerned about their place in the world than about their relations with the United States. Moreover, until Vicente Fox assumed the presidency in December 2000, Mexico was ruled by authoritarian and centralized governments. Recent Brazilian governments, on the other hand, have been more democratic than their Mexican counterparts but also weaker and more susceptible to popular pressure.

A CERTAIN SIMILARITY

BRAZIL AND MEXICO have enough demographic and economic heft to exert real influence in international affairs. At the outset of 2001, Brazil was the fifth most populous country and had the eighth-largest economy in the world. Mexico was the eleventh most populous and had the twelfth-largest economy. (Currency fluctuations in the past year have left the two economies about the same size in dollar terms, although Brazil's domestic purchasing power remains much larger.) On a per capita basis, their ranking falls considerably, but by World Bank standards they are comfortably upper-middle-income countries.

For the past six decades, Brazil and Mexico have also shared a remarkably similar economic history. From 1940 to 1980, they attained some of the highest growth rates in the world, averaging more than six percent a year. At the same time, both nations indulged in massive foreign borrowing. By the early 1980s, crushing debt burdens had pushed them deep into recession. Since 1980, each country's economic growth has dipped to a dismal average of about 2.5 percent a year, although Mexico's growth began accelerating in the latter half of the 1990s.

In this period, both nations initiated economic reforms known as the "Washington consensus," a combination of fiscal discipline, privatization of state-owned businesses, and foreign trade liberalization. Income and expenditures were brought increasingly into balance. In 1994 Brazil finally succeeded in stemming its relentless inflation. The two countries also sold the great bulk of their state companies to private investors, although firms in politically charged sectors such as petroleum and electricity remained in government hands. Average tariffs in Brazil dropped from nearly 50 percent in 1985 to 12 percent in 1995. Mexico dropped its tariffs from an average of 25 percent to 16 percent and aggressively slashed nontariff barriers.

In the past five years, the two countries have shown remarkable economic resilience by recovering briskly from financial crises—Mexico in 1995 and Brazil in 1999. Both crises ran a similar course. Mexico and Brazil each turned to the United States and the International Monetary Fund for large-scale rescue packages that helped avert complete currency collapses and set the stage for a steady recovery. Both also adopted freely floating exchange rates, ending prolonged efforts to peg their currencies to the dollar. Nevertheless, as the global economy has slumped in the past year, the two countries have suffered sharp economic reversals. In addition, energy shortfalls, high domestic debt, and the financial implosion in neighboring Argentina have pushed Brazil dangerously close to another crisis.

Mexico is one of the world's most open economies; Brazil is one of the least.

Finally, democratic governments emerged in both Mexico and Brazil in the past 15 years. Brazil broke sharply with its past in 1985, ending 21 years of military rule with the indirect election of a civilian president. Although long governed by civilian authorities, Mexico moved toward democracy much more gradually. Indeed, it was only a year ago that the inauguration of Vicente Fox—who was elected in the country's first free and fair election contest—finally ended seven decades of one-party rule.

OPEN OR CLOSED?

DESPITE THEIR SIMILARITIES, Mexico and Brazil have pursued their international goals in very different ways. Mexico has made foreign trade the engine of its economy. From 1990 to 2000, its exports catapulted from $45 billion per year to $165 billion, at a dazzling rate of some 15 percent a year. Not only have Mexico's exports quadrupled in the past decade, to the point where they now constitute a third of GDP, but their composition has also shifted dramatically. Manufactured goods now amount to nearly 90 percent of the country's foreign sales, a doubling over ten years. Mexico accounts for almost half of Latin America's foreign trade. Only seven countries worldwide export more.

Brazil's economy, in contrast, remains relatively insular. In the last decade, Brazil's exports grew at a pace less than one-third that of Mexico's, from $32 billion to $58 billion. According to one measure of economic openness—the ratio of exports to GNP—Mexico today ranks second among the world's dozen largest economies, whereas Brazil ranks dead last, although not far behind the United States, India, and Japan. Along with neighboring Argentina, Brazil is also Latin America's least open economy. Exports make up less than ten percent of Brazil's GDP, a number that has hardly changed in the past decade. Moreover, in contrast to Mexico, the share of Brazil's industrial exports has remained unchanged, at about one-half of total exports.

Brazil, however, is not a traditionally closed economy. The country has taken sizeable steps since 1994 to remove barriers to global commerce and investment. Tariffs have been sliced to a quarter of what they were a dozen years ago. Foreign direct investment (FDI) rose from less than $2 billion annually in the early 1990s to more than $30 billion by 2000; in the past five years FDI flows to Brazil have averaged more than three times those to Mexico. Brazil is also pursuing free trade agreements with many nations, and President Fernando Henrique Cardoso declared in August 2001 that Brazil's economic survival depends on expanding exports. Nonetheless, Brazil's trade performance will not come close to Mexican standards anytime soon.

Mexico's emphasis on foreign trade stems from a deliberate strategy to join its economy to that of the United States. This strategy was set in motion in 1990 when Mexican President Carlos Salinas realized that neither Europe nor Japan would do much to strengthen their economic ties with Mexico. Salinas and his advisers concluded that Mexico's best bet was to hitch its economy to that of the United States, and they proposed a U.S.-Mexico free trade agreement to the first Bush administration. This proposal became the tripartite North American Free Trade Agreement (NAFTA) when Canada (which already had a trade pact with the United States) also joined in.

Mexico had always been something of an economic satellite of the United States. The mammoth U.S. economy, 20 times the size of Mexico's, exerts an enormous gravitational pull on Mexican exports and labor. NAFTA has bound the Mexican economy even more tightly to the United States: its share of total U.S.-bound exports has risen from 75 to 90 percent. By buying Mexican goods, dispatching capital and tourists southward, and feeding the flow of workers' remittances (upward of $8 billion annually), the U.S. economy pulled Mexico out of its 1995 crisis and fueled five years of solid growth, averaging 5.5 percent per year from 1996 through 2000. This year the sputtering U.S. economy has dragged Mexico into recession.

Beyond economic and commercial links, a wide range of other issues—migration, drug-trafficking, environmental contamination, energy development, and water rights—link the United States and Mexico. Fox's state visit to Washington in September 2001 was the first by any foreign leader to George W. Bush's White House. It demonstrated the breadth and vitality of U.S.-Mexican ties, which the U.S. president referred to as "our most important relationship in the world." Mexican officials would have preferred greater progress on immigration and other concerns, but they plainly succeeded in refashioning the bilateral agenda. The Fox visit was immediately overshadowed by the events of September 11. But, with Mexico's cooperation on border control and other security issues now so vital, the broader scope of Mexican concerns will likely remain a priority for the United States.

INTO THE WORLD

MEXICO'S FOREIGN POLICY is not focused exclusively on relations with the United States. President Fox and his foreign minister, Jorge Castañeda, have given Mexico a newly active international role. Trading on the nation's close U.S. ties, they have sought to cast Mexico as a bridge between North and South America. Just in the past year, Mexico has sought to bolster peace negotiations in Colombia and to engage Central America more intensely with a plan for joint infrastructure development. Since Fox took charge, Mexico has also become an international advocate for human rights and democracy, themes it had previously shunned due to its long-standing deference to national sovereignty and its own lack of democratic credentials. For the second time since 1946, Mexico last year sought and won a seat on the U.N. Security Council. Nevertheless, Mexico's top priority remains a solid partnership with the United States, and this will shape and circumscribe its entire foreign policy agenda.

Brazil, in contrast, conducts a far more autonomous and diversified foreign policy. Brazil's most important bilateral relationship may also be with the United States, but it is much less consuming and confining than the U.S.-Mexico link. Less than one-quarter of Brazil's trade is with the United States, about the same as with Europe. Moreover, trade with the United States accounts for only 2 percent of Brazil's GDP, whereas it is nearly 30 percent of Mexico's.

Brazil has become more active and assertive in regional and global affairs, especially since Cardoso took office in 1995. On some issues, Brazil has sought to serve as a counterweight to the United States. At times, it has appeared intent on establishing a South American pole of power in the western hemisphere. These new international ambitions were evident when Cardoso convened the first-ever summit meeting of South American heads of state in Brasília in September 2000. The meeting highlighted Brazil's focus on increasing South American integration. A more unified group of nations on that continent would, the logic goes, have greater weight in hemispheric and global negotiations, thereby enhancing Brazil's international influence and strengthening its bargaining power with the United States and other key countries.

Mercosur, a customs union among Brazil, Argentina, Paraguay, and Uruguay that was established in 1991, has formed the core of that strategy. Although badly strained in the past two years by Brazil and Argentina's continu-

ing trade disputes, which are aggravated by their incompatible macroeconomic policies, Mercosur remains a foreign policy priority for Brazilian authorities. Ranking lower is the proposed Free Trade Area of the Americas (FTAA), which would bring Brazil and every other Latin American country into a free trade arrangement with the United States.

Trade is the biggest source of contention between the United States and Brazil. At the April 2001 western-hemisphere summit in Québec City, Cardoso bluntly laid out the conditions under which Brazil would join the FTAA. Many of them involved controversial themes that the United States wants to keep off the negotiating table, such as curbing the use of antidumping measures and subsidies that impede imports of Brazilian steel, soybeans, and orange juice. According to Brazil's ambassador in Washington, Rubens Barbosa, the FTAA talks can succeed, but only if the United States negotiates these issues seriously. Mexico may have been the first Latin American country to establish free trade arrangements with the United States, but Brazil may well be the one that forces the United States to modify its plans for a hemispheric free trade pact. In addition to trade frictions, Brazil and the United States have also clashed in the past several years over election-rigging in Peru and U.S. policy in Colombia.

But Brazil is by no means an adversary of the United States. Indeed, no government in Latin America more effectively demonstrated its solidarity with the United States following the terrorist attacks of September 11. Brazil's swift call to invoke the Rio Treaty—the hemisphere's mutual defense pact that, like NATO, makes an attack against one nation an attack against all—was warmly praised by President Bush and Secretary of State Colin Powell. Although they have wrangled over specific issues, the United States and Brazil maintain largely cooperative relations. The two countries will co-chair FTAA negotiations from 2002 through 2004; despite their differences, both have declared their commitment to reach an agreement. They have worked together to resolve the explosive border dispute between Ecuador and Peru and to support civilian rule in Paraguay. Washington knows that the United States cannot achieve many of its central objectives in the western hemisphere without Brazilian support. Yet Brazil continues to challenge the United States on several fronts and certainly has no plans to follow Mexico's lead and tie its economic or political future to the United States.

LOCATION, LOCATION, LOCATION

IT IS TEMPTING to dismiss the contrasts in policy between Brazil and Mexico as the consequence of geography. Does Mexico—a relatively poor and weak country—have much choice other than to align itself with its powerful neighbor? And surely it should be no surprise that Brazil, a continent away from the United States, would want to pursue independent international policies.

Yet these positions are actually quite new. Until Salinas initiated free trade negotiations with Washington in 1990, Mexico's foreign policy was fixated on how to protect its independence and national integrity from its northern neighbor. Mexican policies reflected anxiety about U.S. corporations taking command of the country's natural resources, about U.S. popular culture overwhelming Mexico's, and about U.S. prodding on democracy and human rights undermining Mexico's political structure. Although the United States vigorously promoted trade liberalization, Mexico maintained a protected, inward-looking economy. Unlike Brazil and other Latin American countries, Mexico steered clear of bilateral treaties and military cooperation with the United States and even rejected U.S. aid. With its profound distrust of the United States, Mexico neither sought nor wanted a close relationship with its northern neighbor and often opposed U.S. positions in international forums. Its voting record in the United Nations was more like Havana's than Washington's.

In contrast, Brazil's policies toward the United States demonstrate far greater continuity. At times, it worked hand in glove with the United States. For example, Brazil helped to legitimize the U.S. intervention in the Dominican Republic in 1965. It was also the largest recipient of aid under President John F. Kennedy's Alliance for Progress. But throughout the 1960s and 1970s, Brasília also opposed Washington on many issues, particularly on economic and financial matters. The U.S.-Brazilian relationship worsened during the Carter years when Brazil's military government took offense at U.S. human rights policies and obstruction of the country's nuclear development. With the return of democracy in 1985, Brazil moved toward its current relationship with the United States, combining cooperation on many fronts with independence and even opposition in some key areas.

In recent years Argentina and Chile have sought to emulate Mexico, not Brazil, in their relationships with the United States. Santiago has begun free trade negotiations with Washington, and it is no secret that Buenos Aires, despite its Mercosur membership, would probably follow suit if it received an invitation. After participating in several overseas peacekeeping initiatives, Argentina has obtained the formal status of "non-NATO ally" of the United States. Thus geography alone cannot adequately explain Brazil's and Mexico's foreign policy orientations.

BIG IDEAS

POLITICS AND IDEAS have also been important in setting the two nations' foreign policies. Mexico's decision to link its future to the United States was a dramatic shift for a country that had long struggled to insulate itself from its powerful northern neighbor. But one important fact remains unchanged: the United States is still the central point of reference for Mexican economic and foreign policy formation. Mexico has altered many of the core tenets and objectives of its international policy, but the driving force of that policy has not varied. Mexico's leaders continue to direct their attention intently toward Washington. Geography determines where Mexico focuses its foreign policy, but not how.

Mexico's authoritarian politics facilitated integration with the United States.

Mexico quietly began integrating with the United States well before NAFTA came into force. By the mid-1980s, Mexico's economy was already heavily dependent on U.S. trade and investment, and emigration north had become a political safety valve. Collaboration between U.S. and Mexican business communities was intensifying, and civil-society groups on both sides of the border were joining forces.

Still, U.S.-Mexican relations have been fundamentally recast in the last decade, thanks in part to the shifting power balance within the ruling Institutional Revolutionary Party (PRI). In the 1980s, as the country struggled with its debt crisis, the party's technocratic wing gained influence. Pragmatic and internationally oriented leaders, many (including Salinas) with advanced degrees from top U.S. universities, took over from more traditional politicians. Although Salinas has since become the nation's most maligned former president, he did more while in office than anyone else to refashion Mexico's economy and foreign policy, and to tilt the nation toward the United States.

The nature of Mexico's authoritarian, highly centralized government, controlled by the PRI since 1929, made Salinas' task a lot easier. For 70 years the PRI not only controlled the executive and legislative branches of government, but also determined who was elected mayor or governor everywhere in the country; dominated labor unions, rural associations, and other institutions; and ensured that the press was passive, if not fully controlled, and the business community docile. As a result, NAFTA was endorsed by nearly every Mexican newspaper and television station and by labor, business, and civic groups as well. Mexican decision-making has now become slower and less tidy in the year since Fox took office with an opposition-dominated legislature. Virtually every presidential initiative, including the response to September 11, is now scrutinized and fiercely debated in the Mexican congress and the media.

The Brazilian approach to foreign relations is very different. Its diplomats, politicians, and commentators write and speak about Brazil as a continental power. Pointing to its size and population, they argue that Brazil should be counted among the world's giant countries, alongside the United States, Russia, China, and India. Indeed, prior to his appointment as foreign minister a year ago, Celso Lafer argued that the interests of Brazil and these other "monster countries" (a term coined by U.S. diplomat George Kennan) go beyond specific issues and outcomes. They have a major stake—and therefore should have a major say—in how global affairs are managed.

Given this self-perception, it is not surprising that Brazil has taken on a more far-reaching international agenda than Mexico has, or that it seeks a permanent seat on the U.N. Security Council, or that it considers itself the natural leader of South America. Brazil's commitment to the Mercosur alliance and its proposals to extend that arrangement to the rest of South America clearly stem from this perspective; so too does Brazil's insistence on a leading role in shaping hemispheric agreements and its resistance to U.S. design or domination of those arrangements. This context also explains why Brazil swiftly offered its support for the U.S. campaign against terrorism and called on Latin American nations to cooperate with the United States under the provisions of the rarely invoked Rio Treaty.

Brazil's limited volume of trade and its domestically oriented economy, however, are out of step with the country's international aspirations. No less than Mexico, Brazil faced the trauma of debt crisis and economic stagnation in the 1980s and had to reform its economy and adjust to a changing global financial system. That Brazil proceeded more slowly and less decisively than Mexico reflected the former's more open and disorderly politics. Since civilian governments took charge starting in 1985, presidential leadership in Brazil has been hobbled by a disorganized party system, an erratic legislature dominated by smaller, more rural states, and by a highly populist constitution.

Mexico's strong, centralized government made economic decisions, such as joining NAFTA, based on technocratic judgments, without serious opposition from Mexico's congress, labor unions, or the press. In Brazil's far more fluid and vibrant democracy, economic policies were shaped by political compromises and constrained by an array of constitutional restrictions. Brazil's powerful labor unions vehemently opposed trade liberalization and most other economic reforms. Unlike the Mexican business community, which developed close links to the United States, most Brazilian companies have had little confidence in their ability to export or compete with foreign firms at home, and they have been loath to open up protected domestic markets. These antitrade biases are reinforced by the popular view of Brazil as a "monster country," big enough to survive on its own.

COSTS AND BENEFITS

MEXICO HAS SUCCEEDED beyond most expectations in reaching its twin objectives: its economy is now joined to that of the United States and is also wide open to foreign trade and investment. Mexico has become the United States' second-largest trading partner and it may soon challenge Canada for the top slot. Exports have grown spectacularly and are today the mainstay of the Mexican economy. Only Brazil and China, among developing nations, have attracted more foreign investment.

Until recently, Mexico was benefiting enormously from its economic partnership with the United States. Even though the country's 1995 financial crisis may have resulted from investor overconfidence, fed by NAFTA's launch a year earlier, the Clinton administration's response to the crisis would have been less quick and less generous without NAFTA in place. As it turned out, Mexico's access to financial markets was restored in five months, and growth was renewed within a year. In contrast, it took Mexico six years to recover from its 1982 debt crisis.

Once growth returned in 1996, it proceeded at a tempo unseen in Mexico since the 1960s and 1970s. The horsepower behind this expansion was the booming U.S. economy. According to NAFTA's detractors, however, the bill for Mexico's decision to tie itself to the U.S. economy came due in 2001, as the U.S. economic downturn quickly pushed Mexico into recession. Nevertheless, most analysts agree that the U.S.-linked Mexican economy is no longer vulnerable to Latin America's frequent crises. This conclusion is supported by the recent stability of the Mexican peso, the uninterrupted flow of FDI into Mexico, and the country's continuing access to international bond markets. In contrast, Brazil's currency is today under intense pressure, FDI is in steep decline, and the country's borrowing capacity has been curtailed by sky-high interest rates.

Despite the harsh impact of the U.S. slump on Mexico's economy, most Mexicans continue to favor close economic ties with the United States. Still, reactions in Mexico to the events of September 11 revealed some national ambivalence about the country's relationship with the United States, along with a streak of anti-Americanism among intellectuals, students, and many politicians. Fox's initially cautious public reaction to the attacks on the United States was conspicuous, occurring just a week after his visit to the White House. In the end, however, Fox did express unequivocal support for the battle against terrorism and traveled to Washington and New York to convey that message. His national security adviser, Adolfo Aguilar Zinser, has also said that Mexico is ready to build a joint security regime with the United States.

Although relations may have been temporarily set back by Mexico's hesitation in joining the U.S. antiterror coalition, the drive toward deeper integration between the two countries is unlikely to flag, since the essential reasons for bilateral cooperation remain unchanged. The most compelling is that U.S. political leaders, Republicans and Democrats alike, now consider the support of Mexican Americans (who make up two-thirds of all U.S. Hispanic voters) vital to their electoral success. Good relations with Mexico are now demanded by U.S. electoral politics.

Brazil is today pursuing a foreign policy agenda more ambitious than Mexico's. Hence it will be much more difficult for Brazil to achieve its international aims. Its pursuit of an independent leadership role faces several obstacles. Although distinguished by the diplomatic and political skill it brings to foreign relations, Brazil's military and financial resources are modest, providing the country with little international or regional leverage.

Even bigger obstacles are Brazil's domestic economic and social conditions. Despite improvements on many fronts, Brazil's economy has underperformed for most of the past two decades. The social picture is even more dismal, although certainly no worse than Mexico's. On crucial issues of international concern, such as poverty and inequality, race relations, education, and environmental management, Brazil is lagging. The importance these internal matters can have is best illustrated by the international acclaim that Brazil has received for its imaginative programs to address HIV and AIDs. Progress in other areas would similarly enhance Brazil's global stature. But, despite some real gains, Brazil remains, as President Cardoso said when he first took office, an unjust society. These handicaps make it hard for Brazil to lead internationally.

In addition, Brazil cannot escape the fact that it shares the same hemisphere with the United States. The gravitational pull of the prodigious U.S. market thwarts Brazil's efforts to establish stable trade arrangements with the rest of South America. Once it received a U.S. invitation to begin free trade talks, for example, Chile lost interest in becoming a full participant in Mercosur. Other South American countries would similarly find it difficult to pass on a free trade deal with the United States. For the Andean nations of Colombia, Peru, Ecuador, and Bolivia, the most urgent trade priority remains continuing the modest trade preferences granted them by the United States.

The United States also plays a central political role in South America. Plan Colombia—Washington's massive, multiyear commitment to assist Bogota in dealing with its drug and guerrilla problems—affects all of Colombia's neighbors. Whenever difficult circumstances arise—when Argentina faces economic collapse, or when Ecuador is threatened by a military takeover, or when Peruvians see their democratic future jeopardized—these countries turn to the United States first. Brazil has made important contributions in many instances, but it cannot match the United States in resources or raw power.

Brazilian leadership confronts yet another hurdle. Most South American countries are still wary about pursuing their external aims collectively. In particular, each has a huge stake in its relationship with the United States, and each wants to represent its own interests. Even

when they support Brazil's policy positions, other South American governments are reluctant to accept Brazilian leadership.

BIPOLAR

WILL BRAZIL and Mexico remain on divergent tracks? Mexico is unlikely to deviate from its strategic commitment to an open economy and close partnership with the United States. It has held a steady course since the mid-1980s, under four successive presidents and two different political regimes. Mexico's first democratically elected president, Vicente Fox, is pursuing this core strategy even more energetically than his predecessors did. And the policies have succeeded, so far, in strengthening Mexico's economy and making it less vulnerable to financial crisis. Moreover, many observers believe that these economic policies themselves helped promote the democratic change in Mexico that made Fox's election possible. The economic interests and structures of the United States and Mexico have become so intertwined that it is hard to imagine backtracking by either side.

Brazil's future course is more difficult to predict. It appears almost inevitable that Brazil will open its economy more widely, discarding impediments to trade and giving new emphasis to exports. Brazil has been heading in that direction for the past decade, albeit more slowly than most other Latin American nations. It is the only way Brazil can be globally competitive.

The United States needs Brazil's backing to achieve its goals in Latin America.

In the short run, however, Brazilian politics will block further economic liberalization. Import restrictions will be loosened only in a negotiated exchange of concessions with other countries—which will require the successful conclusion of hemispheric free trade negotiations or an agreement with the European Union. None of these negotiations will be completed before 2005 at best. A new global round of trade negotiations is likely to take even more time. In the next two or three years, Brazil could well become even more protectionist if the economy continues to stumble or if (although this is still unlikely) Lula da Silva of the populist Workers' Party wins this coming October's presidential elections.

Whatever the election results, Brazil will seek to sustain an active leadership role in regional and global affairs. Although this strategy could remain a source of contention in Brazil's relations with the United States, the two countries might also find more ways to cooperate, a prospect that could enhance Brazil's influence at the regional, hemispheric, and global levels. Brazilian collaboration with the United States has led to some important political and diplomatic successes in recent years, such as the resolution of the Ecuador-Peru border dispute and the avoidance of potential military takeovers in Paraguay and Ecuador. Brazil's adroit and forceful support for the United States in the immediate aftermath of the terrorist assault should now make political cooperation easier.

For its part, the United States should be prepared to work hard to find common ground with Brazil, especially on trade matters. U.S. officials know they need Brazil's backing to make headway on many issues in hemispheric affairs. Brazil may not be powerful enough to fully shape regional policies to its liking, but it has sufficient size and clout to keep the United States from achieving its goals in such crucial areas as the FTAA and Colombia. U.S.-Brazilian collaboration on a variety of international challenges—such as World Trade Organization negotiations and fluctuating oil prices—could advance the interests of both countries.

None of this means that Brazil and the United States have to agree on every issue, nor does it mean that Brazil will ever establish as close or interdependent a relationship as Mexico has with the United States. Geography may not be destiny, but it is not irrelevant, either, and it is not going to go away.

PETER HAKIM is President of the Inter-American Dialogue.

Reprinted by permission of *Foreign Affairs*, January/February 2002, pp. 148-162. © 2002 by the Council of Foreign Relations, Inc.

UNIT 6
Europe

Unit Selections

Key Points to Consider

- Explain why you do or do not expect to see Russia following the Baltic states by seeking membership in NATO and the EU in the near term.

- Who should pay for NATO's activities "out-of-area"?

- Do you believe the differences between the European and American public are large enough to threaten future military and political cooperation between the United States and its European allies?

- Do you agree or disagree with John Hall and Wolfgang Quaisser that expansion eastward by the EU will not harm the future of the EU? Defend your answer.

 **Links: www.dushkin.com/online/**
These sites are annotated in the World Wide Web pages.

Central Europe Online
  http://www.centraleurope.com
Europa: European Union
  http://europa.eu.int
NATO Integrated Data Service
  http://www.nato.int/structur/nids/nids.htm
Social Science Information Gateway
  http://sosig.esrc.bris.ac.uk

After the September 11, 2001, attacks in the United States, European investigators were surprised to learn that an extensive interlocking set of terrorist cells in Italy, Germany, Spain, Britain, France, Belgium, and numerous other countries remained in place. Many of the members of these cells were indoctrinated with combat videos from Chechnya, absorbed into al Qaeda by Osama bin Laden's agents in Europe, and trained in Afghanistan for operations against the West. As authorities continue to identify hidden cell members within their own societies, they coordinate their efforts to rout al Qaeda fighters who had infiltrated the ranks of ethnic Albanian guerrilla forces in Macedonia, Croatia, Bosnia, and Kosovo. Two of the suicide hijackers involved in the September 11 attacks have been traced to a training camp in Bosnia.

The year 2002 saw a remarkable transformation in Europe. As writers for *The Economist* describe in "The Balts and the European Union: Welcome Aboard," the Baltic states of Estonia, Latvia, and Lithuania were offered North Atlantic Treaty Organization (NATO) membership and will be invited to join the European Union, probably in mid-2004. To cope with the upcoming integration, each country is trying to shift the emphasis of its economy to high-tech, "knowledge-based" industrial niches.

Only a few years ago the conventional wisdom was that Russia would never let her former Baltic satellite states negotiate formal ties with NATO. However, NATO has been adept at transforming itself into a collective security organization more like the UN than a collective security defense alliance. Most observers assume that full Russian membership in NATO is only a matter of time now that leaders of Western countries and Russia share a common interest in the fight against terrorism. Deep fissures continue to exist within the alliance, but the recent expansion defies critics who routinely pronounce NATO to be a "dead" organization. In "Reforging the Atlantic Alliance," Philip Gordon provides reasons why it is important to continue supporting NATO and outlines future NATO tasks.

European members of NATO continue to maintain major commitments for peacekeeping operations in Kosovo and other states in the former Yugoslavia Federation. At the end of 2001, the European community assumed the lead role for peacekeeping operations in the former Yugoslavia Federation. However, NATO's military campaign in the region revealed a huge military disparity between the United States and European members of NATO. The United States carried 90 percent of the load in the 47-day air campaign. After the Kosovo experience, leaders of several European countries agreed to create a new regional defense structure, the European Security and Defense Identity. The purpose of the new organization is to make it possible for Europeans to operate in situations in which NATO is not engaged. The dilemma is that most European governments lack the military capability to equip or sustain the new force.

The NATO summit in Prague in 2002 acknowledged this dilemma by adopting the concept of "smart procurement." This idea recognizes that while Europe will never match U.S. defense spending, member states can spend their defense funds more wisely. Discussions also continued on the need for a multinational rapid deployment force of 21,000 troops that would allow NATO to operate quickly and effectively against new enemies far from Europe. The need for such a force became evident again in 2001 when the United States launched a military operation in Afghanistan. Only Great Britain and a few other countries had the military capability and political will necessary to play a role in the military campaign. In contrast, Denmark, even before donor countries met in Tokyo in January 2002 to coordinate reconstruction aid, sent $20,000 (U.S.) worth of office kits to each of 30 Afghani ministries, which included everything that the new government needed to start running the interim government. These very different contributions illustrate the different types of international competencies that European countries are willing to contribute in conflicts outside of Europe.

Immediately after the military campaign in Afghanistan, Germany declined to serve as the lead nation on the ground for peacekeeping due to other international commitments. Instead, Turkey served as the first lead nation of allied forces in Afghanistan. Turkey is scheduled to turn over the reins as the lead nation in February 2003 to Germany and the Netherlands, who will serve as "partners in leadership." This joint leadership is necessary because Germany lacks the capacity to serve as the lead peacekeeping nation-state alone. Neither country has the capabilities to manage the situation if the warlords misbehave or if Afghan president Hamid Karzai falls into real trouble. The lightly armed German Bundeswehr and their Dutch partners can keep the peace, but they are ill-prepared to fight. Moreover, the international peacekeeping mission controls the capital but little more. The shortcomings of the German and Dutch forces in Afghanistan illustrate the larger problems of the limited capacity of European militaries for "out-of-area" operations.

Most analysts agree that NATO's future depends less on its military structures and capacities than on the ability of member states to develop a common political purpose. Tensions among members of the Western alliance may increase in the future since recent public opinion polls indicate that Americans and Europeans do not see eye to eye on many issues. Important differences over such key issues as global leadership, defense spending, and the Middle East are likely to test the seams of the NATO alliance. The Bush administration's rejection of the Kyoto Protocol on Global Warming, the U.S. refusal to ratify the International Criminal Court, and the detention without charge of several European citizens in the United States under the new Patriot Act are three of several issues that are likely to fuel the gap between Americans and Europeans on key political issues.

A month after NATO's historic enlargement, the EU redefined its borders by inviting former Communist nations to join: Poland; the Czech Republic; Hungary; Slovakia; the former Soviet republics of Lithuania, Latvia, and Estonia; and the ex-Yugoslav Republic of Slovenia. Two Mediterranean islands, Cyprus and Malta, will formally join the European Union in May 2004. The historical expansion eastward will put an additional 75 million people under the union's banner and add 23 percent to its territory.

In "Europe's Eastern Enlargement: Who Benefits?" John Hall and Wolfgang Quaisser discuss the fact that the European Union is preparing a quantum leap eastward by adding 100 to 200 million new citizens from central-eastern and southeastern Europe. Hall and Quaisser discuss the issues related to whether this unprecedented expansion eastward will endanger the future of this grand experiment in integration. They conclude that eastern enlargement will create a "win-win" economic situation while also fueling intense frictions between net payers and net receivers of Brussels' budgetary funds.

One immediate loser was Turkey, whose demand to set a date to begin EU negotiations was rejected. Instead, the union's leaders agreed to meet in December 2004 to decide whether the largely Muslim country of 70 million people was democratic enough and respectful enough of human rights to begin negotiations. The EU also failed to reach an agreement in the 28-year-old division of Cyprus, which will become one of the new union members. Talks brokered by the United Nations ended without an agreement between Greek and Turkish Cypriots. That means that only the southern, Greek side of Cyprus—the only government on the island that is recognized internationally—was invited to join the union. The Turkish north could enter later if it agrees on terms to end the island's division.

The decision to expand the membership of the EU may be the easiest step toward further integration. At present there is no political plan for how the organization will govern itself and no economic plan for how the European Central Bank will create a single monetary policy for dramatically different economies. While the EU currently supports a process to draft a political constitution for the new Europe, no one is suggesting that there will be a "United States of Europe" any time soon. With the expansion, the population of the European club will increase by 20 percent, but the average wealth per person will fall by about 13 percent because most of the newcomers are relatively poor. That means that the new union will have to find ways to balance the interests of a country like Luxembourg, with a per capita GDP of nearly $43,000, with a country like Lithuania, with a per capita GDP of $3,200. The new members will also have to adhere to 80,000 pages of European Union laws and regulations.

The expansion also greatly complicates the problems of meeting the monetary and currency requirements of the European Monetary Union (EMU). On January 1, 2002, the 12 EU member countries adopted the new single currency, the euro, as their national currency. Three members—Britain, Denmark, and Sweden—have not yet agreed to adopt the new currency, but the euro is widely expected to quickly creep into their economies as well. However, virtually all European countries are having trouble meeting the EMU criteria, especially those limiting deficit financing in the face of the current international economic slowdown.

Development of a European security force and greater European integration are complicated by the continued multinational peacekeeping missions in Bosnia and Kosovo. Europe's last dictator, Slobodan Milosevic, was forced by popular protests to step down after being defeated in elections in 2000. In 2001 he was charged with 29 counts of war crimes, including genocide, crimes against humanity, and orchestrating a brutal ethnic-cleansing campaign by the Yugoslav army and Bosnia Serb militias that left 200,000 dead and spawned 1 million refugees during the 1992–1995 Bosnian war.

While Milosevic is now on trial for war crimes, several serious political complications to reestablishing stability in the Balkans remain unresolved. Disagreements among Western states about how best to promote stability in the face of periodic tensions between rival factions in Kosovo, Serbia, and Montenegro are among the most serious complicating factors. As the United States focuses on the war on terrorism, the European Union is acquiring greater responsibility for Balkan affairs.

The Balts and the European Union:
Welcome aboard!

The three Baltic states have already had to transform their economies beyond recognition to get into the EU. Now that they are just about in, what next?

TALLINN AND VILNIUS

OF ALL the stunning transformations that have changed the map of Europe since 1989, the Baltic states' shift from Soviet captivity to membership of the top western clubs is among the most remarkable. In the first fragile years of independence, prosperity and security seemed equally distant. Yet in a few short weeks this autumn, Estonia, Latvia and Lithuania have been formally offered NATO membership, and now, at this week's Copenhagen summit, they will be invited to join the European Union, probably in mid-2004. The contrast with the miserable life further east, in Russia, Belarus and Ukraine, has never been bigger.

But making normality work will be a hard slog too. "Small is beautiful when you want to move and change," says an *eu* official. "It isn't so beautiful when you want to attract investments." He is speaking in Tallinn, the small and beautiful capital of Estonia; but his words sum up the challenge facing all three.

In the past few years they have worked wonders to satisfy the nit-picking requirements of the EU. The Balts are so stable and orderly as to be almost boring—which, says Toomas Ilves, a former Estonian foreign minister, is as it should be. Now EU membership brings new opportunities and challenges aplenty. More trade and investment, freedom of movement for their citizens, and the chance to act as bridges between west and east are big pluses. But in many ways reform is still just starting.

Take the Lithuanians. They like to joke that they have the safest Soviet nuclear power plant in the world. The EU has spent millions of euros to prevent the Ignalina plant, which generates four-fifths of the country's electricity, from suffering the same fate as the similarly-designed Chernobyl reactor, in Ukraine, which blew up in 1986. And yet it is still not up to scratch, which is why millions more will be poured into helping Lithuania close it by 2009, a decade early.

Ignalina is symbolic of the Soviet legacy, from which all three Baltic economies continue to suffer. Baltic GDP per head is little more than one-third of the EU average. Business is still generally low tech, particularly in agriculture, where Latvia and Lithuania's workforces have around three times the proportion of farmers and one-third the productivity of the EU average, and that is already a good deal better than ten years ago. EU competition will hurt not help the hundreds of thousands of small, near-subsistence farmers who cannot compete with big, modern producers. Most will have to find other work. Economic development is unevenly spread: 80% of Estonia's booming foreign direct investment, for example, is in Tallinn.

Likewise, say critics, free trade with the EU countries may be good for big businesses, but hard on small ones, and they will all

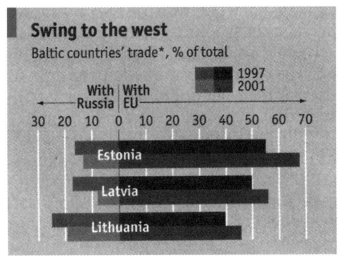

Swing to the west

Baltic countries' trade*, % of total

1997
2001

With Russia | With EU

30 20 10 0 10 20 30 40 50 60 70

Estonia

Latvia

Lithuania

Source: IMF   *Exports and imports

struggle with the extra cost of meeting *eu* rules on such things as workplace safety and environmental protection. When *Verslo Zinios* a Lithuanian business newspaper, polled its readers on which stories interested them most, it found that those on EU accession ranked surprisingly low. "It will come as a shock to some of them to discover that they are not a Lithuanian company but a European company," says Rolandas Barysas, the paper's editor.

Estonia's worry is different. Nordic-mindedly open and transparent, it has virtually no trade barriers, so joining the EU is in some ways a step backwards to more barriers and bureaucracy. Of the three, Estonia is where opinion polls find the lowest support for EU membership; only in the past year has it nudged above 50%.

But petty rules and painful changes, says Mr Ilves, are the price to be paid for belonging to the club. Among the billions of dollars available in restructuring funds, which the countries will keep getting until their GDP per head reaches 75% of the EU average, are portions for agriculture and small businesses. The Eu-

ropean Bank for Reconstruction and Development also lends money more cheaply to local banks if they in turn lend a certain amount to small firms and those outside the capitals. If anything, say officials, the countries' problem has been to come up with enough well-formulated projects to use up all the money on offer.

It will all be no good, however, if the Balts cannot find new sources of employment. Officials in all three countries talk about shifting the emphasis to high-tech, "knowledge-based" industries. These account for only 6% of Lithuania's economy, says Andrius Kubilius, a former prime minister as against around 25% in the EU as it now is.

Each country is developing its niches. Estonia embraced information technology early (people can pay for anything from on-street parking to restaurant bills through their mobile phones), while Latvia plans to make the most of its experience as one of the Soviet Union's main pharmaceutical-development centres. But they will be in tough rivalry with each other, not to mention the rest of the EU.

Which is why there is also a debate about ways to stay competitive. Labour costs are low now, but will not stay that way for ever. One unresolved issue for Europe as a whole is whether countries should be allowed to keep control of their tax policies or have Brussels set them. Many Balts think their relatively low levels of corporate income tax (Estonia cut its rate to zero two years ago) will be one of their few selling points to foreign investors, though there is little evidence as yet that it makes a difference.

Yet despite the challenges, and sizeable Eurosceptic minorities in every Baltic country, there are no concerted anti-EU movements. All three governments are confident that their people will give their nod in referendums next autumn. The alternatives are grim—either a return to Russia's orbit like Belarus, or the in-between status of dirt-poor ex-communist countries like Moldova. As Henrik Hololei, head of Estonia's office for EU integration, puts it, "Either you're in, or you're in… somewhere else."

Reforging the Atlantic Alliance

Philip H. Gordon

LESS THAN 24 hours after the September 11 terrorist attacks on the United States, America's allies in the North Atlantic Treaty Organization (NATO) came together to invoke the Alliance's Article 5 defense guarantee—this "attack on one" was considered an "attack on all." When it came time to implement that guarantee, however—in the form of the military campaign in Afghanistan, Operation Enduring Freedom—NATO was not used. The Americans decided not to ask for a NATO operation for military, political and strategic reasons: only the United States had the right sort of equipment to project military force half way around the world, and Washington did not want political interference from 18 allies in the campaign.

In the wake of these developments—and steadily mounting disagreement and even rancor about a long list of political and strategic issues—some observers have begun to question whether NATO has any enduring role at all. Charles Krauthammer bluntly asserts that "NATO is dead"; it may still have a marginal role to play as an incubator for Russia's integration into Europe and the West, but as a military alliance it is a "hollow shell."[1] Jeffrey Gedmin, director of the Aspen Institute Berlin, once a bastion of Atlanticism, believes that U.S. and European views of security are now so different that "the old Alliance holds little promise of figuring prominently in U.S. global strategic thinking."[2] Robert Kagan urges his readers "to stop pretending that Europeans and Americans share a common view of the world, or even that they occupy the same world."[3] He does not explicitly envisage, let alone call for, NATO's demise, but his views on the growing Euro-American divergence about the use of force—as well as the fact that NATO is hardly mentioned in a 25-page journal article about transatlantic security issues—leaves little doubt about how he sees the future of the Alliance. Even *The Economist*, a strong proponent of transatlantic security cooperation, wonders whether NATO will survive, concluding "it is harder than it used to be to imagine NATO, as it is, advancing far into the 21st century."[4]

NATO-skeptics make serious points; there are good reasons to be concerned about the future of the Alliance—especially if nothing is done to revitalize and adapt it to current circumstances. The Afghanistan campaign re-vealed significant gaps between the war-fighting capabilities of the United States and its allies, most of whom did not have the stealth, all-weather and communications technologies to enable them to take part in the early stages of the operation. The quick success of that operation also reinforced the already strong perception in some quarters in Washington—especially in the Pentagon—that it is easier to fight alone than with allies who have little to offer militarily and who might hamper efficient decision-making. Moreover, the U.S. decision in the wake of the terrorist attacks to increase its defense budget by some $48 billion for 2003—an increase that is larger than any single European country's entire annual defense budget—will only widen the capabilities gap. If the war on terrorism leads the United States to undertake military operations in other distant theaters—such as Iraq—and if the Europeans are unwilling or unable to come along, NATO's centrality will be further reduced. Add to all this the gradually diminishing NATO role in the Balkans (a role that arguably saved NATO from obsolescence in the 1990s), the emergence of a European Security and Defense Policy (ESDP) whose long-term relationship with NATO is uncertain, and a big enlargement of NATO's membership that risks diluting its coherence, and it is not difficult to see why many wonder whether NATO's first actual military missions—in Bosnia and Kosovo—will turn out to have been its last.

Four Reasons to Save NATO

YET TO conclude that NATO no longer has any important roles to play, just because it was not used in a mission for which it was not designed in the first place, would be a mistake. While NATO's roles in the 21st century will be very different from those relevant at the Alliance's founding in 1949, the United States has a strong national interest in preserving and adapting NATO. This is true for four main reasons.

First, NATO remains the primary vehicle for keeping the United States engaged in European security affairs. Perhaps it will turn out to be the case that the wars of the 21st century will be fought without Europeans and far away from Europe. In that case, America's European engagement over the past sixty years could be deemed a job well done and the transatlantic security partnership would no longer be necessary. But after a century that saw two world wars start in Europe and end in horrendous conflict in the Balkans (which, like the world wars, required American intervention to stop), it would seem more prudent for the United States to remain engaged until the continent's future—including that of Russia, Ukraine and eastern Europe—is more clear.

A second enduring role for NATO is to contribute to the continent's integration and stabilization process through enlargement. The incentive of NATO membership has been a powerful force in getting candidates throughout central and eastern Europe to undertake political, economic and military reforms that they would not otherwise have made. Since the enlargement process began in the mid-1990s, NATO aspirants have: resolved border disputes; changed electoral laws to ensure minority rights; discarded old and dangerous weaponry; reduced arms sales to unstable regions; cracked down on anti-Semitism; accelerated privatization; streamlined bloated military-industrial complexes; and promoted the transformation of civil societies. In all of this the desire to meet NATO standards and join the Alliance of democracies was a central motivating factor. Building peace through the development of stable, cooperative allies bears little resemblance to Cold War deterrence, but it represents just as great a contribution to European security.

The enlargement process is now even starting to bear fruit in terms of NATO's relationship with Russia, which under President Vladimir Putin has stopped actively opposing NATO enlargement and is now seeking instead to cooperate with the Alliance. Far from being a factor of division in Europe (as not only its opponents but even some of its supporters reasonably feared it might be), NATO enlargement is making major contributions toward the integration of all European states into the Western security community.

Similarly, through the Partnership for Peace (PfP), the Alliance has promoted military cooperation with partners as far away as Central Asia. These newly independent states—mostly still autocracies—have a long way to go before they would be welcome in an alliance built on democratic values. But the military-military ties, political contacts, institutional links and promise of better relations with the West promoted by the Partnership for Peace are among the West's most promising tools for having a positive impact on the region. Building on their PfP relationships, several of America's Central Asian partners ended up making essential contributions to the campaign in Afghanistan.

A third important role for NATO is ensuring peace in the Balkans, where the Alliance deploys 51,000 troops. Their presence is essential to preventing the region from reverting to the horrible conflicts of the 1990s. Even though the number of troops is gradually being reduced—to around 45,000 by mid-2003—and there is talk of the Europeans under the ESDP eventually taking over the entire mission, NATO's role will be indispensable for at least the next several years. No other organization can effectively plan and coordinate the diverse military forces from all the contributing countries, including the American military presence.

Finally but perhaps most importantly, NATO remains an essential peacetime preparation organization, a "tool box" of interoperable military capabilities that can be drawn upon by allies or groups of allies when needed. It is true that European member-states do not spend enough, and that they spend badly, with many redundancies among national European militaries and too high a proportion of immobile ground forces. Taken all together, the European members of NATO will spend only around $150 billion on defense in 2003, compared with some $380 billion for the United States. But $150 billion is not an insignificant amount of money, even by Pentagon standards. However inefficient they might be, the European members of NATO do have considerable military resources at their disposal, and they are often willing to undertake missions in which the United States does not want to be directly involved. European contributions to NATO's Bosnia and Kosovo campaigns were indispensable to the success of those operations, and today European NATO members (and Partners) are providing the overwhelming majority of Balkan peacekeeping forces—using NATO doctrines, tactics, procedures and interoperable equipment.

It is also worth noting that NATO can make important military contributions even in operations where the Alliance as such is not involved. This was the case, for example, during the Gulf War and in parts of the operation in and around Afghanistan. NATO was not formally involved in either case, but in both cases Allied forces, bases and cooperation among NATO militaries were critical. In the Afghan campaign, most NATO allies were excluded from the initial operations for understandable reasons, but have become more involved over time. This involvement has included combat and special forces missions undertaken by British, Canadian and German soldiers conducting cave-clearing missions in the Afghan mountains; British and French reconnaissance, air-defense, in-air refueling and combat aircraft missions; the deployment of a French carrier-led naval task force alongside American ships patrolling the Indian Ocean; "backfilling" by NATO nations of U.S. military missions in the Balkans and even in the United States (where NATO Airborne Warning and Control Systems were deployed to protect American airspace); and the use of European NATO bases for staging operations during the Afghan campaign.

All of these tasks were facilitated by the existence of common NATO operational doctrines (for example, so that French reconnaissance officers know what kinds of information U.S. pilots need), agreed equipment standards (so that Turkish refueling nozzles fit into U.S. planes), routine multinational exercises (so that allied forces are in the habit of working together and know what problems to expect), and interoperable secure communications (so that various nations' forces can share real-time intelligence and data during an operation). When it came time to organize an international security force for Afghanistan to provide stability once victory was won, European NATO allies provided the vast majority of the forces (first under British, now Turkish, command). By summer 2002 nearly half of the 13,000 foreign troops there came from NATO allies other than the United States. In the long run, NATO itself may prove to be the best option for the maintenance of a long-term, Western-led security force in Afghanistan.

In short, while the war on terrorism does indeed suggest that NATO is no longer the central warfighting institution it was during the Cold War, it would be shortsighted to conclude that its mission is over. As a military machine designed to protect Alliance borders from an external ground attack, the Alliance may indeed be "dead" or dying; that is the benign consequence of its success. But as a community of democracies with common values and interests, and a community that is determined to maintain the political relationships and military tools to protect those interests, it is most certainly worth preserving. Surely, if it may be granted that the United States is wise to seek allies, there is no imaginable substitute for the capacities, military and otherwise, that America's NATO allies can provide.[5]

Five Tasks

INSTEAD of giving up on NATO, the North American and European allies should use the upcoming summit in Prague to adapt the Alliance to the most important security challenges of the day. Just as previous developments— the Soviet deployment of intercontinental ballistic missiles, the end of the Cold War and the conflicts in the Balkans—have obliged the Alliance to adapt, September 11 and the conflict that has followed it require NATO leaders to think boldly and creatively about how to ensure that the Alliance continues to serve European and American security and political interests.

How should NATO adapt at Prague and afterwards? How can the United States best use the Alliance to promote its own interests? Five tasks stand forth.

First, Alliance leaders need to make clear that new threats such as international terrorism are a central concern to NATO member-states and their populations. When the Prague summit takes place in late November, the American public is not going to be focused on NATO enlargement, ESDP or the Balkans, however important those issues remain. They will be wondering whether NATO contributes to the most important security challenge the United States faces, and NATO leaders will have to give them "yes" for an answer.

Already in its 1991 Strategic Concept, NATO recognized that "Alliance security must also take account of the

global context" and that "Alliance security interests can be affected by other risks of a wider nature, including proliferation of weapons of mass destruction, disruption of the flow of vital resources and actions of terrorism and sabotage."[6] NATO repeated the point in its 1999 Strategic Concept, this time moving "acts of terrorism" to the top of the list of "other risks."[7] At Prague, Alliance leaders will need to make vividly clear that NATO is about more than the traditional defense of its borders. This is not to say that any act of terrorism or threat to energy supplies must be treated as an Article 5 contingency for which all Allies are obliged to contribute troops. It does mean, however, that all allies recognize that their common interests and values can be threatened by global developments, a point made dramatically clear by the attacks on Washington and New York. Even if invocations of Article 5 will no longer necessarily mean a formal NATO operation under NATO command, the concept that "an armed attack" from anywhere abroad must trigger solidarity among the member states is an important principle that should be reinforced.

Second, NATO needs to put real substance behind this political commitment by developing its political and organizational capacity to deal with global security threats, including the specific issue of terrorism. European allies who worry about giving the Alliance too great a "global" role have long resisted such new missions. European leaders have been very reluctant to put their countries in a position whereby their soldiers could be dragged off by the United States to fight beyond Europe's borders, particularly since there has traditionally been a good deal of transatlantic disagreement over out-of-area issues. Ironically, today the resistance to expanding NATO's potential mission comes more from the United States, where many in the present administration fear being constrained politically by allies whom they believe have little to contribute militarily.

The United States, however, has every interest in having European allies better prepared to join it on potential global missions. While it is hard to see NATO countries agreeing to use the Alliance for such anti-terrorist matters as law enforcement, immigration, financial control, and domestic intelligence anytime soon, the Alliance should begin immediately to adapt its military structures to contribute more effectively to the war on terrorism. NATO should significantly reform and streamline its command structures to remove redundant and wasteful headquarters, replacing them with a leaner and more efficient structure based on military functions rather than geographical orientation. It should develop a new Force Projection Command specifically responsible for planning out-of-area operations, and increase the number and capability of its rapid reaction forces so that NATO member-states can project forces to distant parts of the world on short notice. Allies should share more information about the proliferation of weapons of mass destruction (WMD) and ballistic missiles, and develop emergency civil defense and consequence management teams that could assist allies subjected to biological, chemical or nuclear weapons attack. NATO should also coordinate the development of joint theater missile defense systems that could protect allies in Europe (including, possibly, Russia) that face a growing ballistic missile and WMD threat.

Finally on this score, NATO should commit itself to a program to coordinate and train various member-state special forces, whose role in the anti-terrorism campaign will be critical. During the Cold War, few could have imagined the need for American and European special forces to travel out-of-area to execute coordinated attacks, but that is now a very real requirement. While NATO was not used for the military response to an attack on the United States, it is unfortunately not difficult to imagine a major terrorist attack on a European city for which a NATO response would be appropriate. The very preparation of new missions for NATO—even if the Alliance is never called upon to conduct them—would be a major contribution in the effort to get European allies and their militaries focused on the new security challenges of the day, most of which are beyond Europe's borders. Such preparation might also buy at least a measure of deterrence against such attacks.

A third and related necessity is for NATO's European allies to accelerate the process of adapting their military capabilities for the most likely new types of missions. At NATO's April 1999 summit, the Allies adopted a Defense Capabilities Initiative (DCI) designed to improve allied forces' deployability, mobility, sustainability, survivability and effectiveness.[8] The DCI process identified 58 areas in which the Allies were asked to make concrete improvements in their forces to fill specific capabilities gaps. But the DCI process never really had political support and few of its goals have been fulfilled. At Prague, European NATO members should reduce this long list to five or six critical categories—to include secure communications, air- and sea-lift capabilities, electronic warfare, precision-guided munitions, defense against weapons of mass destruction, and in-flight refueling—and make real commitments to fulfilling their goals. Not only do the Europeans need to make serious improvements in capabilities if they want to join effectively with the United States in the anti-terrorism campaign, the EU decision process needs to be better integrated with NATO's. If it isn't, current problems with interoperability will only get worse. A division of labor among the Allies, whereby the Americans concentrate mainly on high-end military operations and most Europeans do only peacekeeping and reconstruction, is to a certain extent inevitable and may even have certain advantages.[9] But it should not be allowed to go too far. Dividing military roles in such a stark way would undermine political solidarity within the Alliance and contribute to the growing gap between Americans and Europeans over how and when to use force in the first place. It is a recipe for undermining the political solidarity and sense of shared risk that has been at the core of the Alliance since 1949.

Fourth, NATO should continue the process of enlargement, as a means of developing strong allies capable of contributing to common goals and of consolidating the integration of central and eastern Europe. Barring the unexpected, it now seems clear that the Alliance will take in five to seven new members at Prague: Estonia, Latvia, Lithuania, Slovakia, Slovenia and possibly Bulgaria and Romania as well. Some argue that such a large number of new allies will dilute the Alliance and render it unusable in the future. There is not, however, a fundamental difference between an Alliance at 19 and an Alliance at 24 or 26—NATO never was an alliance of equals and always depended on American leadership in the past, as it will in the future. Moreover, to the degree that NATO is being transformed from a defensive warfighting organization into a force-providing and coalition-facilitating organization, the dilution issue largely goes away.

It is in any case a myth that it is impossible to conduct effective military operations with a large number of countries—the "war by committee" charge leveled against the Kosovo campaign. As General Wesley Clark has argued, if there was a war by committee in Kosovo, the committee was within the Clinton Administration and Joint Chiefs of Staff, not among the NATO allies.[10] Only those few allies actually making a major military contribution—primarily Britain and France—demanded a significant voice in the conduct of the operation. This was not an unreasonable demand, and it was arguably worth the trade-off. If NATO leaders are worried about taking in member-states whose long-term political stability or democratic credentials might be uncertain, they should consider mechanisms that would allow for the temporary suspension of an ally. But it would be a mistake to discard the principle that the Alliance has enunciated for years, that the NATO door is open to all European democracies committed to the Alliance's common values and security interests.

Finally, the Prague summit should be used to promote greater cooperation between NATO and Russia. Significant progress has already been made in this regard, as demonstrated by the May 2002 agreement to set up a new NATO-Russia Council, which will give Russia the opportunity to interact with NATO allies as an equal partner on a range of issues—though without the ability to veto NATO decisions. It is true, as critics of the new relationship with Russia argue, that Russia had the opportunity to cooperate with NATO already through the Permanent Joint Council, created in 1997, and chose not to do so. But the fact is that Russia under Putin has taken a very different approach; the new forum offers an excellent opportunity to advance cooperation in unprecedented ways without sacrificing NATO's ability, when necessary, to act without Russian agreement. Areas for possible joint action could include counter-terrorism, non-proliferation, peacekeeping and crisis management, missile defense cooperation, air defense, civil defense cooperation, collaborative armaments programs and joint work on restructuring Russia's outmoded armed forces and early-warning systems. In the wake of the tragedies of September 11, the prospect that Russia could feel that it is part of the West—rather than threatened by it—is an opportunity that should not be squandered. That NATO can be a vehicle for that opportunity is perhaps the most vivid illustration possible of its continuing utility as an alliance.

NATO IS not dead, nor is it doomed. But it needs to adjust to new realities if it is to serve useful purposes on both sides of the Atlantic. Disagreements among allies are natural; in the case of NATO, certainly, they are nothing new. But allies we remain. The challenge of adjusting the Alliance after the Cold War was already considerable before last September 11; since then it has become both more complex and more urgent. The Prague summit is an opportunity for the Alliance to make that adjustment; it is an opportunity that should not be lost.

Notes

1. Krauthammer, "Re-Imagining NATO," *The Washington Post*, May 24, 2002.

2. Gedmin, "The Alliance is Doomed," *The Washington Post*, May 20, 2002.

3. Kagan, "Power and Weakness," *Policy Review* (June/July 2002), p. 3.

4. "A moment of truth—the future of NATO," *The Economist*, May 4, 2002.

5. Nicely argued by John O'Sullivan, "With Friends Like…Whom?" *National Review*, July 1, 2002.

6. *The Alliance's New Strategic Concept*, North Atlantic Council in Rome, November 7–8, 1991, para. 12.

7. *The Alliance's Strategic Concept*, North Atlantic Council in Washington, DC, April 23–24, 1999, para. 24.

8. See Washington Summit Communiqué, Press Communiqué NAC-S(99)64, Brussels, April 24, 1999, para. 11.

9. See James P. Thomas, *The Military Challenges of Transatlantic Coalitions*, Adelphi Paper no. 333 (London: International Institute for Strategic Studies, May 2000), chapter 3.

10. Wesley Clark, *Waging Modern War* (New York: Public Affairs, 2001).

Philip H. Gordon is a Senior Fellow in Foreign Policy Studies at the Brookings Institution and a former Director for European Affairs at the National Security Council.

The Real
Trans-Atlantic Gap

Americans and Europeans see eye to eye on more issues than one would expect from reading the New York Times *or* Le Monde. *But while elites on both sides of the Atlantic bemoan a largely illusory gap over the use of military force, biotechnology, and global warming, a survey of U.S. and European public opinion highlights sharp differences over global leadership, defense spending, and the Middle East that threaten the future of the last century's most successful alliance.*

By Craig Kennedy and Marshall M. Bouton

Periodic angst about the state of trans-Atlantic ties is perhaps as old as the relationship itself. But the stresses unleashed by the September 11 attacks and their aftermath have brought anxieties about relations between the United States and Europe to a remarkably fevered pitch. Witness the intellectual uproar on both sides of the Atlantic caused by commentator Robert Kagan's article "Power and Weakness," published in the June/July 2002 issue of *Policy Review.* Drawing on growing European complaints about U.S. unilateralism and American counter-charges of European unreliability and weakness, Kagan struck a trans-Atlantic nerve with his assertion that "on major strategic and international questions today, Americans are from Mars and Europeans are from Venus: They agree on little and understand one another less and less."

For the most part, this fierce rhetorical battle has been waged among elites. But these elites reside in democracies and, in the end, the opinions of regular citizens do have some influence on the course of grand policy. What do the American and European publics think about the roles of their respective nations? Do they see a growing trans-Atlantic breach on basic issues of security and terrorism? Are European publics as critical of the United States as are their intellectuals and politicians? Do American voters view Europe as a reliable and favored ally or, like some in Washington would argue, as a barrier and restraint on U.S. efforts to stop terrorism?

To answer these and other questions, the Chicago Council on Foreign Relations (CCFR) and the German Marshall Fund of the United States undertook a detailed study of U.S. and European public opinion on foreign affairs. For the last 28 years, CCFR has conducted a quadrennial survey of U.S. opinion and published the results in FOREIGN POLICY. For the first time, six European countries—Great Britain, France, Germany, Italy, the Netherlands, and Poland—were added this year to provide comparative data on European attitudes.

Based on the bitter harvest of op-ed pieces and news articles during the last years, one would expect dramatically differing views on the goals and means of foreign policy, as well as on specific hot-button issues such as globalization, biotechnology, and the Middle East. For those who fear a permanent split between the United States and Europe, there is good news: Americans and Europeans see eye to eye on more issues than one would expect from reading the *New York Times* or *Le Monde.*

At the same time, areas of disagreement between the European and American publics are significant. On four crucial issues—threat perception, leadership, defense spending, and the Arab-Israeli conflict—both Europeans and Americans harbor the mutual acrimony of their respective elite commentators. Americans and Europeans have always disagreed on these four issues. However, the trans-Atlantic alliance held together during the Cold War because these differences in perception

Average approval for the use of military

	0 5 10 15 20 25 30 35 40 45 50 55 60 65 70 75 80 85
Britain	
France	
Germany	
Netherlands	
Italy	
Poland	
Europe	
US	

CHARTS BY MANUEL BEVIA

and priorities were subordinated to the fight against a common enemy. Now, these points of conflict have moved from the background to center stage and could potentially imperil future cooperation on major strategic challenges in Asia, the Middle East, and elsewhere.

WARM FEELINGS, GOOD INTENTIONS

By almost any standard, the results of our study should soothe the nerves of committed Atlanticists and perhaps deflate the rhetorical excesses of those who have thrived on forecasting a fundamental split between Europe and the United States. First, and most important, Americans and Europeans seem to like each other. When asked to measure their warmth toward various countries on a 100-point thermometer, Americans give Germany a 61 rating and Great Britain a 76. In both cases, American warmth toward these leading European countries is significantly higher than for all other countries except Canada. Europeans give equally warm ratings to the United States, with a high of 68 from Great Britain and a low of 59 from the Netherlands. Even French respondents rate the United States above most of their European neighbors (though below the European Union).

Moreover, Americans and Europeans want their governments to work together as much as possible. Almost 80 percent of Americans want Europeans to exert strong leadership in the world. When asked whether the United States should share more decision making with Europe, even if it means the United States will have to compromise at times, 70 percent of Americans say yes. On the European side, a comfortable majority want to see the European Union (EU) become a superpower like the United States. When asked if the purpose of superpower status should be to compete or cooperate with the United States, 84 percent opt for cooperation. In similar questions, Americans

and Europeans almost always choose trans-Atlantic collaboration over the alternatives.

European critics of the United States have often argued that the United States is isolationist. We asked both sides whether their country should take an active part in the world or stay out of world affairs. A large majority of Americans (71 percent) and Europeans (78 percent) chose an active role. But the United States does have its isolationists. According to this survey, 25 percent would stay out of world affairs. However, that response is only slightly greater than the number for Germany (23 percent) and the Netherlands (24 percent).

Americans and Europeans also have a broadly shared understanding of global threats and countries where they have critical interests [see chart, Critical threats]. Not surprisingly, terrorism ranks high for both sides, as does Islamic fundamentalism. But other threats, such as global warming, elicit very similar reactions from Americans and Europeans. There are exceptions to this general agreement, however. For example, 56 percent of Americans believe that China's rise as a world power poses a critical or extremely important threat, compared with 18 percent of Europeans.

How about support for multilateral institutions? Again, American and European views are quite similar. When asked if the United Nations should be strengthened, 77 percent of Americans say yes, compared with 75 percent of Europeans. On other questions related to the World Bank, the International Monetary Fund, NATO, and other international institutions, no sharp trans-Atlantic differences were evident.

Americans and Europeans respond in very similar ways about the use of military force, belying the argument of Robert Kagan and others that Europe has become a pacifist's paradise.

One of the most interesting results of this study concerns attitudes about the use of military power. When asked a series of hypothetical questions about the use of military force, Americans and Europeans respond in very similar ways, belying the argument of Kagan and others that Europe has become a pacifist's paradise [see chart, Average approval for the use of military]. Strong majorities in both the United States and Europe support using force in many situations. For example, 80 percent of Europeans and 76 percent of Americans agree that military force should be used to uphold international law. Some scenarios concerning the use of force do produce different results. Americans are more supportive of using military action to destroy terrorist camps, while Europeans are much more willing to use force to quell civil wars.

Does that general agreement on the use of force extend to an invasion of Iraq? Yes. Both Americans and Europeans strongly endorse a U.N.-approved invasion that is supported by allies [see chart, Use of military force against Iraq]. Absolute opposition to an American invasion is greater in Europe (26 percent) than in the United States (13 percent) but is still modest.

Other data uphold this finding. For Europe, we experimented with eight scenarios for an American invasion of Iraq. One variable in this comparison proved to be very robust. As long as the United Nations supports U.S. actions, Europeans are willing to participate. In Great Britain, support increases by 30 percentage points when we include the United Nations in the formulation. However, no European country gives majority support for an Iraq invasion without U.N. sanction. German Chancellor Gerhard Schroeder's opposition to U.S. intervention in Iraq makes sense if one looks at the weak German support for military action against Saddam Hussein. In our experiment, German support for military action never rises above 41 percent, while opposition figures range from 49 percent to 72 percent, depending on the scenario.

DANGEROUS DIFFERENCES

Other, more troubling data counterbalance this good-news story and suggest that trans-Atlantic cooperation could become much more difficult. First, Americans and Europeans may have a shared understanding of global threats, but they disagree on their severity. Second, Americans and Europeans have yet to agree on a formula for sharing global leadership. Third, Americans and Europeans have different ideas of appropriate levels of defense spending. And fourth, Americans and Europeans have starkly dissonant views on the Middle East conflict. Each of these areas is critical to the formation, operation, and focus of an effective trans-Atlantic alliance. Disagreement across all four is a cause for significant concern.

Americans and Europeans do rank threats, more or less, in the same order. But Americans find the world much more threatening than do their European counterparts [see chart, Critical threats]. In particular, Americans believe that terrorism and related threats are much more important than do Europeans. Explanations for these pronounced differences start with the impact of September 11 on the American sense of security. Yet whatever their source, these contrasting perspectives do have direct consequences for policymaking. For at least the last year, U.S. foreign policy has been driven by the belief that serious threats require immediate action. By contrast, European policymakers have argued that the United States is acting precipitously and may be overestimating the risks posed by terrorism and other threats. Europeans may think that Saddam Hussein is a threat because he is developing weapons of mass destruction, but they may not have the same sense of urgency about overthrowing his regime as do Americans.

Not only do Europeans feel relatively more secure and less threatened than Americans, they also feel that the United States is in part to blame for its current vulnerability. One of the most disturbing findings of our survey is that a small majority (55 percent) of Europeans think that U.S. policies contributed to the terrorist attacks on New York City and Washington, D.C.

The Cold War alliance held together for more than 40 years because its members largely agreed on the extent and critical

Use of military force against Iraq

	U.S. should not invade	Invade with U.N. and ally support	U.S. should invade alone	Don't know
Britain	20	69	10	1
France	27	63	6	3
Germany	28	56	12	4
Netherlands	18	70	11	2
Italy	33	54	10	4
Poland	26	53	10	11
Europe	26	60	10	4
U.S.	13	65	20	2

nature of the threats posed by the Soviet Union. Given the current differences in public perceptions of the world, it may be much harder to sustain a long-lasting coalition against a broad set of risks—especially those centered outside Europe and the Mediterranean—as long as Europeans do not feel as exposed as Americans.

Resistance to sharing leadership with Europe, as well as European ambivalence toward a subordinate role, may also stymie a sustainable alliance against terrorism. This issue of joint decision making in the trans-Atlantic relationship has been a persistent concern. But especially since September 11, European policymakers and commentators have criticized U.S. "unilateralism" and a perceived American unwillingness to consult with allies before making major decisions. European elites have a pervasive feeling that they are junior partners who are expected to follow without asking a lot of questions. U.S. officials do not completely disavow this position. Driven by a sense of urgency, U.S. policymakers have less patience for the extended consultations and negotiations that Europeans desire. Even before September 11, U.S. officials were tiring of European criticism on the Kyoto Protocol, the International Criminal Court, and other matters.

Europeans would like to do something about the influence gap. When asked whether the United States should remain the sole superpower or if the EU should be a superpower also, 65 percent of Europeans say they would like to see the EU become a peer of the United States. The vast majority of that group believes that EU superpower status will enable Europe to cooperate more effectively, rather than compete, with the United States. The differences within Europe are striking. More than 90 percent of French respondents support EU superpower status (though even in this case, an overwhelming majority see that status as a means for working more closely with the United States). By contrast, only 48 percent of Germans want a European superpower, while 25 percent said that no country, including the United States, should be a dominant force in the world.

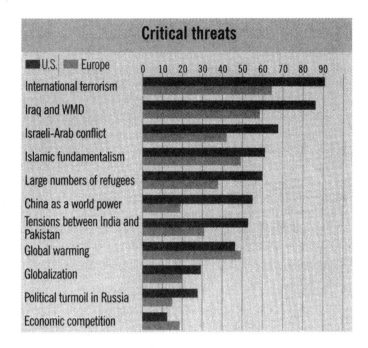

Critical threats

■ U.S. ■ Europe

International terrorism

Iraq and WMD

Israeli-Arab conflict

Islamic fundamentalism

Large numbers of refugees

China as a world power

Tensions between India and Pakistan

Global warming

Globalization

Political turmoil in Russia

Economic competition

Americans, on the other hand, are not interested in sharing their unique position in the world with the EU. In answer to the same question, 52 percent want the United States to be the only major force in the world, and only 33 percent welcome the idea of EU superpower status. As mentioned earlier, Americans want to keep a lock on superpower status, but they are willing to share some decision making with Europeans. Still, it may be easier to agree with the idea of joint decision making when one side perceives the other as weaker and less influential.

This perceived weakness becomes real when one looks at military spending and capabilities. For many years, European governments have reduced or maintained low budgets for the military and defense, arguing that the post–Cold War world poses lesser threats that can be handled more effectively by means other than brute military force. Americans, on the other hand, have argued that Europeans are making a unilateral decision to forgo extensive military cooperation with the United States. Our findings suggest that this tension will not disappear anytime soon.

Europeans may be willing to support the use of force in theory, but they are less interested in spending real money for military purposes. When asked whether defense spending should be expanded, cut, or kept the same, 19 percent of Europeans want to boost military funding, 33 percent want to cut it back, and another 42 percent want to maintain current low defense budgets. By contrast, 44 percent of Americans are willing to spend more on defense, and another 38 percent support current levels of spending, which are already significantly higher than European expenditures.

European proposals for a new form of burden sharing between Europe and the United States have offered new solutions to discrepancies in military power and capabilities. Under this framework, the United States would do the heavy lifting on military matters, while Europe would handle reconstruction, peacekeeping, and other duties that fit European capabilities. We

asked Americans and Europeans if they support this division of labor. A majority of Europeans (52 percent) do support it, as opposed to 39 percent of Americans.

The hesitancy to spend money on defense is so strong in Europe that it dampens the European desire for the EU to become a superpower. As noted, a large majority of Europeans would like to see the EU become a superpower like the United States. However, when we asked those who support that position whether they are willing to spend more on their country's military to achieve that end, public backing for superpower status decreases significantly [see chart, Europeans do not want superpower status].

This startling result can be interpreted two ways. Europeans might say they want to be a superpower, but one that uses means other than military force to influence the world. Other data in our survey support this interpretation. Europeans are quite willing to spend money on foreign aid, for example, and a large majority sees economic aid to poor nations as a key tool in combating terrorism. On the other hand, many Americans might look at these data and conclude that Europe will not be a serious partner in overcoming the biggest challenges facing the world. While some American analysts would disagree, most would argue that this resistance to defense spending consigns Europe to a secondary role in U.S. strategic considerations.

The interplay of these three issues and their corrosive effect on trans-Atlantic cooperation comes to the fore when one looks at the Arab-Israeli conflict. For many years, this issue has been a source of tension. European elites criticize Americans for exaggerating threats to Israel's security and turning their backs on Palestinian suffering. Americans blast European leaders for ignoring the terrorist threats to Israel and for their seeming unwillingness to apply greater accountability to the millions of euros given to Chairman Yasir Arafat and the Palestinian National Authority. European elites express discomfort with American leadership, and Americans wonder why Europe cannot be more supportive of concrete proposals for peace.

When asked to register their "warmth" toward Israel on a 100-point scale, the American response is 55, and the European average is 38. Public support for a Palestinian state indicates another gap in this policy area. A strong majority (72 percent) of Europeans favor such a state, and only 40 percent of Americans do. The European response is fairly consistent among countries, with Italians the most supportive and Poles the least. American respondents also registered the strongest opposition to a Palestinian state by a significant margin, with only Germany coming at all close to the U.S. figure.

Americans have not only quite different opinions about Israel and a proposed Palestinian state but also much stronger feelings about the importance of this conflict to their own security. In our examination of threats, 67 percent of U.S. respondents see military conflict between Israel and its Arab neighbors as a critical threat, while only 42 percent of Europeans share this opinion. These results, in fact, parallel a common U.S. charge that Europeans do not have as much at stake in this conflict as the United States and, hence, can afford to propose strategies that could put Israel at risk.

In this concrete case, one can see the basic problems that plague the trans-Atlantic alliance in many other areas. Ameri-

Closer Than You Think

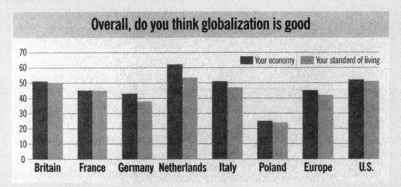

Overall, do you think globalization is good

■ Your economy ▦ Your standard of living

Britain France Germany Netherlands Italy Poland Europe U.S.

Media commentary and reportage often portray the United States and Europe as far apart on a range of social and economic issues. Our survey found unexpected levels of agreement in several areas, including public attitudes toward globalization, global warming, biotechnology, and immigration.

Globalization: Conventional wisdom is that Europeans are much more sensitive to this issue than Americans because of Europe's political history and its hesitancy to embrace lightly constrained market capitalism. Moreover, European elites often suggest that their hands are tied in trade negotiations because of a strong popular resistance to greater liberalization.

Our findings on globalization suggest that Americans are slightly more supportive than Europeans, but the differences are not dramatic [see chart above]. A slight majority of Americans think globalization is good for the United States and for their own standard of living; Europeans are somewhat less positive. Polish respondents are clearly less enthusiastic about globalization as they understand it. When we leave Poland out of the weighted European average, U.S. and Western European attitudes become even closer.

Global warming: We asked respondents to rate different "threats" to their country. Fifty percent of Europeans and 46 percent of Americans say global warming is a high-priority threat. More Americans (17 percent) than Europeans (8 percent) say global warming is not important, but it seems safe to say that public opinion in the United States on this issue is largely in line with that of Europe but out of step with the Bush administration's policies. When asked to rate President George W. Bush's performance on specific

issues, Americans and Europeans both give him their lowest rating for his handling of this topic.

Biotechnology: During the last four years, the European Union has limited greatly the imports of U.S. food that contains genetically modified organisms (GMOs) on the grounds that these biotech products may present unspecified health and environmental risks. We asked respondents if they strongly support, moderately support, moderately oppose, or strongly oppose the use of biotechnology in agriculture and food production: Sixty-six percent of Europeans are opposed to some degree, compared with 45 percent of Americans. However, though a majority of Americans do support biotechnology in agriculture, this level of opposition is somewhat surprising given that U.S. companies dominate this field and that GMOs have been used in the United States for over a decade.

Immigration: Sixty percent of Americans and 38 percent of Europeans rate the level of threat posed by "large number of immigrants and refugees coming into your country" in the highest category. Great Britain (54 percent) and Italy (52 percent) come close to the U.S. levels. By contrast, only 23 percent of German respondents see this threat as extremely important, despite the major debate that immigration policy has provoked in that country. We also asked if respondents favor or oppose immigration restriction as a means for combating terrorism. In this case, 77 percent of Americans and 63 percent of Europeans favor the use of such limits for that specific purpose. Again, Great Britain (74 percent) and Italy (75 percent) have the highest percentages favoring restrictions, and Germans provide the least support for this proposal (44 percent).

cans come to the issues with strong loyalties, a keen sense of the threats posed by the situation, and a willingness to support the use of force if other means prove inadequate. Europeans approach issues with less urgency, a commitment to being even-handed, and a commitment to exhausting all nonmilitary means for solving the problem. Given these differences, U.S. resistance to the idea of sharing leadership with Europeans on this politically and, to Americans, morally, important issue is log-

ical. How can a country work closely with an ally when these significant divisions affect how each analyzes and addresses such a fundamental problem?

CONTAINING THE DAMAGE

Taken together, these four issues have the potential to make the trans-Atlantic alliance much weaker and less effective. Unlike

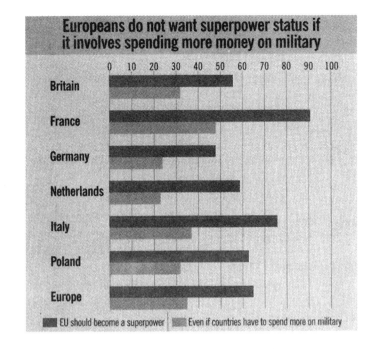

Europeans do not want superpower status if it involves spending more money on military

Britain
France
Germany
Netherlands
Italy
Poland
Europe

■ EU should become a superpower   ■ Even if countries have to spend more on military

many other issues, they also represent areas where the opinions of policy elites and regular citizens are largely similar. We cannot argue that elites are out of touch with their publics, nor can we hope that those in power will try to shield us from populist opinions.

One of the most disturbing findings of our survey is that a small majority (55 percent) of Europeans think that U.S. policies contributed to the terrorist attacks on New York City and Washington, D.C.

Moreover, the September 11 attacks and the changes they produced in the United States have heightened the significance of this trans-Atlantic divergence in perceptions. Americans do feel more at risk now than they did before the attacks. They are more willing now to support military action and defense spending. And Americans may have a stronger appreciation for Israel and its challenges in confronting terrorism and Islamic fundamentalism. Europeans have not responded the same way for understandable reasons. Nonetheless, American expectations for its closest allies have been altered. Europeans also have changed expectations. Many expected the United States to be more demanding of its allies but also more willing to consult and engage in truly joint planning. When it became clear that the United States was on a mission that did not provide much room for consultation on strategic objectives and means, Euro-

peans stepped up their criticisms of U.S. unilateralism and hegemony.

These major disagreements are not easily resolved. Threat perceptions will not change in Europe unless some sort of catastrophic terrorist attack hits Berlin, Paris, or London. Without a greater sense of risk, Europeans will unlikely feel compelled to increase defense spending. But without greater European defense capabilities, Americans will not likely be willing to treat Europe as an equal. Nor do we hold out much hope that attitudes—elite or otherwise—on the Arab-Israeli conflict will change any time soon.

Based on the findings of our survey, significant differences within Europe will also threaten trans-Atlantic cooperation. French aspirations for a stronger European Union with superpower status will almost certainly clash with Germany's and Great Britain's more limited vision of Europe. Britain's willingness to follow the American lead in applying military power will meet resistance from both the French, who are less enamored with U.S. leadership, and the Dutch and Germans, who have a weaker appetite for armed conflict.

These strains, however, need not lead to a gradual dissolution of the trans-Atlantic alliance. As we have emphasized several times, Americans and Europeans want to work together. Americans view Europe as relatively more important than Asia and as a potential ally in addressing a wide range of challenges. Europeans, for their part, show few signs of anti-Americanism. They increasingly question U.S. strategies, as Chancellor Schroeder demonstrated during the recent German electoral campaign. But, based on our data, this questioning comes from a desire for a more equal relationship rather than a desire to cease cooperation.

An ever stronger set of trans-Atlantic economic and cultural ties also supports this commitment to cooperation on both sides. The rise of the U.S.-European corporation formed out of mergers like Daimler and Chrysler is only one sign of how integrated the two business communities are becoming. Trade disputes arise frequently, but these conflicts are small compared with the overall commercial traffic across the Atlantic. On a range of social and cultural issues, Americans and Europeans are also growing closer together. In our study, we found a remarkable degree of agreement, even on controversial issues like global warming and biotechnology.

These common economic interests and shared values may be enough to preserve a strong trans-Atlantic relationship in the face of the differences we have highlighted. Americans and Europeans are not from different planets. Their worldviews differ in important ways, but they also stand on the same ground in many other areas. The challenge for political leaders in the United States and Europe will be to find areas where shared perspectives are strong enough to offset their few—but very strong—differences.

Craig Kennedy is president of the German Marshall Fund of the United States. Marshall M. Bouton is president of the Chicago Council on Foreign Relations.

Want to Know More?

The complete results of this year's survey on trans-Atlantic attitudes by the **Chicago Council on Foreign Relations** and the **German Marshall Fund of the United States** can be found at **www.worldviews.org.** Analyses of the Chicago Council's previous quadrennial polls can be found in former Chicago Council President John E. Rielly's **"The American Mood: A Foreign Policy of Self-Interest"** (FOREIGN POLICY, Spring 1979), **"American Opinion: Continuity, Not Reaganism"** (FOREIGN POLICY, Spring 1983), **"America's State of Mind"** (FOREIGN POLICY, Spring 1987), **"Public Opinion: The Pulse of the '90s"** (FOREIGN POLICY, Spring 1991), **"The Public Mood at Mid-Decade"** (FOREIGN POLICY, Spring 1995), and **"Americans and the World: A Survey at Century's End"** (FOREIGN POLICY, Spring 1999).

Another excellent polling resource is **Americans and Europeans Differ Widely on Foreign Policy Issues,** a survey released on April 17, 2002, by the Pew Research Center for the People & the Press. While the Pew report forecasts a wider divergence in public opinion on the use of force in Iraq, it does not explore the impact on European attitudes of U.N. approval of military action. Useful information on U.S. public opinion on the use of armed force, defense spending, biotechnology, and the Middle East can also be found at the Web site of the **Program on International Policy Attitudes.** One of the best comparative examinations of social and political values is the **World Values Survey,** a series of national surveys in more than 65 societies. See the survey Web site for some results and commentary or consult *Human Values and Beliefs: A Cross-Cultural Sourcebook* (Ann Arbor: The University of Michigan Press, 1998) by Ronald Inglehart, Miguel Basanez, and Alejandro Moreno.

Currently, the biggest skunk at the trans-Atlantic garden party is the Carnegie Endowment for International Peace's (CEIP) Robert Kagan, whose **"Power and Weakness"** (*Policy Review,* June/July 2002) has spurred debate on the extent to which Europe and the United States are growing apart. Antony Blinken, former senior director for Europe at the National Security Council and current chief of staff for the Senate Foreign Relations Committee, argues that "far from diverging, the United States and Europe are converging culturally, economically, and with some effort, strategically" in **"The False Crisis Over the Atlantic"** (*Foreign Affairs,* May/June 2001). CEIP President Jessica Mathews highlights the costs of a trans-Atlantic split in **"Estranged Partners"** (FOREIGN POLICY, November/December 2001).

British diplomat Robert Cooper explains the inherent tensions between "postmodern" entities such as Europe and "modern" nations such as the United States in *The Post-Modern State and the World Order* (London: The Foreign Policy Centre, 2002). In **"America's Imperial Ambition"** (*Foreign Affairs,* September/October 2002), John Ikenberry warns that "America's nascent neoimperial grand strategy threatens to rend the fabric of the international community and political partnerships precisely at a time when that community and those partnerships are urgently needed." And in **"Terms of Engagement: The Paradox of American Power and the Transatlantic Dilemma Post-11 September"** (Paris: European Union Institute for Security Studies, May 2002), Julian Lindley-French argues that "shared interests and values" can no longer ensure a strong trans-Atlantic security alliance.

For links to relevant Web sites, access to the *FP* Archive, and a comprehensive index of related FOREIGN POLICY articles, go to www.foreignpolicy.com.

From *Foreign Policy,* November/December 2002, pp. 66-74. © 2002 by the Carnegie Endowment for International Peace.

Europe's Eastern Enlargement: Who Benefits?

"The historical experience of the East European countries is to dance to the tunes played by regional powers to their west and to their east. Is this just one more round, with Brussels choreographing the show now, rather than Vienna, Berlin, or Moscow?"

JOHN HALL AND WOLFGANG QUAISSER

The European Union is preparing to take a quantum leap eastward. For the EU, previous rounds of enlargement meant adding tens of millions of West Europeans. In contrast, its current plans for enlargement mean adding between 100 million and 200 million new citizens from central, eastern, and southeastern Europe. Will such an unprecedented enlargement make the union that much more perfect? Or might the planned expansion generate such high levels of dissension between and among member countries that the future of this grand experiment in integration is threatened?

Fifteen countries currently compose the EU. Its eastern enlargement means bringing in as many as 12 more countries, likely entering as two or three groups spaced over time. If all goes as scheduled, just after 2010 the population of the EU should approach 500 million. Its population will be significantly larger than that of the United States.

In recent decades the EU has allowed a small number of countries—Ireland, Spain, Portugal, and Greece—with per capita incomes well below the EU average to join. Over time it became clear that membership did allow these countries to escape their historical problems of persistent poverty and industrial backwardness. According to a wide range of statistical indicators, Ireland, Spain, and Portugal have made significant strides in converging toward the EU per capita GDP average.

But eastern enlargement poses a more formidable challenge, one that has generated a set of deep-seated fears mumbled by Western Europe's forgotten public: Could the visions of Europe's elite leaders backfire, threatening and even undermining hard-earned prosperity by extending the EU too far eastward? Citizens to the east ponder: Is the European Union's current scheme for enlargement as virtuous as purported? Is this in reality a quickly spun stop-gap measure to shore up security risks along a newly formed eastern border created by the Soviet Union's unexpected demise? The historical experience of the East European countries is to dance to the tunes played by regional powers to their west and to their east. Is this just one more round, with Brussels choreographing the show now, rather than Vienna, Berlin, or Moscow?

CREATING EUROPE'S NEW DIVIDE?

For a full decade, the Central and East European Countries (CEECs) have undergone transitions from authoritarian, one-party rule to parliamentary democracies, and from planned to market economies. Hungary and Slovenia are judged to have nearly completed their economic transitions and are mostly prepared for their accessions to the EU. Other countries remain farther behind.

The European Commission provides criteria for judging a country's preparedness for accession. The Brussels-based commission annually evaluates countries to determine if each has fulfilled the "Copenhagen Criteria" regarding the establishment of stable institutions guaranteeing pluralistic democracy, the rule of law, and respect for human rights, including protection of national minorities. Its economic criteria presupposes the existence of a functioning market economy. Likewise, candidate countries are evaluated to see if they have implemented the *Acquis Communautaire* (what has to be acquired to belong to the community), the common set of rules, laws, regulatory systems, and institutions that serve as a precondition for fulfilling the EU's common goals.

What makes the EU's eastern enlargement so challenging economically? Foremost are the systemic changes candidate countries confront at home as they abandon legacies of failed communist experiments and embrace the West European model based on democracy and a liberal market economy. The second challenge is the unprecedented scale and speed of the anticipated expansion. In addition, vast differences exist in the levels of economic development and economic performances between the prospective CEECs and the advanced members of the EU.

Levels of capital endowment and technical development are also more extreme than were the differences between the advanced economies in Western Europe and Ireland in 1973, and Portugal, Spain, and Greece at the time of the southern enlargement in the 1980s. Portugal registered at least 30 percent of the EU's per capita output when it joined the EU; Ireland and Spain were roughly at 60 percent of the EU's per capita GDP, and Greece was at 50 percent. However, when measured at current exchange rates, the average level of per capita output in 10 of the CEECs under consideration is only 16 percent of the current EU average. More advanced CEECs, such as the Czech Republic, Poland, and Hungary, are at between 18 and 20 percent of the EU average. Although Slovenia registers at 44 percent of the average EU per capita GDP, it is a small country and its population of only 2 million would increase the EU's size by less than 0.6

percent. Compared to the CEECs, the Balkan countries, as well as Moldova, Ukraine, Belarus, and Russia, are generally poorer, exhibiting significantly lower levels of per capita GDP relative to the EU average. With eastern enlargement, Europe seems to be creating its own new divide.

Which of the CEECs will be invited to join the "rich man's club," and when?[1] The stakes are high. Those invited can expect to prosper, since they will benefit from the EU's agricultural and structural funds—transfers that could be as large as 4 percent of their GDPs. These transfers could continue to flow in for decades, as they have for Ireland. In addition, those chosen will benefit from insider access to the largest single market on earth.

However, as those countries entering the European Union see their economic indicators dramatically accelerate toward EU averages, the countries beyond the EU periphery are expected to fall relatively farther behind. As Darius Rosati has so cogently warned, other countries—farther east in Europe and farther south in the Balkans—surely will never be invited to join the EU, and thus will be left out in the cold: left to manage their economic and associated societal problems without the largesse of massive fund transfers, full market access, and the stability offered by well-funded and competently staffed EU institutions.[2] Even after an eastward expansion, the EU will likely find that the levels of instability, related to high rates of unemployment and comparatively large populations facing persistent poverty just beyond its eastern and southeastern borders, will remain, threatening the EU's clublike atmosphere.

Yet the countries benefiting from the privileges associated with entering the union will face their own share of challenges. Economic preconditions for accession and then full union participation are now higher than they were in the 1970s or 1980s, when countries with lower per capita incomes entered. Brussels introduced additional steps for integration in the 1990s, such as the Single Market Program and the European Monetary Union (EMU). Following EU accession, countries that want to join the EMU must bring into and keep in line rates of inflation, federal budget deficits, and the ratios between total public-sector debt and GDP.

Union membership also means that a member country's firms must integrate into—and then survive in—a single, Europe-wide market. Integration must be achieved in an economic environment in which, even over the long run, levels of labor productivity between members countries will vary significantly. In sum, rising entry standards make joining the EU much more costly and administratively difficult to undertake for the accession hopefuls when compared with what other countries faced when they joined the EU in previous decades.

Entering countries may also not be afforded the same large subsidies to assist farmers and funds for infrastructure improvements that were offered Ireland, Portugal, Spain, and Greece. With comparatively high levels of education and training, the CEEC labor force is lined up and eager to fill higher-paying jobs in the West. However, accession policies will allow individual EU countries to restrict the free movement of labor. Thus, the CEEC workforce will likely face the prospect of sitting at home for up to seven years in what is being euphemistically termed a "transition period" that will allow West European countries to protect their labor markets from the inflow of eastern workers.

If the problems and costs that Europe faces with eastern enlargement outweigh the expected benefits, what keeps the dream of eastern enlargement moving forward?

If the problems and costs that Europe faces with eastern enlargement outweigh the expected benefits, what keeps the dream of eastern enlargement moving forward? The obvious answer is that eastern enlargement can be considered an essential aspect to fulfilling a political vision. Europe's unification was spelled out as a goal in the 1957 Treaty of Rome. But bringing to realization that magnanimous political vision would be difficult—if not impossible—were it not that economic theory is also on the side of enlargement.

At least two major studies suggest that existing EU members can expect modest growth in output resulting from integration. Future growth is estimated as modest since a sizeable portion of potential growth effects were realized shortly after trade liberalization with the CEECs in 1992. The average jump in GDP after the start of accession is expected to range from 0.2 to 0.5 percent of current levels of GDP for most CEECs, although comparatively larger select CEECs can see an increase of between 5 and 7 percent of their current GDP.

On the down side, EU members have started to show signs of divisions that will likely deepen over time as countries estimate their prospects for future gains and losses associated with eastern enlargement. At the December 2000 Nice summit, EU members started to deal with a range of issues related to their voting power relative to other EU members and the countries lining up for accession. With enlargement and an anticipated restructuring of power relations on the horizon, some EU member countries are jockeying to gain positions that will help maximize their gains and minimize their losses.

Austria, Germany, Scandinavia (excluding Denmark), and the Benelux countries (Belgium, the Netherlands, and Luxembourg) are expected to pay more to Brussels but to gain even more through an increase in economic activity with the CEECs. Because of their close geographic proximity, Austria and Germany expect to reap the largest economic benefits from eastern enlargement as CEEC markets expand for their export products. France is expected to pay more but find less returned. A third group, composed of Ireland, Portugal, Spain, and Greece, faces the possibility of losing existing levels of net subsidies, but will remain net beneficiaries of transfers for the foreseeable future. Britain and Denmark should expect neither net gains nor losses.

EU members thus appear to be basing their support (or lack thereof) for eastern enlargement on whether they will benefit or lose when Brussels issues future budgets. In addition, EU countries are calculating how their economies will integrate into an enlarged union.

Trade—as exports from the CEECs to the EU—is estimated to be too small to significantly affect prices in Western Europe commodity markets. Border openings have already affected wages and levels of aggregate employment in the West, but these have been and are expected to remain marginal. CEEC import penetration is expected to continue to exert downward pressures on wages in the apparel and metal industries while also generating additional unemployment. However, a sizeable portion of expected losses has already been absorbed since trade in these goods was liberalized in 1992.

Aggregate employment for existing EU members and entering CEECs is slated to increase. The CEECs are expected to steadily catch up to wage rates, per capita incomes, and levels of productivity, although less-skilled CEEC workers are not expected to share proportionally in these gains.

Increased flows of capital in the form of direct and portfolio investments into the CEECs will not place upward pressure on interest rates in Western Europe since the magnitude of capital transfers to the east will remain too small, relatively, to wield influence over the price of capital in Western Europe. To the east, interest rates stand to fall in most of the CEECs, and the decreasing costs for capital are expected to help foster economic growth in the region generally.

Since trade in goods and services, as well as capital flows, is not likely to lead to an equalization of incomes among the current 15 members of the EU and the CEECs in the short term, gaps in wages between regions are expected to persist for some time. Moreover, as firms and even some CEEC industries are closed and others are expanded with the privatization and restructuring of eastern industry, the EU will be faced with assisting the CEECs through their structural transformations. A special source of funds will be needed to reduce and ameliorate the shocks experienced by individuals and households facing such profound changes.

These changes make it clear that, even with restrictions, labor migration will, over the short term, have greater effects on the EU's labor markets than will trade and investment. Sharing contiguous borders with some of the CEECs, and in relatively close proximity to others, Austria and Germany can expect to absorb the lion's share of eastern migrants. Of the predicted annual net migration of close to 300,000 workers, 200,000 are expected to head for Germany. But estimates suggest that the number of annual migrants should halve within a decade. Even though fears concerning the scope and scale of migration appear unfounded, measures have nevertheless been introduced to slow the free influx of labor during the first several years following accession. Economic growth, welfare gains, and successful structural changes in the accession countries will eventually raise incomes and improve living standards, thereby lessening pressures that promote migration.

THE BALANCE SHEET

From an economic standpoint, eastern enlargement appears to be a win-win situation. CEEC economies are comparatively small, together generating an output that is only 4 percent of the EU-15, and eastern enlargement is not expected to generate significant benefits or costs for Western Europe over the short term. In contrast, the CEECs are expected to gain disproportionately and significantly upon accession.

Friction will occur in the intense power play that will emerge between net payers and net receivers of Brussels's budgetary funds. In addition, the power play could cause the formation of at least two camps firmly at odds with one another. Austria, Germany, Sweden, and the Netherlands apparently will be counting their gains, while France, Spain, Portugal, Greece, and Ireland will be mourning their losses.

The EU faces the task of implementing major institutional and policy reforms to further prepare itself for eastern enlargement—at least by the time of the next Inter-Governmental Conference that is scheduled for 2004. The conference will allow changes to be made to already existing treaties. The exact procedures and objectives of the 2004 conference will be decided upon in December 2001 at the European Council meeting slated to take place in Laeken, Belgium. If at Laeken the EU's leaders and representatives fail to introduce an agenda with new treaties to be considered at the 2004 conference, frictions between and among EU member nations could emerge and diminish substantially the prospects for a broadly inclusive and speedy CEEC accession process.

Notes

1. The selection of countries that will enter in first, second, and even third rounds is not completely settled. Most likely, those countries that have done the most to fulfill the economic criteria will enter first. A "big bang," however, is also likely, with as many as eight countries entering together in 2005 or 2006.
2. Darius K. Rosati, "Economic Disparities in Central and Eastern Europe and the impact of EU Enlargement" (Warsaw: Central European Initiative and the Untied Nations Economic Commission for Europe, 1999).

John Hall *is a professor of economics and international studies at Portland State University in Oregon.* Wolfgang Quaisser *is a senior research fellow at the Osteuropa-Institut in Munich.*

UNIT 7
Former Soviet Union

Unit Selections

24. **George W. Bush and Russia**, James Goldgeier and Michael McFaul
25. **Where Does Europe End?** Nancy Popson

Key Points to Consider

- What do you think about the call for the United States and Russia to form a formal alliance dedicated to fighting terrorism and stopping the proliferation of weapons of mass destruction worldwide?

- What do you think will be Vladimir Putin's main foreign policy priorities over the next few years?

- Given the rapid change in Europe in recent years, do you expect that Ukraine will be invited to join the European Union in the next few years? Why or why not?

 **Links: www.dushkin.com/online/**
These sites are annotated in the World Wide Web pages.

Russia Today
http://www.russiatoday.com

Russian and East European Network Information Center, University of Texas at Austin
http://reenic.utexas.edu/reenic.html

The former USSR is a region composed of 15 independent nation-states, with each state trying to define separate national interests as it experiences severe economic problems. Many ex-Soviet citizens share a sense of disorientation and "pocketbook shock," as their standard of living is lower today than it was under communism. About half of the states are experiencing political instability and growing discontent. Ukraine, the first state to transfer power peacefully from one elected president to another during the early post–USSR era, remains economically highly interdependent with the Russian economy. Despite this dependence, Ukraine has recently signaled its interest in joining the European Union in spite of the fact that many European leaders remain wary about extending the EU borders that far east. The trend of looking to Europe while remaining closely tied to Russia's economy characterizes the situation in many former Soviet Republics.

While uncertainty remains as a principal characteristic of the current Russian political system, the government increasingly appears to have made a permanent transition from a communist state while retaining many features of the old authoritarian regime. The transition was fueled by economic imperatives. After the Russian government failed in attempts to impose austerity measures and collect taxes in the late 1990s, it devalued the ruble, defaulted on domestic bonds, and placed a moratorium on paying overseas creditors. These actions triggered the worst economic crisis in Russia since the collapse of communism.

Russia's parliamentary election at the end of 1999 was widely viewed as a step toward democracy. Optimistic assessments were made despite widespread mudslinging, dirty tricks at the polls, and corruption. The vote was also an endorsement of Prime Minister Vladimir Putin, the sixth prime minister to be appointed to Yeltsin's government during a 2-year period. Boris Yeltsin abruptly stepped down as president in January 2000 after naming Prime Minister Vladimir Putin as his successor.

One of Putin's first acts as president was to grant Yeltsin blanket amnesty from any future investigation of his family finances. After winning the 2000 presidential elections, Putin displayed a willingness to use military force in Chechnya, to change the reliance on nuclear weapons in the military doctrine, and to implement reductions in conventional armed forces and the size of the defense budget. Putin also built bridges to rival politicians and implemented measures designed to rein in the power of the country's economic oligarchs. After the bungled rescue attempt of the Kursk submarine that cost 118 sailors their lives and became a national scandal, Putin formulated a plan to transform Russia's internal balance of power in his favor by reorganizing the country into seven "super-regions" with Kremlin loyalists in charge. Putin argued that only with a strong state can Russia establish the rule of law and the foundations of a civil society. Provincial elections at the end of 2000 suggested that Putin's approach had wide support with Russian voters. The overwhelming majority of Russians in recent public opinion polls indicated a willingness to trade away some democratic freedoms if it was necessary to achieve order. While progress was made on the political front, funds for conventional military forces have continued to decline. The dismal state of the Russian Army was indicated by reports that desertions and suicides were on the rise again during 2002.

Most Russians also support Putin's hard-line approach to dealing with Chechen rebels. The long-simmering conflict had received little news coverage until it burst into the national consciousness during October 2002 after Chechen rebels seized 700 patrons in a Moscow theater. President Putin refused to negotiate with the terrorists and authorized Russian special forces to pump gas into the theater before storming the building in a dramatic pre-dawn raid. The raid left 50 militants and over 90 hostages dead. The Russian government refused to identify the type of gas used by security forces to secure the building. The gas is widely thought to be a BZ incapacitating agent and it remains unclear whether the use of the gas violated international conventions.

Even before the September 11 attacks in the United States, Putin was moving to implement far-reaching changes in Russia's foreign policy by proposing greater cooperation between Europe and the United States on security matters. The Bush administration also dropped its campaign rhetoric about Russia after the attacks and moved quickly as well to reengage Russia. James Goldgeier and Michael McFaul in "George W. Bush and Russia" discuss how the bond between the two leaders grew stronger after September 11 as the war on terrorism, rather than the pace of Russia's internal reform or continued fighting in Chechnya, came to dominate the new Russian-American relationship. Putin expressed his interest in working with NATO and the United States in the war on terrorism. During 2002, representatives of NATO and Russia met to form the NATO-Russia Council that gives Russia a formal but limited role in the alliance. Russia and NATO continue to explore other ways for their armed forces to cooperate, including an agreement to sign a Military Cooperation Pact in 2003 to cooperate on rescue missions at

sea. Recently, NATO secretary general George Robertson said the alliance is also prepared to assist Russia in modernizing, downsizing, and professionalizing its armed forces.

On the eve of Putin's first summit meeting with President Bush in the United States, the Russian leader, in a television interview on a U.S. news show, emphasized that the most serious proliferation threat that Russia and the West shared was the threat that terrorists might be able to procure and use tactical nuclear weapons. At his meetings with President Bush in Crawford, Texas, Putin has allied himself with the West in the war on terrorism. Both countries agreed to make deep cuts in their arsenals of nuclear weapons over the next decade. President Putin acknowledged publicly that the Anti-Ballistic Missile Treaty was probably an outdated relic of the cold war. In response, President Bush announced support for Russia to gain most-favored-nation trading status and admission into the World Trade Organization.

Despite dramatic signs of an emerging strategic realignment in geopolitical relations, it is still too soon to know the longer-term domestic and international consequences of the current period of instability on Russia's future leaders or foreign policy. Since 1991, Russia has been trying to fashion a national-security policy to fit its changed status in a new era. Today, Russian strategists are deeply concerned by the failure to secure Russia's links to CIS states and by the paucity of promising options. Most Russian foreign-policy elite consider themselves *derzhavniki*—believers in strong central government and Russia as a nuclear superpower and great power in Eurasia and East Asia. Whether this elite can accept the tenets of a more modest Russian foreign policy remains uncertain.

Russia, the United States, and China are all interested in countering the rising influence of Islamic fundamentalism in Central Asian states. The U.S. military had negotiated military pacts with several states in the region prior to the September 11 attacks. These agreements paid off for the United States in the war to oust the Taliban in Afghanistan. The existence of these agreements permitted U.S. military personnel to operate from several states in the Caucuses during the war, including Tajikistan, Uzbekistan, and Kyrgyzstan.

Western states, along with Russia and China, are also anxious to maintain good relations with the Central Asian states bordering the Caspian Sea in part because all of these states are poised to undergo dramatic changes due to their oil and gas reserves. When developed, these reserves are estimated to be worth between $2.5 and $3 trillion dollars. The untapped resources have stimulated a growing web of recent deals in Central Asia and increasingly have made the area appear to be a new kind of post–cold war zone of competition where the interests of three former military rivals—China, Russia, and the United States—and a variety of multinational corporations now intersect.

George W. Bush and Russia

"Why the major reversal in Bush's thinking on Russia? Most have attributed this amazing transformation to September 11.... But September 11 is only part of the story."

JAMES GOLDGEIER AND MICHAEL MCFAUL

The group of foreign policy officials that advised Texas Governor George W. Bush during the 2000 presidential campaign declared the Clinton–Gore approach to Russia a total failure. Their central criticism, especially when expressed privately, was not that President Bill Clinton had done too little to promote markets and democracy in Russia and the subsequent integration of Russia into the Western community of democratic states. Instead, they argued that Clinton and his team had devoted too much time and too many resources to trying to change Russia internally. As for integrating Russia into Western international institutions, Bush's foreign policy advisers expressed indifference during the campaign, and instead emphasized the need to strengthen the "core"—that is, American allies in Europe and Asia—rather than expanding the core to peripheral places such as Russia.

Bush's team did not advocate neglect of Russia. The candidate's foreign policy team—a group called the Vulcans and headed by Stanford professor Condoleezza Rice—believed the best way to repair United States–Russian relations was to begin to treat Russia like an international power. They advocated greater focus on the "national interest." This implied, in their view, more attention on the balance of power in the international system and among the great powers, such as Russia and China, and less attention to "humanitarian interests" and lesser powers, such as Haiti, Somalia, Bosnia, and Kosovo. Bush campaign adviser Robert Blackwill explained that Bush planned to focus on Russia and China and "not Haiti, not Somalia" because these were countries that could threaten American national security interests. "The reality is that a few big powers can radically affect international peace, stability, and prosperity," Rice wrote in a January–February 2000 *Foreign Affairs* essay during the campaign. Rice also recognized the importance of promoting American values in foreign affairs, which she described as "universal."

Greater attention to great powers with nonliberal values did not mean a softer line. On the contrary, in reference to both Russia and China, Bush and his campaign

officials promised to depart from the Clinton strategy of accommodation and adhere instead to "tough realism." The Bush team promised to end the romanticism that Clinton's Russia team held for Russia. For Bush's advisers, Russia was still a great power, but one in decline, which made it erratic and dangerous. As Rice wrote in *Foreign Affairs*, "Moscow is determined to assert itself in the world and often does so in ways that are at once haphazard and threatening to American interests." Bush advisers promised to end the "happy talk" and discontinue the overpersonalized approach they claimed Clinton practiced with Russian President Boris Yeltsin. "The problem for U.S. policy is that the Clinton administration's embrace of Yeltsin and those who were thought to be reformers around him has failed," Rice stated bluntly and emphatically throughout the campaign. Clinton's team, in Rice's view, mistakenly let its Russia policy become "synonymous with the agenda of the President of Russia." Bush advisers also threatened sanctions if Russia continued to supply Iran with nuclear technologies and pledged not to consider Russian interests in dealing with European security matters or American strategic interests more generally. In particular, candidate Bush made clear that he planned to withdraw from the Anti-Ballistic Missile (ABM) treaty or attempt to amend it to allow for the building of national missile defense, no matter what the Russian position on the issue. He also pledged to unilaterally lower the United States nuclear arsenal to levels dictated by American interests alone, which meant there was no need to consult the Russians or slow the process by signing treaties with them.

Because members of the Bush foreign policy team were realists, they tended to downplay the importance of regime type and internal politics generally and instead focused on the external behavior of states, which they believed were influenced first and foremost by the balance of power in the international system. Generally, they pledged a similar approach to Russia. According to Rice, "the United States needs to recognize that Russia is a great power, and that we will always have interests that conflict as well as coincide." During the campaign, Rice recommended that the United States should not get bogged down in Russian internal developments, but instead "must concentrate on the important security agenda with Russia." In one presidential debate, Bush stated even more bluntly: "The only people that are going to reform Russia are Russia [sic]. They're going to have to make the decision themselves."

This said, two internal problems in Russia—corruption and Chechnya—were simply too juicy politically to ignore. Candidate Bush and his advisers repeatedly touched on these issues and blamed the Clinton administration for not doing enough in responding to them. In his one major foreign policy speech of the campaign, Bush described Russia as a power "in transition" whose final regime type was still unknown. Echoing Wilsonian themes, Bush argued that "dealing with Russia on essential issues will be far easier if we are dealing with a democratic and free Russia." In supporting reforms in Russia, however, Bush stated emphatically that "we cannot excuse Russian brutality. When the Russian government attacks civilians—killing women and children, leaving orphans and refugees—it can no longer expect aid from international lending institutions. The Russian government will discover that it cannot build a stable and unified nation on the ruins of human rights. That it cannot learn the lessons of democracy from the textbook of tyranny. We want to cooperate with Russia on its concern with terrorism, but that is impossible unless Moscow operates with civilized restraint."[1]

Four months later, as the following exchange on February 16, 2000 between candidate Bush and journalist Jim Lehrer during *The NewsHour with Jim Lehrer* demonstrates, Bush promised to do more regarding Chechnya.

JL: On Chechnya and Russia, the U.S. and the rest of the Western world had been raising Cain with Russia from the beginning, saying "You are killing innocent civilians." The Russians have said essentially, "We're fighting terrorism, and, by the way, mind your own business." What else—what else, if anything, could be done by the United States?

GWB: Well, we could cut off IMF aid and export/import loans to Russia until they heard the message loud and clear, and we should do that. It's going to be a very interesting issue to see how Russia emerges, Jim. This guy, Putin, who is now the temporary president, has come to power as a result of Chechnya. He kind of rode the great wave of popularity as the Russian military looked like they were gaining strength in kind of handling the Chechnya situation in a way that's not acceptable to peaceful nations....

JL: But on Chechnya, specifically, you think we should not—we should hold up International Monetary Fund aid. Anything else we should do?

GWB: Export/import loans.

JL: And just cut them off?

GWB: Yes, sir, I think we should.

JL: Until they do what?

GWB: Until they understand they need to resolve the dispute peacefully and not be bombing women and children and causing huge numbers of refugees to flee Chechnya.

JL: And do you think that would work?

GWB: Well, it certainly worked better than what the Clinton administration has tried.

JL: You mean, just using words, you mean?

GWB: Yes.

Candidate Bush, his campaign, and his campaign supporters also tried to make corruption in Russia—and Clinton and Gore's inattention to it—a campaign issue. Vice President Al Gore and Russian Prime Minister Viktor Chernomyrdin had developed a close personal relationship during their years of service as cochairs of the United States–Russian Joint Commission on Economic and Technological Cooperation, often referred to as the "Gore–Chernomyrdin Commission." According to Republican critics of Gore, the vice president had allowed his personal relationship with Chernomyrdin to blind him to the corrupt practices of the Russian prime minister. In 1995, when Gore was said to have received a CIA report linking Chernomyrdin with corrupt practices, he allegedly returned the report with the word "bullshit" written across the top of the document.[2] Five years later, Bush and others reminded the American voters of Gore's intimate relationship with this reputedly corrupt Russian official. In a 2000 presidential debate, Bush claimed that Clinton policies fueled corruption: "We went to Russia, we said, 'Here's some IMF money,' and it ended up in Viktor Chernomyrdin's pockets and others'. And yet we played like there was reform." At the same time, reporters revealed that Gore had signed a secret agreement with Chernomyrdin in 1995 that allowed Russia to continue to sell conventional weapons to Iran until December 31, 1999, a deadline the Russians had ignored.

The Cox report, a Republican investigation of Clinton's Russia policy by a committee chaired by Representative Christopher Cox (R.-Calif.) and published on the eve of the 2000 election, went even further, accusing Gore of deliberately trying to cover up widespread corruption within the Russian government. Commenting on the epithet allegedly scrawled on the CIA document, the Cox report concluded, "It is difficult to imagine a more dangerously intemperate reaction by the vice president to official corruption in Russia."[3] Based on other testimonials, the Cox report suggested that "[i]t is therefore clear that the vice president rejected not an initial report unsupported by other evidence, but rather a detailed report built on extensive earlier work by the CIA of which Gore must have been aware. Moreover, the allegations against Chernomyrdin were made in the context of numerous charges against other senior Russian leaders—suggesting widespread corruption at the top levels of government." The Cox report also accused the Clinton team of failing to "mount an aggressive challenge to organized crime in Russia." Candidate Bush never embraced a causal connection between Russian crime and American foreign policy, but he and his campaign staff did endorse the general characterization of Russia as a lost cause, burdened by imperial proclivities from its past and criminal undertows in its new present.

> *Candidate Bush's statements on Russia were not the product of strategic thinking, but oversimplified campaign slogans.*

Despite the efforts of the Cox team and others, Russia—and foreign policy in general—never became a major issue in the 2000 campaign. Candidate Bush's statements on Russia were not the product of strategic thinking, but oversimplified campaign slogans. Before becoming president, in fact, little evidence suggests that Bush had devoted a great deal of time to thinking systematically about foreign policy. Only the actual practice of policy revealed his true intentions toward Russia.

FROM CONFRONTATION TO REENGAGEMENT

In the first weeks of his administration, President Bush and his new foreign policy team signaled their intention to maintain a tough line on Russia and Chechnya in particular. After being named national security adviser but before taking office, Condoleezza Rice wrote an opinion piece for the *Chicago Tribune* in which she restated many themes of her *Foreign Affairs* essay from a year earlier. In the December 31, 2000 newspaper column, Rice emphasized again that "the United States needs to recognize that Russia is a great power" and therefore "U.S. policy must concentrate on the important security agenda with Russia." At the same time, she also reiterated many of Russia's domestic ills, including weak democratic institutions, halfhearted economic reforms, and corruption. She devoted special attention to the ill effects of the Chechen war and Putin's role in it: "As prime minister, Vladimir Putin used the Chechnya war to stir nationalism at home while fueling his political fortunes. The Russian military has been uncharacteristically blunt and vocal in asserting its duty to defend the integrity of the Russian Federation—an unwelcome development in civil-military relations. The long-term effect of the war on Russia's political culture

should not be underestimated. This war has affected the relations between Russia and its neighbors in the Caucasus, as the Kremlin has been hurling charges of harboring and abetting Chechen terrorists against states as diverse as Saudi Arabia, Georgia, and Azerbaijan. The war is a reminder of the vulnerability of the small, new states around Russia and of America's interest in their independence."

Rice hoped that this blunt statement about Russia's problems and their impact on United States interests would stand in contrast to the sugar-coated rhetoric of the Clinton years, which, in her opinion, greatly damaged United States national security: "Frustrated expectations and 'Russia fatigue' are direct consequences of the 'happy talk' in which the Clinton administration engaged."

In spring 2001, Bush and his foreign policy team did seem determined to end the "happy talk." In March his administration ordered the expulsion of nearly 50 Russian diplomats from the United States, who were accused of being spies. Bush personally did not make any statements about Chechnya in his first months in power, but his State Department sent a loud signal of support for the Chechen cause by arranging a meeting between the Chechen foreign minister in exile, Ilyas Akhmadov, and the acting head of the State Department's Bureau of Newly Independent States, John Beyrle, the highest-level meeting ever with a Chechen government official. In this early period, Bush officials also seemed poised to maintain a tough line on Russia's relations with rogue states. Secretary of Defense Donald Rumsfeld called Russia an "active proliferator" while Deputy Secretary of Defense Paul Wolfowitz described the Russians as immoral proliferators, who "seem to be willing to sell anything to anyone for money." As Wolfowitz explained, "My view is that they have to be confronted with a choice. You can't have your cake and eat it too. You can't do billions' worth of business and aid and all that with the United States and its allies, and then turn around and do small quantities of obnoxious stuff that threatens our people and our pilots and our sailors."

Leaks from the White House suggested that assistance to Russia would be reduced. A new, more confrontational approach to relations with Russia seemed to be emerging. As Jane Perlez concluded in her review of Russia policy in the March 23, 2001 *New York Times*, "The Bush administration has not articulated a broad policy toward Russia, but in thoughts and deeds it has taken a sharp departure from the engagement policies of its predecessor, moving toward isolating Russia and its president, Vladimir V. Putin."

More generally, Bush also hinted that the promotion of democracy might be a major component of his foreign policy agenda. For instance, in introducing his future secretary of state, Colin Powell, Bush stated clearly that "our stand for human freedom is not an empty formality of diplomacy but a founding and guiding principle of this great land. By promoting democracy we lay the foundation for a better and more stable world." If applied to Russia, this statement implied a greater focus on, not neglect of Russia's internal problems.

At the same time, Bush and his team suggested that addressing the bilateral relationship with Russia was not a top priority. Adopting a policy line that contradicted its earlier statements on the need to focus on great powers, the new administration said that engaging and strengthening relations with America's allies were a greater and more immediate concern. Only at the insistence of American allies in Europe did Bush agree to schedule a meeting with Putin as a final stop on his first trip to Europe in summer 2001. Symbolically, the Bush administration also downgraded Russia's place within the foreign policy bureaucracy by dismantling the Bureau of Newly Independent States within the State Department that Clinton had created. In the new organizational chart at the State Department, Russia was one of 54 countries in the new Bureau of European and Eurasian Affairs. Rice initiated a parallel reorganization at the National Security Council, folding the directorate on Russia, Ukraine, and Eurasia affairs into a new one encompassing Europe and Eurasia.

THE DEMISE OF "TOUGH REALISM"

"Tough realism" toward Russia did not last long. Like his father in 1989, Bush ordered a major review of United States policy with Russia. Even before this was completed, however, Bush's new approach toward Russia emerged. Instead of confrontation and neglect, Bush decided to reembrace the policy of engagement with Russia and its president, Vladimir Putin. Rather than departure, this decision signaled continuity with Clinton's Russia strategy.

> *Bush deliberately tried to forge a personal bond with his Russian counterpart.*

As the June meeting with Putin in Slovenia approached, Bush began to become personally involved in his Russia policy for the first time. In spring he made a strategic decision that he would not confront Putin with a laundry list of American concerns. Instead, he wanted to establish a personal rapport with the Russian leader as a necessary first step in developing a partnership with his Russian counterpart. It was a businessman's approach to foreign policy.

With this decision, Bush was pursuing a strategy similar to Clinton's, but for different ends. Clinton had embraced Yeltsin because he believed Yeltsin the best hope for Russian reform. In helping his Russian friend, Clinton believed he was also aiding Russian internal reform. Bush's objective in reaching out to Putin had little to do with Russian reform because he had a different set of for-

eign policy goals. The new American president wanted to avoid long discussions or arguments about Russian internal politics. Instead, the security agenda trumped all other concerns. In particular, Bush wanted to establish a relationship with Putin to secure Russia's acquiescence to American withdrawal from the ABM treaty. At the time, many critics of Bush, in Europe especially but also in the United States, warned that United States withdrawal from the treaty would produce a cataclysmic break in United States–Russian relations. Bush and his foreign policy team were determined to abrogate the treaty without derailing United States–Russian relations. Doing so required less focus on Russian internal flaws and more on the security agenda between the two countries. Now that the presidential campaign was over, Bush was returning to the realist inclinations of his closest foreign policy advisers—inclinations that were also shared by his father.

At their first meeting in Slovenia in June 2001, Bush went out of his way to praise Putin. Instead of depersonalizing relations with Russia, Bush deliberately tried to forge a personal bond with his Russian counterpart. At this meeting, Bush reported, "I looked the man in the eye. I found him to be very straightforward and trustworthy…. I was able to get a sense of his soul"; he liked what he saw and sensed. According to White House staffers, Bush and Putin did discuss Chechnya privately, but almost no mention was made of the issue publicly.

Nor did Bush mention publicly the issue of press freedoms, which Putin had done so much to limit in his first year in office. Instead of a public rebuke on the press issue, the Bush administration decided to work on this concern privately. While in Moscow the following month, Rice took time away from her main agenda—the death of the ABM treaty—to meet with representatives of the Russian press. The roundtable was not a press conference but a frank discussion of the future of the independent media in Russia. According to participants in this meeting, Rice expressed understanding of the issues and sympathy with the "opposition" representatives. But no concrete policy changes resulted from this meeting or any other between United States officials and representatives from the Russian opposition media. Eventually, the Bush administration did establish a media initiative, an exchange between American and Russian press executives whose noble aim was to foster the political independence of the Russian press by securing financial independence for independent media outlets. No concrete projects of assistance—rhetorical or otherwise—have resulted yet from this program.

SEPTEMBER 11

The bond between Bush and Putin grew even stronger after September 11. Putin was one of the first foreign leaders to call Bush that day to communicate his full support for the United States and the American people. Putin expressed sympathy as a leader of a country that also had suffered from acts of terrorism against civilians in the capital. Putin then followed his words of support with policies of assistance. On September 24, 2001, Putin announced a five-point plan to support the American war against terrorism. He pledged that his government would share intelligence with its American counterparts, open Russian airspace for flights providing humanitarian assistance, cooperate with Russia's Central Asian allies to offer similar airspace access to American flights, participate in international search-and-rescue efforts, and increase direct assistance—humanitarian as well as military—to the Northern Alliance, the guerrilla army opposed to the ruling Taliban in Afghanistan.

Putin's agreement to an American military presence in Central Asia represented a historic change in Russian foreign policy. Before September 11, President Putin had vacillated between pro-Western and anti-Western foreign policy stances. Putin had pushed through the Russian parliament ratification of the Comprehensive Test Ban Treaty and the second Strategic Arms Reduction Treaty; expressed a clear desire for Russia to become a fully integrated member of the Group of Eight (G-8) Western industrial nations, the World Trade Organization, and, more generally, Europe; and stressed in his new foreign policy doctrine that "Russia shall actively work to attract foreign investments" and will endeavor to "ensure favorable external conditions for forming a market-oriented economy in our country." At the same time, Putin had also reached out to North Korea, Cuba, and China, and signed a major arms deal with Iran.

> ## In Bush's view, Russia has become a partner, a friend, and even an ally of the United States in the global struggle against terrorism.

Putin's personal dual impulses of seeking to integrate into the West while also trying to balance against the West reflect Russia's longstanding love–hate relationship with the West. In the wake of September 11, however, Putin seemed to lean much farther toward the West and especially the United States. His foreign minister, Igor Ivanov, compared the new situation to the alliance between the United States and the Soviet Union during World War II, only now "we are joined by common democratic values, and it is even more obvious that a struggle against a world threat requires the cooperation of our countries and the entire world community."

Since September 11, Bush has articulated a much clearer, simpler vision of his foreign policy. Obviously, the new defining issue before American foreign policymakers is the "war on terrorism." In Bush's view, this war has divided the world into two groups—those countries supporting the United States and those that do not. In this black-and-white world, Russia is clearly a supporter of the American war on terrorism. Therefore, in Bush's view, Russia has become a partner, a friend, and even an ally of the United States in the global struggle against terrorism. United States Ambassador to Russia Alexander Vershbow even went so far as to declare in February 2002 that "the United States and Russia are closer today—politically, economically, and militarily—than at any time in our history."

THE DISAPPEARANCE OF RUSSIA'S INTERNAL PROBLEMS

Bush rewarded Putin for his immediate embrace of the right side after September 11 by changing the way he spoke about Russia's "war against terrorism." On September 26, 2001, White House Press Secretary Ari Fleischer communicated Bush's appreciation for Putin's statement. Fleischer also stated that the "Chechnya leadership, like all responsible political leaders in the world, must immediately and conditionally cut all contacts with international terrorist groups, such as Osama bin Laden and the Al Qaeda organization." The Clinton administration had previously connected some Chechen fighters to bin Laden's network; the Bush administration had not. Bush radically changed his views about the Chechen war from his campaigning days, eventually accepting Russia's definition of the war on terrorism to include Chechnya. Meetings between the Bush administration and the Chechen government-in-exile have since been downgraded. When visiting Washington in spring 2002, on the eve of Bush's trip to Moscow, the Chechen foreign minister, Ilyas Akhmadov, could not secure an official meeting with any senior United States government representative.

President Bush's statement did not give Putin a green light to do what he wanted in Chechnya: the Russian armed forces already were doing that, with little or no reference to American opinions. The statement of support did underscore the notion that the United States and Russia faced a common enemy. Putin had been pushing this theme for years with his American counterparts. In November 1999, then Prime Minister Putin even published an opinion piece in *The New York Times* in which he asked Americans to "imagine ordinary New Yorkers or Washingtonians asleep in their homes. Then, in a flash, hundreds perish at the Watergate or at an apartment on Manhattan's West Side." Putin therefore was pleased to hear that Bush finally recognized their common cause publicly.

In subsequent meetings between Bush and Putin, the war in Chechnya has not been a major agenda item. Journalist Jamie Dettmer summed up the attitude in the August 27, 2002 *Insight on the News*: "Bush has shown remarkable discipline in ignoring Russia's increasingly brutal campaign against separatists in the rebel republic—a campaign dubbed by Yelena Bonner, widow of Nobel Prize-winning human rights activist Andrei Sakharov, as the 'political genocide' of the Chechen people." Before meetings between the Russian and American presidents, Bush administration officials repeatedly have stressed that the issue of Chechnya is covered at length behind closed doors. When Bush has alluded to the Chechen situation publicly, however, he and senior government officials have often adopted Putin's portrayal of the Russian military operation as part of the war on terrorism. As Bush reaffirmed at the G-8 meeting in Canada in summer 2002, "President Putin has been a stalwart in the fight against terror. He understands the threat of terror, because he has lived through terror. He's seen terror first-hand and he knows the threat of terrorism.... He understands what I understand, that there won't be peace if terrorists are allowed to kill and take innocent life. And, therefore, I view President Putin as an ally, strong ally, in the war against terrorism." Even Secretary of State Colin Powell changed how he described the Chechen conflict, stating soon after the Moscow 2002 summit that, "Russia is fighting terrorists in Chechnya, there is no question about that, and we understand that." These remarks suggest that the references to Chechnya behind closed doors may not be as hard-hitting as United States officials have claimed.

Although little in the conduct of the war has changed since candidate Bush pledged to sanction Russia until it stops bombing "women and children" and causing "huge numbers of refugees to flee Chechnya," no sanctions have been applied.

The Bush administration has not always spoken with a unified voice about Chechnya. Although the president himself has not criticized the Chechen war since the 2000 presidential campaign, members of his administration have condemned the conduct of the Russian military operation. When pressed to talk about Chechnya, Condoleezza Rice has continued to express a nuanced view of the war: "[W]e clearly have differences with the Russian government about Chechnya. We've said to them that we fully agree that the Chechen leadership should not involve itself with terrorist elements in the region, and there are terrorist elements in the region. But [we have also noted] that not every Chechen is a terrorist and that the Chechens' legitimate aspiration for a political solution should be pursued by the Russian government. And we have been very actively pressing the Russian government to move on the political front with Chechnya."

The United States ambassador to Russia, Alexander Vershbow, has been particularly vocal in condemning the methods of the campaign, urging a political solution and distinguishing between international terrorists fighting in Chechnya and local Chechen fighters whose aim is independence. In public statements, Deputy Assistant Secretary for European and Eurasian Affairs Steven Pifer also has stressed the need to differentiate freedom fighters from international terrorists. Pifer stated bluntly that the "danger to civilians in Chechnya remains our greatest concern. The human rights situation is poor, with a history of abuses by all sides."

If the rhetoric of the Bush administration has changed considerably over its first two years in office—from the critical to the supportive, but with dissident voices continuing to highlight the negative—actual policy has changed very little from the Clinton era. When asked in his confirmation hearings how the Bush approach to Chechnya would differ from the Clinton policy, Powell answered, "I don't know that I can answer that." Subsequent statements by Bush administration officials suggest that the actual policy on Chechnya has changed very little. On the basic issues concerning Chechnya, State Department spokesman Richard Boucher said, "To reiterate, our policy has not changed. We recognize Chechnya as part of Russia." He also added that "they [the Russians] need to take steps to bring the violence to an end... [T]here is no military solution to the problem, and they need—both sides need to find ways to begin a dialogue and reach a political settlement." Under Bush, the United States has continued to provide humanitarian assistance to the region. At the same time, administration officials have refrained from pursuing new policy initiatives regarding Chechnya. They have not embraced a more activist role in the region such as those proposed by former national security adviser Zbigniew Brzezinski, or offered American mediating services to the Russians and Chechens.

Although little in the conduct of the war has changed since candidate Bush pledged to sanction Russia until it stops bombing "women and children" and causing "huge numbers of refugees to flee Chechnya," no sanctions have been applied. The only significant policy change is rhetorical. If Clinton begrudgingly added statements critical of the Chechen war to his talking points on Russia, Bush has eliminated them.

ON NOT MENTIONING THE "D" WORD

Other issues regarding Russian internal reform also have assumed a marginal position in United States–Russian relations. Under Putin, nearly every democratic institution has become weaker, not stronger. In the Putin era, criticism of the state has proved costly. The State Security Service has stepped up harassment of investigative journalists, human rights activists, environmental leaders, and Western nongovernmental organizations and religious groups and their Russian affiliates. Putin and his surrogates have gone on the offensive against critical in-dependent media outlets, seizing control of NTV and then TV-6—the country's only opposition networks—and fir-ing nonconformist journalists at publications such as the popular weekly, *Itogi*.[4] Putin also has weakened alterna-tive power centers within the state. His "reform" of the upper house of parliament, the Federation Council, has emasculated this once-powerful check on presidential power. Perhaps most disturbingly, as demonstrated in the 2002 presidential election in the Russian republic of Ingushetia, Moscow has begun to actively engage in se-lecting regional governors by disqualifying candidates deemed unacceptable to the Kremlin.

Despite these assaults on Russian democratic institu-tions—already fragile before Putin came to power—Pres-ident Bush has rarely mentioned the "d" word in public during his meetings with Putin. Instead, United States of-ficials have explained that Bush has decided to discuss is-sues of democracy with Putin privately. Even concerns about freedom of the press are not discussed with the press. As with Chechnya, lower-level officials have tried to speak more publicly and critically about the antidemo-cratic trends in Russia. In Moscow, United States Ambas-sador Vershbow has spoken out repeatedly on the state's abusive use of power against independent media outlets, human rights activists, and environmentalists. Vershbow was especially vocal during the state's campaign against TV-6 in winter 2001 and spring 2002. It is true that in Washington, national security adviser Rice has made a point of meeting with democracy activists and indepen-dent journalists from Russia. And in its annual report on human rights around the world, the State Department documented in detail the scope and scale of abuses in Russia. Despite these symbolic gestures and words, how-ever, democratic erosion in Russia has not been a top agenda issue for Bush in his execution of Russia policy.

According to human rights activists and democracy proponents in Russia, this change in policy has had nega-tive consequences for their causes. At the beginning of the Bush administration, these groups were optimistic about the return of a Republican to the White House. Bush said the right things and appeared willing to be tough with the Kremlin authorities. Since September 11, however, Rus-sian democracy activists have noted the change in tone in Bush's statements about Russia. Tatiana Kasatkina, exec-utive director of the human rights group Memorial, noted that she and her associates "were not satisfied" with Bush's comments about human rights during his first visit to Russia in May 2002. "He spoke about Chechnya and human rights only in passing. There was nothing in the speech like what he said during the elections cam-paign." These same groups now feel abandoned. As Lyudmila Alexeyeva, head of the Moscow Helsinki Group, explained, "The integration of Russia into the an-titerror coalition became a pardon of violations by West-ern democracies. This ally that we [the Russian human rights movement] had in Western governments, the

United States, the European Union, and Canada, is im-measurably less of an ally now." Other Russian human rights activists have complained that Bush's references to the joint American-Russian war on terrorism have given the Russian military in Chechnya even more leeway to act as it pleases. As Valentina Melnikova, an activist with the Soldier's Mothers Committee, stated in reaction to Bush's comments on Chechnya during the May 2002 Moscow summit, "We know for sure that the way he spoke about it [the war on terrorism] gives more freedom to the Rus-sian military."

Regarding Russian economic reform, the Bush admin-istration has had to make few policy decisions because re-formers within Putin's government have proceeded with extensive economic reforms without asking for major ex-ternal financial or technical assistance. Putin's team has pushed through major tax reforms, produced trade sur-pluses, maintained balanced budgets, and quelled infla-tion.[5] In 1999, Russia recovered from the 1998 financial meltdown and recorded 5.4 percent growth in GDP. In 2000, the Russian economy grew 8.3 percent, the highest annual rate of growth in decades, but this tapered to 5.2 percent in 2001. Russia's success made decisions about aid to Russia easy. The Bush administration did not sup-port new IMF loans to Russia during these years, because the Russian government did not request them.

Some analysts called for the forgiveness of Russian debt owed to the United States, but the Russian govern-ment has never asked for debt forgiveness, and so the Bush administration has never offered it. Although with smaller budgets, bilateral economic assistance programs have continued under the Bush administration, they are not the focus of policy in either capital. The Bush admin-istration dissolved the much-criticized Gore–Chernomyr-din commission and has given its blessing and support to a set of private bilateral organizations dedicated to the same set of issues pursued by Gore–Chernomyrdin. Rus-sian economic reform, however, is not, as it was during much of the 1990s, a major issue in United States–Russian relations. Especially after September 11, Russia's internal problems disappeared from the United States–Russian diplomatic agenda.

A NEW RUSSIAN-AMERICAN SECURITY AGENDA

Putin and his immediate circle of foreign policy advis-ers welcomed the return of realpolitik as the philosophy guiding United States–Russian relations. Even before September 11, this new realism embraced by both presi-dents helped reverse the perceived setbacks in United States–Russian relations in the latter half of the 1990s. The simple fact that both presidents were new also created a sense of optimism in the bilateral relationship. After Sep-tember 11, the personal bonds between Putin and Bush and the positive ambience surrounding United States–Russian relations grew even stronger.

The new "happy talk" between presidents produced policy results. On the war on terrorism, Secretary of State Powell and other Bush officials have praised Russia as a "key member of the antiterrorist coalition." Powell asserted, "Russia has played a crucial role in our success in Afghanistan, by providing intelligence, bolstering the Northern Alliance, and assisting our entry into Central Asia. As a result, we have seriously eroded the capabilities of a terrorist network that posed a direct threat to both of our countries." United States and Russian officials have continued to echo similar cooperative themes well beyond the efforts established during the military campaign against the Taliban in Afghanistan. At the June 2002 G-8 meeting, Bush praised Putin as a "man of action when it comes to fighting terror."

Beyond Afghanistan, American and Russian actions in fighting the war on terrorism have occurred in parallel and not in conjunction with each other. In spring 2002, Russian and American officials discussed a joint operation in Georgia to root out Al Qaeda operatives allegedly camped in the Pankisi Gorge area. Eventually, American armed forces were deployed in Georgia, but not accompanied by their Russian counterparts, who are not perceived as allies by the Georgian government. Russian and American officials differ on how long American troops should stay in Central Asia. Russian parliamentary speaker Gennady Seleznov has warned that "Russia will not approve of permanent U.S. military bases in Central Asia." The Bush and Putin administrations also have not agreed on a definition of terrorism or a course of action when discussing Iran or Iraq. Despite its new alliance with the United States, Russia has refused to stop selling nuclear technology to Iran. Officially, Russia has remained opposed to a military attack against Iraq. And Russian officials have accused the United States of "double standards" for pushing for democracy in Iraq and Iran but not in Saudi Arabia or Egypt. Nonetheless, the degree of understanding about security threats to both the United States and Russia from third parties—whether states or organizations such as Al Qaeda—was never greater than in the wake of September 11.[6]

In addition to cooperation (or at least a shared vision) on terrorism, Bush and Putin took advantage of the new warm relationship between their two countries to finally complete unresolved security issues from the previous decade. On arms control, the Bush administration failed to secure Russian approval for amendments to the ABM treaty, which would have allowed the United States to deploy national missile defense. Instead, on December 13, 2001, Bush officially notified Moscow of his intention to withdraw from the ABM treaty, which occurred six months later. Putin as well as many other Russian officials repeatedly stated their disapproval. But the withdrawal announcement had no discernible negative consequences for the bilateral relationship. Five months later, the two presidents signed the Moscow Treaty on nuclear weapons during Bush's first visit to Moscow. The two-page treaty committed both countries to reduce their nuclear warheads to between 1,700 and 2,200 by December 31, 2012 (Bush originally did not want to sign a treaty, proposing instead a handshake or a memorandum of understanding; the signing of an actual treaty that had to be ratified by American and Russian legislators was a concession to Putin). Although the treaty represented the largest reduction in strategic nuclear weapons ever codified in an international agreement, critics of the treaty have rightly noted that the treaty does not obligate either country to actually destroy the nuclear warheads. Rather, the treaty requires both sides to remove these warheads from their delivery vehicles. The Bush administration has acknowledged that the Pentagon intends to store 4,600 warheads in a "responsive force," since the American capacity to reproduce these warheads is limited. But most treaties signed between the United States and the Soviet Union/Russia were more illusion than real reductions.

Bush and Putin also have initiated several important steps that should eventually resolve lingering security concerns regarding Europe. Bush's courtship of Putin helped make the second round of NATO expansion a nonevent. Even as a presidential candidate, Bush had stated consistently that he intended to continue to enlarge NATO and was not going to allow Russia a veto over candidate countries. At the same time, the Bush administration reestablished a special relationship between NATO and Russia by creating the NATO–Russia Council, a new institutional arrangement ratified by NATO members in Rome in July 2002. Russian membership into NATO, once considered a subject too ridiculous even to mention, is now discussed as a real possibility in the distant future. Disagreements—most notably about Belarus—still exist, and competition—most strikingly over the future orientation of Ukraine—remain. But after September 11, the once-distant goal of a Europe "peaceful, undivided, and democratic" no longer seems so remote.

Finally, Bush and Putin have cooperated to push the agenda of Russian integration into Western international institutions beyond NATO. In summer 2002, the G-8 leaders agreed that in 2006, Russia would assume the presidency and host the group's annual summit. According to the White House, "The decision reflects Russia's economic and democratic transition in recent years under President Putin." President Bush also has stated emphatically that he wants to facilitate Russia's speedy entry into the World Trade Organization. To promote membership, his Department of Commerce declared Russia a market economy in spring 2002. To aid Russian economic integration, Bush also called for Russia's graduation from the Jackson–Vanik Amendment, although his administration has failed to convince or cajole Congress to change the legislation.[7] In some economic sectors such as steel and poultry, Bush policy actions impeding United States–Russian trade have trumped these rhetorical pledges in support of Russian economic integration. But overall,

Bush's basic economic and security policy impulse toward Russia has been one of integration.

INTEGRATION WITHOUT TRANSFORMATION

Midway through his first term in office, Bush's Russia policy looks very similar to Clinton's. Both American presidents sought closer relations with Russia. Both pushed for Russian integration into the West. During his first two years as president, Bush may have joined greater results from pursuing this policy—Russia seems more integrated into the West and closer to the United States in 2002 than in 1999—but the basic strategy pursued by Bush was not distinct from Clinton's, or from President George H. W. Bush's. Beginning with the elder Bush and continuing with Clinton, United States foreign policymakers in the 1990s embraced Russian integration into the West as the objective, and engagement of Russia as the strategy to achieve this. George W. Bush's strategy toward Russia does not signal a qualitatively new approach. Rather, his policy represents the continuation of Clinton's basic strategy. After all, Clinton talked with Yeltsin about reducing strategic nuclear weapons to below 2,000 warheads, but never closed the deal. It is a deal, however, that Clinton or Gore would have signed in a heartbeat. Likewise, the new Russia–NATO pact looks like the old Russia–NATO Charter from 1997.

The approaches of Bill Clinton and George W. Bush differ in one important way. Whereas Clinton and his foreign policy team recognized democratization and marketization within Russia as preconditions for full-scale integration into the Western community of democratic states, Bush does not. Like Clinton and his father, George W. Bush pushed for integration of Russia into the West. Unlike Clinton but similar to his father, Bush does not believe that Russia's internal transformation must precede Russia's external integration into the Western clubs. Bush does not emphasize Russian internal reforms, and defines a very small role for the United States in facilitating the process of Russian internal reform. Regarding Russian democracy, which has eroded considerably during his first two years in office, Bush rarely has mentioned Russia's problems, let alone proposed policies that might address them.

Why the major reversal in Bush's thinking on Russia? Most have attributed this amazing transformation to September 11. This is partly correct. In Bush's view, the battle against terrorism is a black-and-white issue. Putin made the clear choice to wear a white hat, and Bush appreciated his unequivocal decision to join the right side. Some in Washington even believe that the Russian reaction to September 11 has been more sympathetic to the American cause than those of some NATO allies.

But September 11 is only part of the story. Well before the terrorist attacks, Bush already had decided that he would abandon the policy of "tough realism" and instead work to cultivate a personal relationship with Putin and a cooperative relationship with Russia. This new approach was clearly apparent at the first Bush–Putin summit in Slovenia.

At the beginning of all new administrations, whatever the previous administration did is deemed wrong. This was true for the Bush team in the first months of 2001. Yet as it began to think more strategically about American interests with Russia, the Bush administration realized that a Russia inside the Western community of states was better for the United States than a Russia outside this community. Bush also realized that his own security agenda—including first and foremost the abrogation of the ABM treaty—could be achieved more easily with a cooperative rather than confrontational relationship with Russia.

Putin's own thinking and behavior must be included as part of the explanation for this shift. A different leader in the Kremlin might have reacted more negatively to an American president who abrogated the ABM treaty, pushed for the expansion of NATO to include former Soviet republics, and stationed American troops in Central Asia. Often in history, great powers—and especially great powers in decline—have banded together to balance against an expanding hegemon. Putin might have pursued the formation of a new anti-American coalition among Russia, China, and even Europe to act as a counterweight to the United States. Instead of pursuing this balance-of-power strategy, Putin opted to stay the course of integration. Instead of balancing against the United States and its allies, Putin's Russia has tried to join with the Western community of states.

For nearly two decades, Kremlin leaders have been pursuing the same basic strategy toward the West: integration. Soviet President Mikhail Gorbachev started this new trajectory (remember his phrase, a "common European home"), Yeltsin deepened it, and Putin has continued it. From Moscow's perspective, there have been hiccups along the way—NATO expansion, the August 1998 financial crisis, Kosovo—but these challenges to integration have all been temporary. Integration into the West changed from a slogan to a policy after the Soviet and Russian leaders initiated fundamental transformations of Russian political and economic institutions. These internal changes triggered a new foreign policy. On first coming to office, Putin toyed with a different approach. Some in his entourage actively encouraged a less Westerncentric direction to foreign policy. After some initial hesitation, Putin recommitted Russia to a policy of integration into the West. September 11 further reaffirmed his strategic choice. Although Putin still faces many domestic critics of this pro-Western orientation, Bush's decision not to make Russian internal reform a precondition for Russia's Western integration or for closer ties with the United States has made Putin's decision to lean Westward much easier.

NOTES

1. Governor George W. Bush, "A Distinctly American Internationalism," Ronald Reagan Presidential Library, Simi Valley, Calif., November 19, 1999.

2. Although refusing to comment on this revelation, Gore advisers maintain that the vice president generally complained about the vagueness of the charges against Chernomyrdin.

3. *Russia's Road to Corruption: How the Clinton Administration Exported Government Instead of Free Enterprise and Failed the Russian People* (Washington: U.S. House of Representatives, September 2000), p. 79.

4. Masha Lipman and Michael McFaul, "'Managed Democracy' in Russia: Putin and the Press," *Harvard International Journal of Press/Politics*, vol. 6, no. 3 (Summer 2001), pp. 117–128.

5. For an overview of positive and negative trends in the Russian economy, see especially the set of articles published by the Joint Economic Committee, Congress of the United States, *Russia's Uncertain Economic Future: Compendium of Papers* (Washington: Government Printing Office, December 2001).

6. Although never discussed publicly, China constitutes another shared threat—or more precisely "future" threat—that has helped strengthen the bond between the Russian and American presidents.

7. Thirty years ago, Senator Henry "Scoop" Jackson (D.-Wash.) and Congressman Charles Vanik (D.-Ohio) co-sponsored an amendment to the 1974 Trade Act that linked the Soviet Union's trading status to levels of Jewish emigration. Russia eliminated state controls on Jewish emigration over a decade ago, but the American legislation has not yet been amended.

JAMES GOLDGEIER *is director of the Institute for European, Russian and Eurasian Studies, George Washington University, and adjunct senior fellow, Council on Foreign Relations.* MICHAEL MCFAUL *is the Peter and Helen Bing Research Fellow at the Hoover Institution, associate professor of political science at Stanford University, and senior associate at the Carnegie Endowment for International Peace. They are completing a book on United States policy toward Russia after the cold war.*

From *Current History*, October 2002, pp. 313-324. © 2002 by Current History, Inc. Reprinted by permission.

Where Does Europe End?

Throughout its history, Ukraine has straddled the border between East and West. Now, barely a decade after breaking away from the crumbling Soviet Union, it is leaning strongly toward Europe. But Europe is wary.

by Nancy Popson

Even the tiniest of the 33 parties competing in Ukraine's parliamentary elections this past spring boasted all the ephemera of the modern American-style political campaign, from catchy logos to slick television ads. A few members of Ukraine's burgeoning homegrown public relations elite snatched some of the business from even the dominant Russian and Western imagemakers. One 30-second television spot perfectly distilled the choices facing Ukrainians. It opened with a black-and-white animated line drawing of an old train filled with elderly people. The passengers sit tiredly in the compartments, dressed in peasant garb that hangs loosely on their sturdy frames. Their faces are gaunt. The train moves slowly, and the viewer soon sees that the tracks lead to a cliff, where the rails are mangled and broken. The scene then changes to a

color animation of a modern high-speed train filled with young people enjoying themselves. The passengers—good-looking, thin, happy—are dressed in European-style clothes. The spot ends with the declaration that it is time for a new generation to take the reins of power in Ukraine.

The ad failed to win the New Generation Party a single seat in the Rada, or parliament, but it put the choices clearly: What kind of train will Ukraine be, and in which direction will it head?

These are questions that Ukrainians have been trying to answer for hundreds of years. Since the 15th century, Ukrainian leaders have struggled to carve out a space for themselves between East and West, between Russia (and later the Soviet Union) to the east and a succession of other powers to the west—Lithuania, Poland, the Austro-Hungarian

Empire, and now the European Union (EU). Ukraine's very name means "borderland."

Twice before in the 11 years since the country achieved its independence from the Soviet Union, voters gave a relatively clear answer to the question of the nation's future, saying, in effect, *ni dyakuyu* (no thanks) to a distinctly Western orientation. Awarding the Communist Party of Ukraine the most seats in the Rada, they chose to pursue a glacial pace of reform and to maintain very close ties with Russia. But on March 31, Ukrainians chose a somewhat different course.

This time, the Communists came in second in the party-list contest. A plurality of seats in the new parliament will be held by groups that back either President Leonid Kuchma, a canny ex-apparatchik and self-proclaimed reformer, or some dozen economic oligarchs who, for the most part, support him. These groups are generally pro-Western. Far more significantly, for the first time in Ukraine's brief democratic history, voters put a notable pro-reform opposition in parliament. The top vote-getter in the party-list contest was Our Ukraine, a bloc led by the 48-year-old former prime minister Viktor Yushchenko. It was joined by the reform-minded Bloc for Julia Tymoshenko, led by the charismatic former vice prime minister for energy issues, and Oleksandr Moroz's Socialist Party. Together, the three blocs, which run from the center-right to the center-left, control about one-third of the seats in parliament. (The exact balance of power is difficult to determine, because only half the 450 deputies are selected in the national party-list vote, while half come from single-member districts where the party identities and loyalties of those elected are often unclear.) If these three blocs are able to work together and attract unaffiliated deputies, they may be able to nudge the Ukrainian train toward higher speeds and, working with the pro-presidential forces, in a distinctly westward direction. A few weeks after the election, the presidentially appointed foreign minister, Anatoliy Zlenko, declared, "Ukraine chooses the union it prefers. This is the EU."

The West, however, may not choose Ukraine.

For generations, it was said derisively that "Europe ends at the Pyrenees." Now it appears that Europe's leaders may be drawing another line across the landscape. They have met Ukraine's inquiries about eventual membership in the EU with studied cool. The EU is already preoccupied with plans for an enlargement that could boost membership from the current 15 countries to 27 by the end of the decade, including four of Ukraine's western neighbors—Poland, Hungary, Slovakia, and Romania. Ukraine's appeal to the Europeans is further limited by

political and economic institutions (especially its legal system) that fall far short of European standards. Perhaps just as damning in the EU's eyes is the fact that Ukraine's main exports are items already overproduced by important EU countries, notably farm products and steel. European officials encourage Ukrainian cooperation and compliance with European legal, democratic, and economic standards, but despite regular entreaties from Kyiv, they refuse to speculate about a schedule for Ukraine's accession to the EU.

As if the cold shoulder were not bad enough, the EU's expansion is likely to measurably harm Ukraine. Because EU rules require members to tighten visa requirements for visitors from non-EU countries, Ukrainians will have difficulty crossing borders into Poland and other countries where they have prospered as traders, and where many have relatives. The border could become, in effect, a new cliff lying in front of the Ukrainian train.

The past decade has not been kind to Ukraine's dreams. When the country declared its independence from the Soviet Union in 1991, and later agreed to give up the hundreds of formerly Soviet nuclear weapons on its soil, many observers thought it would quickly become a success story. Larger than any country in the EU and with a population of almost 50 million, Ukraine has abundant natural resources. Its "black soil" farmland made it the breadbasket of the Soviet Union. It contains major industries, concentrated in the east, and like other former Soviet republics it boasts a highly educated population. The eastern city of Kharkiv alone is home to more than 25 universities.

But the reality has fallen dismayingly short of expectations, thanks largely to the Communists' power in parliament. The country's official gross domestic product (GDP) shrank more than 60 percent in the first nine years of independence

(though the large, unofficial shadow economy cushioned the fall). Privatization, especially land reform, progressed slowly. The transformation of Ukraine's large collective farms into joint stock companies, peasant associations, cooperatives, and the like has been completed, yet little has really changed. Smaller private farms remain rare. Ukraine's GDP per capita was only $3,850 (in purchasing-power parity) in 2000—about the same as El Salvador's. According to a 1999 U.S. government estimate, 50 percent of the population lives in poverty. Many workers are paid intermittently, if at all.

The electoral success of Our Ukraine owes much to Viktor Yushchenko's engineering of a significant economic turnaround during his stint as prime minister, from 1999 to 2001. Yushchenko insisted on transparency in economic transactions, particularly in the energy industry, where barter and the process of holding long-term debts on official books (in the full knowledge that the government would bail out enterprises in dire straits) had become common practice. Government budgets were kept in check. Ukraine worked closely with the International Monetary Fund, securing credit and implementing IMF-mandated reforms. But many of these changes hurt the interests of the country's dozen or so powerful economic oligarchs, and in April 2001, just as the economy was beginning to pick up, the oligarchs and Communists in the Rada dumped Yushchenko's government in a vote of no-confidence.

While Ukraine struggled economically, its democratic development took an encouraging path, at least through the 1990s. Parliamentary and presidential elections were considered free and fair. A new constitution, ratified in 1996, provides for both a strong president and a vigorous parliament. (Some observers argue that the difficulty of pushing economic reforms through the strengthened Rada is a testament to the strength of the new constitu-

tional system.) And unlike in Russia, the president has not resorted to tanks and mortar shells to mold the parliament to his wishes.

But more recent developments have been less encouraging. In April 2000, a Kuchma-backed national referendum on proposals that would give the president much greater power over parliament won the approval, according to the official tally, of more than 80 percent of the voters. But there were widespread reports of fraud and other irregularities, and the Rada has refused to implement the measures. An even more ominous sign came with the release in November 2000 of audiotapes allegedly made in Kuchma's inner office. The tapes—whose authenticity has not been established—include excerpts of conversations between Kuchma and his aides that cast doubt on the legitimacy of the voting in the 1999 presidential election and the 2000 referendum. Other conversations allegedly document Kuchma's ap-

proval of the sale to Iraq of advanced air defense systems capable of detecting stealth bombers. They also suggest that Kuchma or his highest aides were involved in the disappearance of journalist Hryhoryi Gongadze, an outspoken opponent of the president. An official review of the investigation inspired no confidence in Kuchma's government. When Gongadze's headless body was found in a ditch outside Kyiv, his hands and torso marred by acid, a series of DNA tests by Russian and Ukrainian authorities purportedly showed that the body was not that of the missing man. A Western test proved that it was.

Kuchma soon cracked down on his critics. Julia Tymoshenko was jailed in February 2001 on corruption charges, which raised eyebrows not so much because the charges were implausible—few Ukrainian politicians could pass Western tests of political hygiene—but because of the timing. After spending six weeks in jail, Tymoshenko was released for

lack of evidence by a Kyiv court. Two months later, Kuchma supported parliament's dismissal of then-prime minister Yushchenko, a potential rival.

The elections in March 2002 probably put an end to Kuchma's hopes of implementing the referendum measures, but he remains a powerful force, especially with the uncertain balance of power in the Rada. There is no guarantee that Kuchma will not take advantage of future divisions to strengthen his presidential powers and challenge the legislation that bars him from seeking a third term in the 2004 election.

To a certain extent, Ukraine's political divisions reflect the deep cultural, ethnic, and linguistic differences that even a casual visitor can see etched in the country's landscape. It requires only a short drive from the Polish border to reach the regional capital of L'viv, a city of picturesque cobblestone streets whose life revolves around its grand old opera house and the tree-lined pedestrian park that lies before it. The rolling countryside is dotted with crumbling palaces of the Austro-Hungarian elite and farms that would look at home in Île-de-France. But some 500 miles to the east, the city of Kharkiv offers a blunt contrast, its grandiose boulevards lined with monumental buildings in the Stalinist and post-Stalinist style and a vast central square—one of the largest in Europe—still dominated by an imposing statue of Lenin.

Almost in the middle of the country's east-west axis, appropriately enough, sits the capital city of Kyiv. It is no Moscow—life moves a bit more slowly here, skyscrapers are nowhere to be seen, and the streetscape is muted by many trees and parks. Kyiv too has kept its monument to Lenin, and a hulking metal statue of a redoubtable female comrade defending the city with upraised sword (called the "Baba" by locals) dominates the view of its bluffs from the river below. Yet the real center of the city is at Independence Square, the site of a substantial underground shopping mall and a monument to independence, which, in an ambiguous testament to the country's modernizing impulses, occupy space once graced by an array of European-style fountains.

U kraine's oldest ties are to Russia. Both countries trace their origins to a single ancient society, Kyivan Rus, and its capital, Kyiv. The Orthodox religious tradition dates back to Kyivan Rus's acceptance of Christianity in the 10th century, and the modern Russian and Ukrainian languages both descend from old church Slavonic.

For centuries after the collapse of Kyivan Rus in a 13th-century Mongol invasion, the territory that is now Ukraine was divided. The western principalities found themselves under Lithuanian and later Polish rule, while those to the east fell under what would become the Russian Empire.

After World War II the Soviet Union reunited the regions, and in 1954 it transferred Crimea, which had been an autonomous republic within the Russian Soviet Federated Socialist Republic since 1921, to Ukraine.

While eastern areas of Ukraine have more deeply rooted ties to Russia, decades of Soviet rule strengthened the entire country's web of connections to its former master. After World War II, the Soviets forcibly resettled most of Ukraine's Poles and Hungarians, leaving a population that was 73 percent Ukrainian and 22 percent ethnic Russian, according to the last Soviet census. Russians form a majority in the Crimea, and they are especially numerous in other areas that were part of the Russian Empire. Then there is a linguistic split, which follows slightly different lines from the ethnic divide. Because Russian was the language of social mobility during the Soviet period, many ethnic Ukrainians—especially in the cities in the south and east—are more comfortable speaking Russian than Ukrainian. However, the two languages remain mutually intelligible, at least to people raised in the bilingual atmosphere of Ukraine.

C ulturally, Ukraine straddles the divide that defines what political scientist Samuel Huntington famously called "the clash of civilizations." More than 97 percent of its religious congregations are Christian, but most are Orthodox and trace their history to Kyivan Rus. More "Western" strands of Christian faith are also strong, notably the Ukrainian Greek Catholic Church in the west and the Roman Catholic Church in the central part of the country. Some of the country's cultural-religious fault lines were exposed last summer when Pope John Paul II's visit to Ukraine stirred protests by Orthodox leaders who were alarmed by alleged Latin-rite encroachments on their turf. However, the Orthodox believers themselves are not united. Most of the Orthodox communities remain loyal to the pa-

triarch in Moscow, but many now proclaim their allegiance to an independent Ukrainian patriarch in Kyiv or to the smaller Autocephalous (independent) Orthodox Church.

Ukraine's history of division and heterogeneity goes a long way toward explaining its post-independence "multivector" foreign policy. Opinion polls show that the public is equally willing to support closer ties with Europe and with Russia, depending on how the question is worded. The voting patterns of the March 31 elections reflected some of these divisions. Opposition candidates fared best in the western regions, while the Communists (who are seen as pro-Russian) did very well in the south and east. This is a somewhat simplistic picture of the cultural-regional divide, but it underlines the difficulty Ukraine—and the new generation in parliament—will have in charting the future.

No such ambiguities hamper Russia. Many Russians cannot understand why Ukraine wants to be independent, or why Ukrainian patriots choose to emphasize the differences between their close cultures and languages. Russians saw Ukraine's decision to separate from the Soviet Union in 1991 as a pragmatic vote for freedom from communism and for what seemed a rosy economic future, not as a break from the historical relationship between the two countries. While many Ukrainian voters may have shared that view, the first post-communist politicians used their mandate to rebuild the country as an independent nation-state. The sharp focus on nation-building throughout the 1990s led to a pervasive discourse of Ukraine as "other" than Russia. In textbooks and the popular media, Ukrainian history was recast as a long, winding path out of oppression toward the ultimate goal of becoming a united, independent state. While that ideal state is viewed as multicultural and inclusive, it is built on a foundation of Ukrainian ethnic distinctiveness.

All of this is unfathomable to many Russians. The very fact that

Kyiv—the historical center of Kyivan Rus, and thus a very strong part of Russian national identity—is now located in another country is baffling to them. Russian business has close connections with Ukrainian factories and mines that were built when Ukraine was part of the Soviet Union, and Russian nationalists are acutely aware of Ukraine's large Russian-speaking population. Although no official claims have been made, popular Russian politicians such as Moscow's mayor Yuriy Luzhkov often speak of Crimea as rightfully Russian territory. Even the most liberal Russian thinkers cannot conceive of that traditional playground of the tsars and their Soviet successors as part of another state. The Russians maintain a distinctly proprietary demeanor. When a Ukrainian foreign ministry official spoke recently of the country's commitment to EU membership and declared that Ukraine could not simultaneously integrate with the Eurasian Economic Community (Russia, Kazakhstan, Kyrgyzstan, Tajikistan, and Belarus), the Russian ambassador undiplomatically declared him an "obtuse man" and re-minded Ukrainians that the EU had not issued them any invitations.

Politicians in Kyiv have openly courted Europe. Ukraine has been active in the North Atlantic Treaty Organization's Partnership for Peace—it recently announced plans to apply for NATO membership—and since 1992 has sent more than 8,000 soldiers to serve in the former Yugoslavia under the United Nations flag. And Ukraine's western regions cooperate with their Central European neighbors in regional economic and environmental initiatives such as the Carpathian Euroregion.

This is not to say that Ukrainian politicians are anti-Russian. Russia is the country's biggest trading partner and a vital source of oil and gas. The two countries are bound by strong cultural, linguistic, and familial ties. Ukrainians may savor their political independence, but the prospect of severing all links is something few can imagine.

Which way will the Ukrainian train go? While the majority of the Rada's deputies now look to Europe as Ukraine's future partner, the tracks leading toward Russia, though fraught with perils for Ukrainian independence, remain alluringly open. Ukraine's leaders are likely to turn back to brother Russia if faced with too many obstacles on the track to Europe.

With Ukraine occupying an important position on the new frontier between "the West and the rest," the country's domestic political squabbles and shifting coalitions take on far more international importance than they otherwise might. A Ukraine that embraced Western standards of law, business, and politics would be in a position to enhance European stability, either within a larger Europe or as a strong, democratic, economically stable neighboring state. Ukraine needs the support of the international community if it is to move in that direction. Building more restricted borders between Ukraine and Europe will only weaken the reform impulse and strengthen Russia's influence, leaving a frontier that in the future could require far more attentive guarding.

NANCY POPSON *is deputy director of the Wilson Center's Kennan Institute.*

From *The Wilson Quarterly*, Summer 2002, pp. 13-19. © 2002 by Nancy Popson.

UNIT 8
The Pacific Basin

Unit Selections

26. **Japan's Slow-Moving Economic Avalanche**, Scott B. MacDonald and Jonathan Lemco
27. **India, Pakistan, and the Prospect of War**, Alexander Evans
28. **Great Leap Backwards**, Ahmed Rashid

Key Points to Consider

- What should the United States, South Korea, China, Japan, and Russia do now that North Korea has unilaterally left the NPT Treaty and announced its intention of reopening a nuclear fuel reactor capable of producing weapons-grade nuclear fuel?

- How would the collapse of the Japanese financial system affect international relations in the region?

- Do you think there will be a limited nuclear war between Pakistan and India in your lifetime? Defend your answer.

- Explain why you believe that the United States will or will not remain engaged in Afghanistan over the next decade.

 **Links: www.dushkin.com/online/**
These sites are annotated in the World Wide Web pages.

ASEAN Web
   http://www.asean.or.id
Inside China Today
   http://www.insidechina.com
Japan Ministry of Foreign Affairs
   http://www.mofa.go.jp

Conventional wisdom about economic development was shaken by the economic instability experienced by the Asian "tigers" and by prolonged recession in Japan. Analysts now worry about the political fallout from the economic downturn on domestic political stability. Many analysts remain concerned that the financial crisis will continue to have a lingering impact on several countries in the region at the same time that China becomes the center of growth by becoming both the major importer and exporter in the region. The economic crisis occurred at the same time as a relatively unstable Pacific Basin security environment experienced a series of shocks, including the continued development of missiles by North Korea and the mobilization of Indian and Pakistani troops on their shared border as the war on terrorism in Afghanistan was winding down.

After U.S. intelligence sources made public their assessment that North Korea may have already developed two nuclear warheads, the Bush administration abandoned a declared policy of preemption and opted instead to pursue diplomatic means to resolve the conflict. North Korea's policy shift occurred at the same time that the United States was mobilizing for war in Iraq. South Korea was in the midst of a presidential transition. The new South Korean president, Roh Moo Hyun, had run on a platform of continued engagement with North Korea and greater separation from the United States. North Korea's move to reopen its nuclear reactor in violation of an international agreement and escalating rhetoric about a "nuclear fight to the finish" was widely interpreted as a bluff by a dictator facing desperate economic conditions at home. After North Korea announced that it was withdrawing from the Nuclear Nonproliferation Treaty, some analysts argued that North Korea had decided that its best defense against a U.S. preemptive attack was to go ahead and develop a strong nuclear deterrent. While experts debated North Korea's motives, the escalating situation had immediate ramifications as these moves shook up the geostrategic balance in the region. Alarms were sounded in several capitals that the crisis, if not properly managed, could trigger a nuclear and conventional arms race among Japan, Russia, China, North Korea, and South Korea.

These internation conflicts flared up at the same time that groups allied with al Qaeda stepped up their activities in such key countries in the region as Indonesia and the Philippines. In the summer of 2002, bombing of a resort in Bali, popular with Australian and other Western tourists, underscored the fact that al Qaeda's terror campaign continued to have global reach. Many analysts now worry that continuing poverty and rebellions in such populous countries as Indonesia and the Philippines will provide both large numbers of new recruits and strongholds for radical Islamic terrorist groups.

The economic slowdown in Asia started during 1997 in Thailand's supercharged economy, when a run on the local currency burst the speculative bubble. Thailand was the first of several Asian tigers to devalue its currency and to accept harsh austerity measures as a condition for an International Money Fund (IMF) bailout package. The Thai crisis triggered a wider regional slowdown. By early 1998, local stock markets had lost two-thirds of their value, unemployment was rising, prices were skyrocketing, living standards were falling, and recession spread throughout the region.

Some analysts now view the Southeast Asian currency crises of 1997 as part of a pattern of financial instability that often accompanies rapid economic growth. These analysts predict that Asian economies will continue to grow and account for over half the world's income by 2025. Economists and financial analysts continue to debate whether recent economic recovery is a meaningful upswing fueled by consumer demand in the United States or a temporary recovery. Most analysts agree that sustained economic recovery will require growth in the economies of both Japan and China.

While its economy also slowed, China managed to avoid a currency devaluation. The transfer of control of Hong Kong from Great Britain to China went smoothly, despite a dramatic fall in tourism and a sell-off of property. Instead, China's most serious economic problems stem from dislocations caused by government efforts to reform its "iron rice bowl" economy based on socialistic principles. Economic reforms are eliminating many of the subsidies for food, housing, and jobs that had been promised to all citizens under socialism. China is also eliminating nonperforming state entities. The result is massive unemployment, growing resentment over increased disparites among urban and rural areas and across regions, and a fundamental change in the relationship between the state, the Communist Party, and workers. Many analysts have speculated about the probable domestic and international changes that would be made after Jiang Zemin relinquished the top position in the ruling Communist Party in November 2001. However, Jiang Zemin retained his position as head of the Central Military Commission. This partial leadership change raises questions about the real extent of the power held by his successor, Hu Jintao, and whether Mr. Jiang has actually gone into retirement.

None of the recent political changes are expected to trigger major changes in China's foreign policy. Instead, China's changing economic policies are expected to continue to drive changes in its foreign policy. Taiwan remains China's number one foreign policy interest. At the same time, China is increasing its influence and presence throughout the region. When the United States tried to build a coalition in Central Asia in 2001, it was following in the footsteps of China who had forged a regional pact against terrorism, drug running, and Islamic radicalism during 2000. Bold new foreign policy moves are transforming China's relations with its neighbors. The changes are little noticed because China is using quiet diplomacy and modest pressure. China's inroads with its 14 territorial neighbors mark an important and rapid shift in its foreign policy. The broader mission underlying the recent moves is to assert the country's place as the dominant power in the region.

Many observers hope that China will increasingly act like a status quo power as it increases its economic integration into the world capitalist economic system. This belief and the potential new business offered by China as a market were the two main reasons for the U.S. push to ensure that China gained admission to the World Trade Organization (WTO) in 2001. China's entry into the World Trade Organization means that for the first time since the 1920s, Asia's largest nation is participating in helping to define the important issues for the rule-based trade and investment regime.

China's admission to the WTO was greeted with a great deal of trepidation in Japan. Many Japanese fear that the continuing recession means that Japan will be unable to compete with China in the economic or political realm. Sino-Japanese relations are a vital national interest for Japan because half of Japan's largest companies are planning to increase production overseas—and 71 percent of those planned to do it in China.

Japanese fears are also fueled by a growing sense that the country will never be able to climb out of the current recession until the government and major banks acknowledge billions of dollars worth of bad debt that they have not yet written off. In 2002, the government was embarrassed by its inability to receive enough bids to sell 10-year government bonds. After the Japanese government paid for a third bailout in 4 years of major financial groups, several senior officials resolved to require banks and major financial groups to write off bad debt and change their basic way of doing business before seeking more government aid. Scott MacDonald and Jonathan Lemco explain the current economic morass in "Japan's Slow-Moving Economic Avalanche." They conclude that for now Japan is too integrated into the world financial system and too big to fall, but that a time of reckoning will come if changes are not made.

The continuing depressed state of Japan's economy is spreading economic problems throughout Southeast Asia that threaten three decades of political stability. The economic crisis in Asia wiped out the gains of a large proportion of middle-class citizens. The economic upheaval that has occurred to date is fueling major political changes in Thailand, Malaysia, Singapore, Vietnam, Indonesia, and the Philippines. Many analysts believe that major political change has only just begun. The reaction of a government-backed militia to an independence vote in East Timor in 1999 may confirm the predictions.

Economic slowdown in Asia may also have the unintended consequence of speeding up efforts to reduce the influence of international financial institutions that are dominated by the major industrialized countries. New regional arrangements are being fashioned in East Asia by Japan, China, South Korea, and the 10 members of the Association of South-East Asian Nations (ASEAN). The ASEAN+3 group illustrates how a growing number of states in different regions are turning to subregional preferential trade pacts and cooperative financial arrangements.

Future events in Afghanistan, India, and Pakistan are also likely to determine future regional relationships. As Alexander Evans and Sumit Ganguly explain in "India, Pakistan, and the Prospect of War," the important question about the prospects for future stability in South Asia is whether the United States will avoid abandoning Afghanistan once the al Qaeda threat is eliminated while remaining sufficiently engaged to help prevent tensions between India and Pakistan from escalating into war. Ahmed Rashid in "Great Leap Backwards" claims that what the United States will do in the future is uncertain because it has no long-term strategy for implementing political or economic change in Afghanistan. He shares the fear of many that without a long-term strategy, ethnic divisions and Islamic fundamentalism will reassert themselves and divide Afghanistan, and possibly other countries, in the region.

Japan's Slow-Moving Economic Avalanche

"For now Japan is too integrated into the financial system and too big to fail.... But a time of reckoning is coming if changes are not made."

SCOTT B. MACDONALD AND JONATHAN LEMCO

Ten years ago, a favorite topic for American media pundits was the perceived economic threat from a mighty Japan. Japanese investors, whose financial resources seemed limitless, had bought American icons such as Rockefeller Center, the Pebble Beach Golf Course, and Universal Studios. Representative Helen Bentley (R.-Md.) could muse that the United States was "rapidly becoming a colony of Japan." Japan's strong political leadership, close government–private sector coordination, and high-tech export-led growth were pronounced as the new model for economic growth in Asia, the source of the "Asian Miracle."

A decade later the United States is the economic powerhouse. Rockefeller Center and Pebble Beach are again in American hands. More important, the Japanese economy, although still the second largest in the world, is in decline.

Does this decline signal depression, mass unemployment, and catastrophe? In the short term it does not. For now Japan is too integrated into the world financial system and too big to fail. Instead, we should expect an acceleration of unemployment, widening economic inequality, unprecedented domestic debt levels, deflationary pressures, and feeble political efforts to address these problems (Prime Minister Junichiro Koizumi is the first unabashed reformer in high office in recent memory, although his popularity has declined recently). But the good news is that Japan's foreign currency reserve capacity is so substantial (over $200 billion) that it can withstand internal and external shocks for two or more years before it is forced to act.[1] Japan also maintains substantial savings. Equally important, most of the country's debt is held by local investors, who traditionally have rarely pursued other options than investing in what has proved to be safe (thus far): Japanese government bonds. But a time of reckoning is coming if changes are not made.

SUMMING UP JAPAN'S CHALLENGES

What are Japan's major challenges?

• Too much debt (public debt is 135 percent of GDP—$5.6 trillion—and rising). This is up from 61 percent 10 years ago. The United States peak was 76 percent in 1993. Debt servicing alone consumes 35 percent of all budgetary outlays. Many private-sector analysts expect that public-sector debt will surpass 200 percent of GDP within the next two years, imposing an even greater drag on the Japanese economy.

• Negative economic growth (down 2.1 percent as of September 2001). The Organization for Economic Cooperation and Development and the International Monetary Fund both expect that the Japanese economy will continue to contract in 2002 and 2003 and show only marginal growth in 2004.

• Unemployment is on the rise and will stay that way. Reflecting the structural problems of the Japanese economy, unemployment reached a post-war high of 5.6 percent in February 2002.

• Deflation is becoming pervasive in all sectors of the economy. It is eroding the government's revenue base, which is contributing to the fiscal imbalance. Consumer prices fell 0.7 percent in 2001, the fastest decline in 30 years. There is a strong sense of stagnation in economic policy. Indeed, as the international rating agency Moody's Investors Service stated in February 2002: "The longer it takes for the government to fashion an effective policy response to deflation, the more complicated solving other economic problems becomes."

• Weak and technically insolvent banks. Many banks have so many nonperforming loans (NPLs) that they just will not lend. The government conservatively puts NPLs at 4.3 trillion yen or

8 percent of GDP. Private-sector analysts believe this number is much higher. The NPL problem has crippled conventional monetary policy and limited the benefits of aggressive government spending.

• An ineffective and inequitable taxation system. The country's corporate and individual taxes are high; even the ministry of economy, trade, and industry acknowledges that heavy taxes are the largest factor prompting major companies to shift manufacturing offshore.

• Many corporations have suspect financial strength but are sustained by government subsidies. The government continues to be reluctant to allow companies to fail, especially those in the agricultural, construction, and retail sectors. These "walking dead" or "zombie" companies are another substantial drag on the economy.

• The government tends to prefer state-guided solutions to business problems rather than market-driven resolutions. In early 2002, Snow Brands, a struggling dairy and food company, was unable to overcome a mislabeling scandal. As the company went through its third restructuring plan, it came under pressure for considering a friendly takeover bid from Swiss multinational Nestlé, with which it has a joint venture. Instead, the government guided three of the country's agricultural cooperatives to help Snow Brands, while it was suggested that its banks forgive or generously restructure debt. Although the ministry of agriculture was to later reverse its position about foreign equity involvement, the tendency remains for government guidance in finding a solution.

• The government seems resigned to a stock market slide, although in a measured fashion. The downturn in the economy, a lack of investor confidence, and poor (albeit improving) transparency and disclosure in the corporate world clearly are negative influences on the stock market. The government has established a fund to buy shares on the market to slow the process of stock market decline and help provide support for banks that sell stock to improve their financial situations. While this creates a brake for a stock market collapse, it still does not address the fundamental structural problems facing the Japanese economy.

• The Japanese currency, the yen, may be the world's strongest, but it also has a more dubious distinction: it may also buy the least amount of goods. Japanese consumers live in the world's most expensive country. Labor and other costs are extremely high, which provides a further disincentive for international investment in Japan. Indeed, even Japanese companies are leaving Japan for cheaper labor costs in countries such as China and Mexico. Although the yen is likely to weaken (possibly to Y150 to US$1 by the end of 2002) as the Japanese government tries to help slow the decline of the current account surplus and rekindle export growth, it will remain one of the world's strongest currencies.

The many challenges facing Japan are daunting. Yet the government apparently will not make any dramatic policy moves until the economy or the financial system actually reaches what it considers the "danger point." Officials have repeatedly pledged radical action in the case of a crisis while denying that a crisis exists at present. In fact, senior public servants frequently see it as in their interest to stifle the reform effort. Consequently, the policy track is characterized by tinkering, while the macroeconomic fundamentals steadily grow worse.

The long list of challenges Japan faces is slowly eroding confidence inside the country. A new twist is that the Japanese people are now second-guessing the government. The price of gold, the favored instrument of those fearing downward volatility, was at a two-year high in February 2002. According to stories in the Japanese press, growing numbers of Japanese are buying gold and taking it home as a hedge against possible bank failures. Many Japanese believe that gold stored under the mattress is probably a safer investment than money or gold in a bank, which could fail in the not-too-distant future. The collapse of the mid-sized Ishikawa Bank in December 2001 clearly spooked many Japanese and made gold appear a safe option.

The international investment community is well aware of Japan's problems. In February 2002, Moody's Investors Service put Japan's Aa3 yen-denominated debt on watch for a possible downgrade, possibly to as low as A2. If this were to occur, the world's second-largest economy would be rated at the same level as such second- or third-tier economies as Cyprus, Greece, Latvia, and Mauritius.[2]

THE CORE PROBLEM

Japan's many challenges reflect a core problem: the Gordian knot of special-interest politics at the center of a system based on coordination and stability. The Liberal Democratic Party's long period of rule from the 1950s to the present with only a few short interruptions has stemmed from its ability to provide stability based on policy coordination between the public and private sectors. This has appealed to many Japanese, especially those in parts of the economy that sought and received protection from outside competition: retail companies, construction firms, and agriculture. In all fairness, the system initially worked well, providing strong economic growth from which everyone benefited. It was difficult to argue that the 9.3 percent annual growth rate in the 1956–1973 period and the 4.1 percent growth rate in the 1975–1991 period were bad.

Yet, while some economic sectors were pushed to compete in the rough-and-tumble international markets, the protected areas were allowed to plod along, never developing competitive instincts. As Japan became a world leader in automobiles and electronics, it maintained an inefficient retail sector, a high-cost and consumer-unfriendly agricultural sector, and a bloated construction sector, highly dependent on the public sector for ongoing largesse. The banks were an important component of this system, providing loans and keeping many weaker companies afloat by loose accounting standards. Banks were willing to keep the "zombie" companies from bankruptcy by providing even more loans or not categorizing the companies as troubled. These government-guided bailouts continue today because Tokyo is unwilling to handle the alternative: mass joblessness in a nation with a mediocre social safety net. Cleaning up all the bad borrowers would swell the ranks of the unemployed by 3 million to 4 million people, or 6 percent of the total workforce.

Of course, the current policy of "muddling through" has already cost 2 million jobs since 1997.

The LDP's ability to raise considerable cash from its key supporters was another key element in the coordination and stability system. As political historian Richard L. Sims notes, the "LDP, either as a party or through its various factions and Diet members, acquired enormous funds—far beyond the amounts which rival parties could raise—notably from big business, the construction industry, and numerous small businesses which owed gratitude to LDP politicians or sought their aid."[3]

Today we are seeing the gradual fragmentation of the coordination and stability system. The end of the bubble economy (the period during the 1980s that witnessed a massive rise in stock market and real estate prices) left the Japanese economy in a weakened condition, increasingly less able to handle the demands of the protected economy. At the same time, international competition has grown, demands for less protection of the Japanese domestic market have increased (including a Sino-Japanese trade dispute in late 2001 and early 2002 over the access of Chinese agricultural goods), and the banking system threatens to collapse unless propped up by government assistance.

KOIZUMI'S PROMISE

In April 2001, Prime Minister Koizumi took office with strong public support, a call for reform, and a hopeful international audience. His reform program aimed to remodel Japan into a more powerful, competitive, and taxpayer-friendly economy. Along these lines he advocated fiscal rehabilitation (addressing the budget deficit and growing public-sector debt), dealing with the nonperforming loans in the banking system, and public corporation reform. The last included the abolition or privatization of 62 public corporations out of a total of 163. It also entailed reform of the postal banking system, long one of the major repositories for Japanese savers.

Koizumi and his team of reformers faced considerable opposition from within the LDP, the bureaucracy, and those businesses that would be affected by the reduced government largesse incorporated in fiscal reform. Consequently, Koizumi was forced to make a number of compromises with the conservatives and party fence-sitters to pass legislation, which subsequently was watered down and less effective than initially envisioned. For example, only 17 corporations are to be abolished and 45 others privatized, but over lengthy periods of time. At the same time, the banking sector continued to have massive problems with bad loans. Despite an improvement in dealing with old bad debt, new bad loans multiplied as the economy shifted in late 2001 into its third recession in five years (that is, for the third time in five years, Japan experienced two consecutive quarters of negative economic growth).

While the reform effort was to restructure the economy, the central bank pledged on February 10, 2002 to loosen monetary policy. But because the discount rate is zero percent and the Bank of Japan has already extended so much cheap credit, the banking sector cannot absorb any more. The banks, saddled with many troubled loans, have little appetite to lend. Moreover,

potential customers have other options: foreign banks (for the large and still financially sound companies), local finance companies, and new non-bank finance institutions.

THINKING POST-KOIZUMI?

Prime Minister Koizumi sought to break with Japan's system of coordination and stability and move the country to a more open, more market-driven economy. This also implied a shift in the country's political life to greater transparency and careful weighing of public opinion. But Koizumi increasingly appears boxed in by conservative LDP factions that oppose his reform plan. He can expect little support from his coalition partners, New Komieto and the New Conservative Party, or from the largely ineffectual opposition.

It was because of the efforts of conservative LDP members that Foreign Minister Makiko Tanaka left Koizumi's cabinet earlier this year, dealing another blow to the fragile unity of the Koizumi team. Highly popular with women voters and in her electoral region, Tanaka, the daughter of one of Japans' former prime ministers and long-dominant political players, was brought to the foreign ministry to refashion Japans's foreign policy and clean it up. The foreign ministry since January 2001 has been the focus of several investigations, initiated by a scandal involving the long-running embezzlement of discretionary funds by a ministry official. Other scandals had followed. Tanaka, however, was disliked by foreign ministry bureaucrats, who had little intention of lending their support to a cleanup of the ministry. Moreover, powerful members of the LDP, in particular, Muneo Suzuki, thought little of interfering in the ministry and attempting to impose their own stamp on foreign policy. Hence Tanaka was constantly upended by the bureaucrats. Offhand comments were leaked to the press, causing an embarrassment on more than one occasion. The media, with its allies in the LDP, was more than happy to comply in seeking to bring down the foreign minister. Finally, on January 29, 2002, Koizumi forced Tanaka to resign over the trivial issue of two NGOs not being allowed to attend a summit held in Tokyo on the reconstruction of Afghanistan.

Tanaka's ouster at the hands of LDP conservatives was immediately reflected in a downward shift in Koizumi's popularity in opinion polls from 85.6 percent in December 2001 to 55.6 percent in early February 2002. Opinion polls taken shortly after Tanaka's sacking also showed that 62 percent disagreed with the action. For Koizumi, the fall in public opinion indicated a loss of leverage over LDP conservatives. As Minoru Morita, a political analyst, noted in the February 4, 2002 *Nikkei Weekly,* "Koizumi won't be able to control the Liberal Democratic Party's Old Guard, who oppose his reform plans, if his popularity continues to slide. The extremely high public approval ratings have been his strongest weapon to support his fragile power based with the LDP."

Prime Minister Koizumi remains committed to his reform agenda, despite the formidable opposition he faces. The nosedive in his approval ratings does not derive from public dissatisfaction with the reform agenda, but is the popular reaction to the firing of the foreign minister. (Tanaka has, however, expressed her continued support for Koizumi since her resignation.)

Koizumi will likely be sustained in office through 2002 and possibly beyond because the economy has not yet collapsed. Although public-sector debt is expected to climb to 140 percent of GDP by the end of the 2002 fiscal year (March 31), the economy will continue to contract and unemployment will climb further upward. Undoubtedly, the government will resort to tinkering to stop a stock market and bank meltdown. The government's stock market buying fund may buy stocks dumped into the Nikkei by the banks as they seek to raise cash. The government also is likely to inject capital into major banks that are threatening to collapse. And major corporations are likely to receive some form of government-guided assistance to postpone bankruptcies. This is geared in particular to help fend off further collapses in the retail and construction sectors, in sharp contrast to the United States, which recently let major retailed Kmart file for Chapter 11 bankruptcy.

Despite the gloom and doom hanging over the Japanese landscape, the private sector offers a glimmer of hope.

Koizumi thus will continue in office for the foreseeable future. This is especially the case since no major rival appears to be on the political landscape, with the possible exception of former Prime Minister Ryutaro Hashimoto. Hashimoto, however, lacks popular support, which has become increasingly important to hold office. The public hardly puts much more trust in LDP conservatives or the opposition Liberal Party, which may be on the verge of falling apart due to internal conflicts. The bottom line to understanding the economic crisis facing Japan is the lack of urgency on the part of the Japanese political elite to deal with it. It requires too much work and reordering of the political economy.

Koizumi does have one additional card to play to prolong his tenure as prime minister—nationalism. Although both the reform and the conservative factions of the LDP have different views of economic policy, they share a growing sense of Japanese nationalism. Moreover, many around Koizumi espouse a "new nationalism," which can be broadly defined as economically progressive with regard to the required structural reforms, a need for a more entrepreneurial business culture and a more open economy, and a desire to see Japan take steps to convert its economic strengths into political and military power. This does not imply the creation of a militaristic culture similar to that of the 1930s, but it does mean a more assertive Japan in regional and international affairs, complete with a real modern army as opposed to the Self Defense Forces. This group also would like to see an end to the American military presence on Japanese soil and a stronger policy line toward China and North Korea. Without any other glue to keep a government together, the LDP may opt for a stronger form of nationalism, which would probably play well at home, but complicate relations with China, the Koreas, and the United States.

HOPE FROM THE PRIVATE SECTOR

Despite the gloom and doom hanging over the Japanese landscape, the private sector offers a glimmer of hope. The wrenching economic downturn is gradually leading to three developments: consolidation, a shifting of operations overseas, and greater domestic pressure for deregulation. All three developments are largely positive, although there is some angst over the "hollowing out" of Japanese industry to China and other parts of Asia.

The consolidation of various sectors of the economy is a painful yet necessary process. While the agricultural, construction, and retail sectors continue to maintain some clout with the government, other parts of the economy are carrying out structural adjustments. In the auto, electronics, and high-tech sectors, capital expenditures and workforces are being reduced. In the brewing, chemical, insurance, and steel sectors, mergers are occurring more rapidly. In many cases, companies have decided that mergers are preferable to bankruptcy. Even in the banking sector, the number of banks is shrinking, especially in areas outside the major cities.

The consolidation process does offer some hope but it must be allowed to run its course, which should include the downsizing of bloated workforces, the reduction of government assistance, and the introduction of more measures to deregulate the business environment. In this area, legislation has been passed to improve transparency and the disclosure of corporate information, and to provide a better corporate governance and a more clearly defined and market-friendly path to bankruptcy. Yet considerably more must be done in these areas, especially since a gap continues to divide what is passed as law and how it is put into practice.

Japan must also do more to encourage the creation of a vibrant sector of small and medium-sized enterprises that can compete globally. This has been a major factor in the long economic boom in the United States during the 1990s and a source of considerable technological innovation, which continues to positively impact productivity trends. Currently the small and medium-sized business sector in Japan is being hit hard by the recession, the lack of cost-efficiencies due to longstanding protectionism, and a lack of credit needed to augment productivity. Without further development of this sector, a Japanese recovery will take that much longer.

THE SLOW-MOVING AVALANCHE

Japan has been defined in the postwar era by its economic development—development that overcame the trauma of defeat and resulted in one of the world's most affluent societies. It also made Japan the leading economic power in Asia, dominating regional trade and finance. The danger for Japan is that when the unprecedented economic crisis that has gripped the country for nearly a decade really hits home sometime in the near future, all claims to past success will ring hollow. Much can be said for policies that promote coordination and stability, but not if they result in economic collapse. Japan remains prosperous. But when Japan's sense of economic security is truly threatened, the

Japanese people will demand concerted action. This may halt the slow-moving avalanche. If not, the world will feel Japan's collapse.

Notes

1. For an exhaustive historical analysis of how Japan found itself in its current predicament, see Richard Katz, *Japan: The System that Soured—The Rise and Fall of the Japanese Economic Miracle* (Armonk: N.Y.: M. E. Sharpe, 1998).

2. Moody's singled out the yen rating and not the foreign currency rating, which is Aal and stable, because the bulk of Japan's debt is in yen and held by Japanese investors. This is an important distinction, especially since Japanese investors have remained willing to buy Japanese government bonds despite the economic deterioration, as opposed to foreigners, who have in the past more easily panicked and dumped troubled securities.

3. Richard L. Sims, *Japanese Political History since the Meiji Renovation, 1868–2000* (London: Hurst and Company, 2000), p. 337.

SCOTT B. MACDONALD, *the director of research at Aladdin Capital,* and JONATHAN LEMCO, *an analyst working in the financial services industry, are the authors of the forthcoming* Soldiers, Holy Men, and Mandarins in Twenty-First Century Asia.

From *Current History,* April 2002, pp. 172-176. © 2002 by Current History, Inc. Reprinted by permission.

India, Pakistan, and the Prospect of War

"India seized an opportunity in December 2001. In escalating a crisis into a global drama, Prime Minister Vajpayee and his colleagues took a calculated risk. Has it worked?"

ALEXANDER EVANS

On Thursday, December 13, 2001, five militants armed with automatic weapons and grenades stormed the Indian parliament building in New Delhi. Equipped with a false security pass and an official car of the kind often used by high-ranking politicians, they managed to make their way into the parliament courtyard. Once inside, they dashed out of the vehicle and moved toward the main parliament entrance, firing as they went. Security forces managed to restore order, but not before 14 people—including all five assailants—were killed. It was one of the most serious terrorist attacks to take place in the Indian capital. Although no group claimed responsibility, the Indians blamed two extremist militant organizations based in Pakistan, the Jaish-e-Muhammad (Army of the Prophet) and Lashkar-e-Taiba (Army of the Pure) for the attack. Within weeks it would become the critical event that could lead to war.

The Indian prime minister, Atal Vajpayee, spoke to the Indian nation live on television. "This was not just an attack on the building, it was a warning to the entire nation," he said. The Indian media responded in kind. The next day's newspapers were full of horror at the attack—and calls for Pakistan to end support to militants, once and for all. Pointed references were made to Israel and the United States. If these two countries could combat terrorism and take on the countries behind it, why couldn't India? India's hard-line home minister, L. K. Advani, set the tone when he said: "We will liquidate the terrorists and their sponsors whoever they are, wherever they are."

Within hours Pakistani President Pervez Musharraf had condemned the attack. He added, "I would like to convey our sympathies to the government and people of India as well as our deep condolences to the bereaved families." The attack on the Indian parliament was also condemned by United States President George W. Bush, British Foreign Secretary Jack Straw, and many other world leaders.

As the days passed, the crisis deepened. To India, it was the final straw in a series of terrorist attacks—including a suicide attack on the local Kashmir State Assembly on October 1, 2001. By December 20 the atmosphere was heated. India turned down Pakistan's request for evidence backing up New Delhi's assertion that Pakistani-backed militants were responsible for the parliament attack. This was also a rebuff to the United States, which had suggested that releasing this evidence would help reduce tension. This suggestion was not taken well in New Delhi—with Indian officials pointing out (privately) that the Americans had been equally unforthcoming in their campaign against terror.

There were unconfirmed reports of Indian troop movements close to the Pakistan border in Rajasthan. Meanwhile, President Musharraf flew to China, where he held meetings with Chinese President Jiang Zemin. China has long had warm relations with Pakistan, but in two previous Indo-Pakistani wars—in 1965 and 1971—had chosen not to intervene. While there, Musharraf slipped into combative language, accusing India of "arrogance" and engaging in knee-jerk reactions.

On the border, tensions were growing. Two Indian border guards were killed on December 22, allegedly by Pakistani fire. In New Delhi, Mohammad Sharif Khan, a Pakistani diplomat, was allegedly detained and beaten by Indian security officials. Khan was accused of spying, but whether he was spying (or not) or was beaten (or not), this incident further undermined already weak diplomatic channels between India and Pakistan. The next two days saw moves by Pakistan to stave off Indian action. On December 24, the State Bank of Pakistan froze Lashkar-e-Taiba bank accounts (at the instruction of the Pakistani government).

This was followed quickly the next day by the detention by Pakistani security forces of Jaish-e-Muhammad chief Maulana Masood Azhar.

But by December 25 war looked inevitable. Heavy Indian troop deployments along the Pakistani border were now accompanied by mass evacuations of civilians from adjoining areas. Indian and Pakistani forces secured their positions, laying extensive minefields in recently vacated fields and villages. Accidents on both sides linked to the transport of mines and munitions began to claim military and civilian lives.

On December 26, the United States again tried to reduce the political temperature when Secretary of State Colin Powell announced that the two militant groups India blamed for the parliament attack had been formally placed on the United States list of banned terrorist organizations. At the same time, Pakistani intelligence suggested that India was now poised to invade Pakistan; most of its army, and almost all its air force, was deployed in an offensive formation aimed at Pakistan.

India continued to apply pressure and took a series of steps on December 27, including announcing that it would halve its diplomatic representation in Pakistan, forbid Pakistani planes to enter Indian airspace beginning January 1, and close down transport links between both countries. Pakistan reciprocated.

India also prepared a list of 20 people it accused of involvement in acts of terrorism on Indian soil—and believed to be in Pakistan. On December 31, Arun Kumar Singh, a senior Indian external affairs ministry official, called in Pakistan's deputy high commissioner, Jalil Abbas Jeelani, to present him with the list. Singh then demanded that Pakistan hand over to India for trial those named on the list.

A CRISIS SLOWLY DEFUSED

As the new year rolled in, feverish diplomatic activity was taking place in Washington D.C. and London. British Prime Minister Tony Blair, it was announced, would soon travel to South Asia to meet directly with Indian and Pakistani leaders. While Britain and the United States were worried about the threat of direct conflict, they were also keen to head off the impact the crisis was beginning to have on the American-led coalition against terrorism. With work still to be done in Afghanistan, and unconfirmed reports that senior Taliban and Al Qaeda members might be slipping away into Pakistan itself, keeping Pakistan focused on supporting the war on terror was an important foreign policy priority. Already Pakistan had moved forces to the border with India that could have been used to intercept and detain suspected terrorists entering from Afghanistan.

The onus was on Pakistan to make concessions. On January 4, 2002, the Pakistani police raided a number of locations, mainly in Punjab province, detaining militants from the Lashkar-e-Taiba and the Jaish-e-Muhammad. But Colin Powell continued to apply pressure from the American side, saying that he expected Musharraf to do more. American officials were worried that India was determined to see major Pakistani concessions, and that nothing less would assuage New Delhi's leaders.

The Indian and Pakistani leaders themselves had assembled in the Nepalese capital, Kathmandu. A long-planned regional summit of leaders from the South Asian Association for Regional Cooperation, the weak South Asian regional body established in 1985, had begun a few days before. Musharraf and Vajpayee arrived on January 4, and considerable discussion ensued as to whether they would exchange words—or even a glance—during the summit. Musharraf arrived late, leading to speculation in the Indian press that he had no intention of taking the summit seriously.

Musharraf seized the diplomatic high ground—and the photo opportunity—when he walked over to Vajpayee and offered the startled Indian prime minister his hand. Vajpayee took it, and the summit handshake added weight to Musharraf's offer, made a few hours before, of a Pakistan "hand of friendship" to India. When Vajpayee addressed the summit, however, he made it clear that India stood by its position that only concrete action from Pakistan would pave the way for normalization of relations. The only positive sign came, once more, from the media advisers, who ensured that Vajpayee reciprocated Musharraf's visual gesture at the close of the summit. Two handshakes—but no serious talks—later, both leaders headed for home.

In both cases their next international engagement was with the British prime minister. Blair had a difficult role to play. He needed to affirm British support for India in cracking down on terrorism and soothe Indian concerns about the links between Washington and Islamabad that had been renewed with the United States–led coalition attack on Afghanistan. Even before he touched down on Indian soil, Blair clarified that he would not be telling either country how to run its affairs. Blair told reporters that, while Britain had no magic formula for peace, both he and President Bush were determined to prevent war from breaking out.

Blair flew into New Delhi the following day and met with the Indian prime minister on January 6. They signed a joint declaration condemning terrorism and those who support it. And in a joint press conference the following day, Blair was careful not to endorse the specifics of Indian demands on Pakistan—although he again used language that endeared him to his Indian hosts. "The terrorist attacks of eleventh September, first October, and thirteenth December were deliberate attempts to shatter the peace of our peoples and to undermine democratic values. The attack on the Indian parliament was an attack on democracy worldwide" read the joint declaration. India's wounded national pride was given its due by the visiting British prime minister.

On January 7, Blair turned to playing to a Pakistani audience. His meeting with President Musharraf was private, and officials unofficially suggested that Blair had been blunt about what Pakistan needed to do—although warm in his gratitude for Pakistan's support in the war in Afghanistan. The press conference afterward said it all. A quiet Musharraf and a tired Blair fielded questions from the world's media. Musharraf condemned terrorism, but avoided commenting on Indian demands. He said that he had stressed to Blair Pakistan's "policy of restraint and

INDIA AND PAKISTAN: ENEMIES SINCE BIRTH

INDIA AND PAKISTAN were the two independent states that emerged when the British decolonized South Asia in August 1947. The two states were established under very different ideologies. India was a secular state, inheriting most Indian Hindus and many Indian Muslims. Pakistan was a state created for the Muslims of South Asia. The partition of British India to form these two successor states was riven by communal violence: hundreds of thousands of civilians were murdered as they fled their homes to join India or Pakistan. Within months, India and Pakistan went to war over the disputed mountain state of Kashmir. It has been split between the two countries ever since, with war once again breaking out over the former principality in 1965 (India and Pakistan also went to war in 1971 when East Pakistan seceded to become the new state of Bangladesh).

During the 1990s, hostility between India and Pakistan deepened. Pakistan covertly supported a guerrilla war in Kashmir and, as it faltered, imported Islamist extremists to replace Kashmiri fighters. And India, increasingly confident, felt anger that it could not strike back directly against Pakistan. In 1999 a crisis developed as Pakistan unilaterally occupied strategically important mountaintops in Kashmir, with major fighting breaking out between troops from the two countries in these mountains. By 2001 India had grown increasingly impatient with what it saw as Pakistan-sponsored terrorism in Kashmir. Pakistan disagreed, arguing that Kashmiris were simply fighting for their right to self-determination. Then came the terror attacks on the Indian parliament building in December 2001.

A. E.

responsibility." Blair made his views clear. There was, he said, no likelihood of international intervention to solve the Kashmir dispute—a blunt remark that did not appear on either the official Pakistani- or British-edited transcripts. And the same day India and Pakistan were again trading diplomatic brickbats, with India claiming it had shot down a Pakistani drone in Indian airspace (Pakistan denied the charge).

What did the Blair mission achieve? The British press was critical, perhaps following accusations by the opposition Conservative Party that Blair was neglecting domestic priorities. One or two commentators acerbically noted that Blair was also encouraging India to buy British-manufactured Hawk jets, a role that sat uneasily with his mission for peace. But Blair helped convey an important message from New Delhi (and Washington) to Pakistan—that the regime in Islamabad needed to respond substantively to Indian demands—while keeping India informed. Both the United States and Britain tilted toward India throughout the crisis to keep India from military action (Indian action could have provoked a nuclear exchange between the two recently declared nuclear powers, especially if Pakistan felt close to collapse following an Indian assault).

Following Blair's departure from Pakistan, the Pakistani administration again was keen to keep the United States involved. On January 8 it became public knowledge that the government of Pakistan would allow American forces to enter Pakistani territory in "hot pursuit" of escaping terrorist suspects. Pakistan also formally detained the Lashkar-e-Taiba's supreme leader, Hafiz Muhammad Saeed (although he had already been seized by security forces a week or so before). India expected more—but would Pakistan make any further concessions?

MUSHARRAF'S NEW PAKISTAN?

On Saturday January 12, 2002, newsrooms across the world waited for President Musharraf to speak to his nation—and India—in a live televised broadcast. Nobody quite knew what he might say, although Pakistani diplomats had been at pains to indicate that it would be a major speech. And whatever it contained, they said in a series of hurried briefings with commentators in the United States and Britain, no further concessions would be made.

It did prove to be a remarkable speech. Musharraf spoke for over an hour, mainly in Urdu, but switching into English for the crucial section that dealt with India. It began as a vibrant defense of Pakistan's founding principles but quickly turned to why Pakistan is foundering today. Topping his list was sectarianism and education. Pakistanis are sick of a "Kalashnikov culture," Musharraf said, where sectarian violence rules. He spoke of how Islam, the foundation stone of Pakistan, had been manipulated for sectarian ends. An "extremist minority," he said, was engaging in fratricidal killings.

In education, he explained, the traditions of Islam—which include achievements in the fields of mathematics, science, medicine, and astronomy—had been replaced by the current woeful state of affairs. Again, he accused extremists of responsibility for this, putting sectarianism before Islam. They had abused the concept of jihad. What Pakistan needed, Musharraf suggested, was a jihad against "illiteracy, poverty, backwardness, and hunger."

Pakistan's education system would be reformed. The madrassa (religious schooling) system could no longer be abused by sectarian interests. All mosques and madrassas would be regulated in the future.

It took the president over 30 minutes to mention the critical subject of Kashmir. Kashmir, he said, "runs in our blood." He promised that Pakistan would never budge an inch from its support for the Kashmir cause. And after warning the Indian leader that the Pakistani armed forces were ready for anything, he asked the international community to intervene to protect Kashmiris from human rights abuses.

Musharraf has not been an old-school dictator, replete with dubious dress-sense and an insatiable appetite for power.

In an important move, Musharraf said he had banned two radical sectarian groups held responsible for violence within Pakistan itself. Then he said he had banned the Jaish-e-Muhammad and the Lashkar-e-Taiba—the two groups India holds responsible for the December 13 attack on parliament. He cloaked the ban as an action against sectarianism.

Next, he announced new regulations for Pakistani madrassas, mosques, and foreign students (some of whom have been linked to militancy). The new regulations would include compulsory registration of all religious institutions, as well as individual registration of foreign students. He also promised that no terrorism would be conducted from Pakistani soil. As he made these announcements, the camera slowly panned in to his face, somberly filling the screen.

But he also drew a line in the sand. India, he explained, had provided a list of 20 people it wanted Pakistan to hand over. Pakistan would never extradite Pakistani citizens to India—although non-Pakistani nationals would have their cases investigated. On this demand, at least, Musharraf was unwilling to compromise.

Musharraf closed with a vision of a stronger Pakistan: a country that could take its place in the international community with honor—and act as a beacon for Islam.

Musharraf's January 12 speech was groundbreaking, but not unexpected. The government had alerted observers to expect a significant statement—in this case the key concession to India (the banning of the Jaish-e-Muhammad and the Lashkar-e-Taiba). And it also established a vision for Pakistan's future—one in which extremist rhetoric and violence would be curtailed.

The United States welcomes Musharraf's position, with warm statements from both President Bush and Secretary of State Powell. Tony Blair also endorsed his stand. And the government of India, while cautious about what Pakistan would do in practice, slowly welcomed Musharraf's commitments in the days that followed. The doubters persisted, however, in asking whether Musharraf would really crack down on Pakistani support for militancy. In February 2002, there were reports that Musharraf had closed down the ISI's Kashmir directorate. Whether this is cosmetic surgery, a specific move to delink extremist groups from former ISI sponsors, or a sweeping reform continues to be unknown.

INDIA'S GAMBLE

India seized an opportunity in December 2001. In escalating a crisis into a global drama, Prime Minister Vajpayee and his colleagues took a calculated risk. A sharp deterioration in Indo-Pakistani relations was to be expected, but the massive military buildup that followed was optimal. Indian officials sensed a

brief window of opportunity, and put together a strategy to make the most of it. The Indian public came on board, not least because the direct nature of the attack on the Indian parliament resonated with American shock at the September 11 terror attacks. Yet the policy was not driven by Indian public opinion; differences between elements in the ruling Bhartiya Janata Party–led coalition were more important.

Has it worked? The jury is still out. Musharraf did ban the Lashkar-e-Taiba and the Jaish-e-Muhammad, but his carefully phrased speech on January 12 did not mention Kashmiri terrorism and studiously avoided any mention of the Hezb-ul Mujahedeen, the main Kashmiri militant group. For Pakistan, the Hezb-ul Mujahedeen are freedom fighters, not terrorists. The Hezb-ul Mujahedeen do not engage in suicide attacks and have strong political links with Kashmiri separatist politicians.

Musharraf has accepted Indian demands that he act against extremist militant groups operating from Pakistan. His country has been portrayed as a safe harbor for terrorists, and one in which action is only belatedly being taken. Even so, Musharraf has managed to take back much of the presentational territory lost in the December 13 suicide attack on the Indian parliament. By boldly setting out a fresh path for Pakistan—with support from his fellow generals—he may achieve more than many of his elected, civilian predecessors. Ironically, Musharraf may have seized a series of small victories from an apparent diplomatic defeat in January. He has earned United States praise for responding to Indian demands. He has traded in extremist groups (who opposed him anyway), but has kept open lines with the Hezb-ul Mujahedeen. In Pakistan itself, he has renewed a national vision. It is not enough, though, for him to attempt to eliminate sectarianism and regulate foreign students (who are sometimes militant) resident in Pakistan. He faces several major challenges, all of which will need sustained action, not just words, to overcome.

Pakistan's economy and institutions are in a poor state. The additional aid, both bilateral and multilateral, that has flowed into the country since it joined the international coalition against terror—$1 billion from the United States alone—is only a stop-gap. The country's creditors agreed to restructure $12.5 billion of the country's external debt in December, and fresh loans have been promised.

But funds alone cannot solve Pakistan's crisis of governance; only a strengthening of Pakistan's institutions, action against corruption, and a collective commitment from the nation's elite will alter the trend of previous decades. If Pakistan is to prosper, Musharraf must offer more than words. And democracy, barely mentioned since Pakistan recovered its position as a significant United States ally, must form a part of the equation. To be fair, Musharraf has not been an old-school dictator, replete with dubious dress-sense and an insatiable appetite for power. Instead, he took on the reins of government almost reluctantly in October 1999, displacing former Prime Minister Nawaz Sharif's disintegrating and corrupt administration. Musharraf says he will return Pakistan to democracy—and it looks as if he means it, unlike Pakistan's last military dictator, General Zia ul-Haq, whose rule only ended with his death in 1988. Pakistan's new friendship with the United States—assuming it can last—can help steer it into safe waters as it pursues a return to democracy.

AMERICA'S ROLE

The December 2001 crisis showed how critical the American role is in South Asia. United Nations efforts to forge a peaceful settlement foundered; while India welcomes functional UN bodies (like UNCTAD and UNESCO), it is directly opposed to a UN role in settling South Asian disputes. Direct intervention by the United States or Britain is also rejected—but Washington can use good offices to help tamp down tensions.

How did the United States intervene during this latest crisis? Washington conducted an open and a private campaign to encourage India to back down from open conflict, all the time encouraging Pakistan to take steps against its own militants. President Bush also personally announced the banning of the Lashkar-e-Taiba on December 21, 2001, calling it a "stateless sponsor of terrorism." His statement signaled America's commitment to take a stand against groups determined to exacerbate Indo-Pakistani hostility. And when Musharraf finally conceded to some of India's demands, American leaders were quick to praise him.

America has intensified its efforts to reduce regional tension by restraining India and encouraging concessions from Pakistan. In the future it needs to focus on Kashmir, which is the proximate cause of Indo-Pakistani tension. The Kashmir issue must be solved—or at least salved. The United States has tried to do so before, each time failing to deliver a peaceful compromise acceptable to India and Pakistan. In the 1950s, American efforts were largely directed through the UN, failing mainly due to the Indians. In the early 1960s, an intensive bilateral effort involving six rounds of Indo-Pakistani talks yielded little, largely in the face of Pakistani obstruction. Since the 1965 Indo-Pakistani war the United States has shied away from active attempts to solve the Kashmir dispute, while keeping open the offer of its good offices should India and Pakistan jointly seek to call on them.

FUTURE TENSIONS, FUTURE HOPES

Another South Asian crisis has apparently subsided. But with no clear sign of improved relations between India and Pakistan, tensions are bound to bubble to the surface once more.

Three facts give cause for optimism. First, the crisis has not become a war—as it easily could have on December 29. Second, in late February 2002 the border between India and Pakistan was remarkably quiet. After a surfeit of cross-border shelling and occasional displays of machismo, the message on both sides appears to be restraint. Third, Pakistan has reviewed its Kashmir policy by banning the two major militant groups, the Lashkar-e-Taiba and the Jaish-e-Muhammad, and by showing a willingness to address sectarian strife—and international militants. Pakistan thus appears to have prevented Indian military action.

There are also three reasons for pessimism. First, India and Pakistan have not resolved their differences—which remain vast. The organizing principles of Pakistan's Kashmir policy continue to challenge Indian claims to sovereignty in Kashmir. And, on a regular basis, senior Indians continue to use strong language—for example, talking of an "axis of terror" (to echo President Bush's State of the Union address regarding Iraq, Iran, and North Korea) based in Pakistan. Reduced diplomatic links make misperception and renewed sources of tension likely.

Second, both countries remain at a high level of military mobilization—an expensive and possibly dangerous state of affairs. High concentrations of military forces, mines, and borders do not mix well. The lack of significant demobilization points to continuing concerns in New Delhi that Indian diplomatic objectives have not yet been met. And third, there is a wildcard. Most militants have fallen into line behind Musharraf, but a radical tail retains the capacity to strike at Indian targets—and lacks the constraint of Pakistani support to hold them back from attacks like that on the Indian parliament. For example, militants from the banned Lashkar-e-Taiba have reformed themselves, mounting a campaign to disprove their terrorist status and stating that they will not target Westerners in an attempt to gain Islamabad's favor, but elements in the Jaish-e-Muhammad have vowed to continue their war—with or without Pakistani support.

Can violence—and the threat of violence—ever be removed from Indo-Pakistani relations?

One member of this radical tail is surely the suspected terrorist Ahmed Sheikh. A one-time student at the London School of Economics, he has already been involved in one kidnapping (of American and British backpackers in India in 1994). Although he was arrested by India in connection with that case, he was then released as part of a deal with the hijackers of an Indian Airlines flight in December 1999. Slipping back into Pakistan, he went deep underground.

On January 23, 2002 Daniel Pearl, an American journalist with the *Wall Street Journal,* disappeared while pursuing a story in Karachi, Pakistan. He was kidnapped by unknown militants, probably connected to Ahmed Sheikh. Although Sheikh was arrested by Pakistani authorities on February 12, at some stage Pearl was murdered—and his videotaped death was confirmed on February 21, 2002. Pearl was a victim of the same sectarianism that has ripped apart Pakistan since the early 1980s—a sectarianism that makes simplistic assertions about religion and politics (one of the putative reasons given for Pearl's kidnap was that he was a Jew).

An equally deformed politics shapes relations between India and Pakistan. When thinking about each other, both nations are obsessed with the past and blind to their own current domestic problems. In his January speech, President Musharraf made a start by promising to take on Pakistan's extremists. He will be judged by his actions. In India, the festering problem of Kashmir requires political as well as military attention—and it is not clear whether the Indian government has the will or the desire to rectify past wrongs and engage with ordinary Kashmiris. As one sign of progress, India has appointed a new representative to advance a process of dialogue in Kashmir. The official, Wajahat Habibullah, a Muslim Indian bureaucrat, is well respected on all sides and could make headway, despite expectations to the contrary.

There continues to be ample scope for tension between these two hostile neighbors. India still harbors doubts about Pakistan's commitment to peace, and Pakistani policymakers remain anxious about Indian policies, wondering whether the threat of war may be used again. Unfortunately, the precedent suggests that a measure of military threat helped deliver Indian diplomatic objectives. Fortunately, Pakistan's response to the crisis suggests that its covert war on India in Kashmir may soon be reined in. Can violence—and the threat of violence—ever be removed from Indo-Pakistani relations?

ALEXANDER EVANS *is a research associate at the Center for Defense Studies, King's College, London.*

Reprinted from *Current History*, April 2002, pp. 160-165. © 2002 by Current History, Inc. Reprinted by permission.

AFGHANISTAN

Great Leap Backwards

The U.S. has no long-term political strategy for Afghanistan. This raises the danger that ethnic divisions and Islamic fundamentalism could again divide the country

By Ahmed Rashid/WASHINGTON AND KABUL

AS EARTHQUAKES, warlordism and coup reports continue to sweep through Afghanistan, there is mounting criticism within the Bush administration in Washington and the interim rulers in Kabul that the United States military and Central Intelligence Agency's continuing control of U.S. policy is preventing creation of a U.S. political and economic strategy that would help strengthen the interim government.

"The war in Afghanistan is over and what is left is a mopping-up operation in a few provinces, but the Department of Defence is still controlling policy," says a U.S. official in Washington involved in Afghanistan affairs. Adds Edmund McWilliams, a retired U.S. diplomat who served in Afghanistan in the 1980s: "Policymakers in Washington have failed to recognize that the key challenges are no longer simply military, but instead increasingly political."

Since the fall of the Taliban in December, U.S. military operations have taken place in only three of 32 provinces in the country. However, Secretary of Defence Donald Rumsfeld does not let a day pass without reminding journalists that the war is continuing.

Although the Al Qaeda terrorist network still poses a major international threat, its influence in Afghanistan is now small. "For 95% of the population there is no war and what people want now is greater security and reconstruction," says a senior aide to Hamid Karzai, chairman of Kabul's interim government. "But the U.S. doesn't get it."

A U.S. political strategy could strengthen the interim government through reconstruction aid and support its efforts in calling the *loya jirga*, the grand tribal council that is to convene in June to establish a new transitional government. The lack of such a strategy has been cause for concern in the interim government and the offices of the United Nations Special Representative for Afghanistan.

Some U.S. officials say that the fault lies in the same small group of senior officials who have run the war. This group is drawn from the Pentagon, the CIA and the National Security Council. Secretary of State Colin Powell has not made a single statement on U.S. policy towards Afghanistan since February, while other arms of the government such as the U.S. Agency for International Development, or USAID, have little or no say in setting policy direction for reconstruction.

The lack of a political strategy is having its effect on the ground. The Pentagon and CIA's support for Pashtun warlords in southern and eastern Afghanistan in order to mop up Al Qaeda members has led to the arming and financing of some 45,000 Afghan mercenaries. These "American warlords," as some officials in Kabul call them, often battle with each other, and pay scant attention to Karzai's government. Some have tricked U.S. forces into bombing their rivals rather than Al Qaeda forces. Others are overseeing harvesting of a new poppy crop, which is refined into heroin near U.S. bases around Kandahar.

Zalmay Khalilzad, the U.S. special envoy to Afghanistan and a National Security Council official, told reporters in Kabul on March 26 that there was "no contradiction" in the U.S. policy of trying to discourage warlordism by arming new warlords.

Meanwhile USAID, which has nearly $300 million to spend this year for the reconstruction of Afghanistan, has been unable to establish a single reconstruction project from its strategic programme. It has farmed out some support through other aid organizations and small grants. But several USAID officials are reportedly in the process of resigning, disillusioned with their agency's inability to adequately contribute to reconstruction.

The U.S. has also rejected demands to expand the International Security Assistance Force (ISAF) outside Kabul. This means there is little security for the elections to the loya jirga, which start on April 23. The Pentagon says that foreign peacekeepers would disrupt the war against terrorism.

In Kabul, many Afghan ministers are frustrated that the U.S. military has made no attempt to rein in the growing power of the Panjshiri Tajiks, who were the key U.S. ally in the war against the Taliban and run the three most powerful ministries in Kabul—Defence, Interior and Foreign Affairs. "The Panjshiris are alienating the Pashtuns and undermining Karzai's ability to extend the writ of the government," says a minister in the interim government. "Karzai cannot rein them in as long as the Americans say nothing to them."

From April 3-5 the Panjshiris, who also run the intelligence service, arrested some 300 people, mostly Pashtuns, without informing the UN, the ISAF or the interim cabinet, saying there was a plot by Afghan extremists to destabilize the government. Although the majority were quickly freed, it has widened the rift between the Pashtuns and the Tajiks.

Washington appears partly cognizant of the problems, but its remedies are questionable. On April 3 at a meeting of 35 donor countries in Geneva the U.S. pledged to lead the effort to fund a new 60,000-strong Afghan army at a cost of $235 million, while Germany promised to help rebuild a 70,000-strong police force. But the creation of a new army is still several years away and it is not clear who will lead it—the Panjshiris or a new, neutral and professional Afghan officer corps.

Meanwhile, the urgent need for reconstruction has prompted the CIA to fund "quick impact projects" using its Afghan mercenaries, according to officials in Washington and Kabul. Such projects bypass the interim government and UN agencies, and are likely to further strengthen the warlords and alienate the Karzai government. Western relief agencies are already highly critical of the U.S. military carrying out relief operations in uniform.

There is the danger that ethnic divisions could again split the country and Islamic extremism take root. For Rumsfeld, the war against terrorism remains a war, but for most Afghans it is now all about how to build the peace. Says McWilliams: "Failure to develop an independent political strategy and insensitivity to human rights are strongly reminiscent of the mistakes U.S. policymakers made in the 1982–92 period."

UNIT 9
Middle East and Africa

Unit Selections

Key Points to Consider

- Why do so many Arabs in the Middle East say that the United States is to blame for the terrorist bombings of September 11, 2001?

- Explain why you believe that Osama bin Laden's message is so popular with young Arabs around the world.

- What actions could be taken to facilitate resumption of a peace process between Israel and the Palestinians?

- Why do you think al Qaeda operatives targeted a tourist resort in Kenya for an attack at the end of 2002?

- Do you agree with William Thom's prediction that criminal actions and war will become virtually indistinguishable in the future?

 **Links: www.dushkin.com/online/**
These sites are annotated in the World Wide Web pages.

Africa News Online
http://www.africanews.org

ArabNet
http://www.arab.net

Columbia International Affairs Online
http://www.ciaonet.org/cbr/cbr00/video/cbr_v/cbr_v.html

ei: Electronic Intifada
http://electronicintifada.net/new.shtml

IslamiCity
http://islamicity.com

Israel Information Center
http://www.accessv.com/~yehuda/

Mail and Guardian
htp://www.mg.co.za

MEMRI: The Middle East Research Institute
http://www.memri.org/video

The initial phase of the U.S.–led military campaign against al Qaeda and its Taliban hosts in Afghanistan was surprisingly brief. The campaign began in October with massive air attacks. The response began only a few weeks after the terrorist attacks. There was little press coverage of the conflict on the ground so few details are available about that aspect of the military campaign. Most of the fighting was conducted by the Northern Alliance and other Afghani groups with military aid, advice, and support from U.S. special forces. The Taliban surrendered at the end of November after the fall of its last northern stronghold of Kunduz. The second phase of the war on terrorism involves an intensive search for Osama bin Laden and remaining Taliban leaders and armed fighters in the region, along with additional efforts to identify and destroy remaining components of al Qaeda's network throughout the world.

An interim Afghani government that included representatives from the major ethnic groups of the country was quickly formed. The government started to function with large amounts of Western aid. Media coverage paid little attention to a massive airlift of food that averted widespread starvation in Afghanistan during the winter. Instead, media attention shifted to a rapidly escalating crisis on the Pakistani-India border after the two countries engaged in a tit-for-tat military mobilization of forces. The incident triggering the crisis was an attack by Pakistan-based guerrillas on the parliament building in New Delhi, India. Several outside countries, including the United States, pressured the two sides to find a peaceful resolution to the crisis. By the end of 2001, President Pervez Musharraf had arrested dozen of leaders of Islamic militant groups in Pakistan whom India had blamed for the parliamentary attack.

President Bush echoed a question on the minds of many Americans when he asked in a speech to Congress why those 19 men choose to wreck the icons of U.S. military and economic power? Peter Ford in "Why Do They Hate Us?" uses interviews with a number of Muslims to explain why, asking Muslims and Arabs, including those who are sympathetic toward the United States and those who are not, to explore their feelings. Dale Eickelman explains in "Bin Laden, the Arab 'Street,' and the Middle East's Democracy Deficit," how part of bin Laden's success in recruiting young Arabs to his cause is because he speaks in the "vivid language of popular Islamic preachers, and builds on a deep and widespread resentment against the West and local ruling elites identified with it."

For many in the Arab world the carnage of September 11 was retribution for 50 years of U.S. policies in the region. Many Arabs believe that America's policy toward Israel has been excessively one-sided, supporting Israel financially as it expands its borders, while ignoring the Palestinian victims of Israeli policy. After the violent collapse of the Israeli-Palestinian peace negotiations in 2000, the Bush administration has not opted to take a lead role in efforts to promote multinational peacekeeping efforts. Instead, the United States remains supportive of Israel as the country struggles to cope with growing unrest. Mouin Rabbani, in "The Costs of Chaos in Palestine," describes the goals of Ariel Sharon's hard-line approach to the intifada and the use of suicide bombings. According to Rabbani, Israel has now launched a comprehensive war of attrition in the occupied territories, whose objective is a decisive military victory leading to prolonged interim arrangements dictated by Israel. Facing these overwhelming odds, the Palestinians remain plagued by the crisis of leadership that has already exacted a high price. Rabbani warns that if Israel's current policies continue, a "two-nation" peace settlement may not be possible.

Since 2000, Western intelligence sources have warned that Saddam Hussein of Iraq had rebuilt a covert chemical and biological weapons operation. Reports that Hussein offered to launch weapons of mass destruction against Israel if a neighboring state would permit him to place missiles on their territory after the resumption of fighting between Israel and the Palestinians, added credence to intelligence warnings. As the Bush administration mobilizes for a second military intervention in Iraq, growing pressures by U.S. allies and support for the UN's IAEA weapons inspectors pushed back the date.

Largely overlooked in Eurocentric analyses of international relations is the fact that Africa, rather than the Middle East, will soon become the continent with the largest number of Muslims. This demographic trend helps to explain why so many bin Laden foot soldiers were from Africa. Many analysts now call Africa the "soft underbelly" of international relations. For the past several decades, the number of Africans adopting fundamentalist Islamic (and Christian) beliefs has steadily risen. Al Qaeda and other radical Islamic movements have found many recruits among mostly unemployed youth in urban African centers. After hosting bin Laden for several years in the 1990s, the radical Islamic government of Sudan asked him to leave in 1996. While the Sudanese cooperated with the United States during the military campaign in Afghanistan by sharing intelligence, many citizens in Sudan and other African countries continue to treat bin Laden as an icon. Growing concerns about terrorist networks prompted representatives from 30 African countries to launch an "African pact against terrorism."

The bombings of an Israeli-owned tourist resort and the failed attempt in Kenya at the end of 2002 to shoot down an airplane bound for Tel Aviv served as a reminder of the fact that the war on terrorism is also being waged in Africa. In "Kenya's Porous Border Lies Open to Arms Smugglers," Dexter Filkins and Marc Lacey note that the recent terrorist attacks in Kenya shed light on the continuing flow of illegal goods—lobsters, drugs, people, even missiles—across the Somali-Kenyan border. The illegal traffic flows across historic smuggling routes that form a 2,000-mile arc from Pakistan down the eastern coast of Africa to southern Africa. Western officials suspect that al Qaeda operatives move people and guns around the region using these routes.

Sweeping generalizations about political and economic trends in sub-Saharan Africa are difficult to make because conditions vary widely among countries and between sectors in the same country. Most countries in sub-Saharan Africa are neither on the verge of widespread anarchy nor at the dawn of democratic and economic renewal. Economic growth in Africa is stronger today than in the disastrous 1980s, but most countries were adversely affected by the downturn in many commodity prices during the late 1990s. Talk of "a new African renaissance" at the end of the 1990s was premature. African states, however, are taking a number of steps that may eventually lead to greater economic integration and prosperity.

During 2001, the Organization for African Union was renamed the African Union. With the backing of such diverse leaders as Muammar Kadaffi of Libya and Thabo Mbeki of South Africa, the goal of the new continent wide organization is to establish a common market and common political institutions along the lines of the European Union. Two regional organizations, the Economic Community of West African States (ECOWAS) and the Southern African Development Community (SADC) have negotiated far-reaching agreements in recent years that are designed to create viable regional peacekeeping capabilities, and several elder African statesmen have met with success in mediation efforts in several conflicts between states.

The pace of African integration is slow for several reasons. One of the most important reasons is that most African economies remain highly fragile, dependent on world commodity prices and other forces beyond the control of national policymakers. Another reason is that conflict resolution in several African states is likely to require a substantial modification of the territorial and political status quo as well as extensive involvement by UN peacekeepers. Eritrea and Ethiopia finally reached agreement on the terms of a lasting peace settlement in 2000, but most observers warn that renewed hostilities could flare up again. Similar warnings are being made in Sierra Leone where an extensive UN peacekeeping force and a small British contingent are implementing a disarmament process and attempting to stop the sale of "conflict diamonds" that fuels rebel activities in Sierra Leone, Liberia, and Guinea.

The limitations of regional efforts to maintain security were underscored by the fact that France felt compelled to send its largest contingent of troops in years to Ivory Coast to protect French citizens and help the government prevent northern rebels from taking more territory. Reports that bin Laden supporters were among those benefiting from the mining and selling of "conflict diamonds" from the area underscore the fact that African nation-states, like nation-states worldwide, must cope simultaneously with security threats that are tied to domestic and external sources. The recent revelation of the ability of al Qaeda operatives to exist and profit in zones of unrest in West Africa also underscores the dangers that should be expected when outside actors ignore conditions on the ground in collapsed states.

The diplomatic success of South Africa and other members of the Southern African Development Organization in facilitating another cease-fire signed by the government and leaders of major rebel groups in the Democratic Republic of the Congo (DRC) at the end of 2001 was quickly diminished by reports of renewed fighting in the eastern portion by rebel factions who are not controlled by representatives of the major groups. A series of interrelated clashes throughout the Great Lakes region since the mid-1990s led many observers to label the situation as the first continental war in modern Africa. The UN has encountered difficulties in its efforts to send peacekeepers to monitor the cease-fire along the Eritrea-Ethiopian border, to maintain order in Sierra Leone, or to enforce a multilateral cease-fire in the Congo and Burundi. Similar problems in negotiating a lasting cease-fire agreement were recently encountered by France, despite the fact that France's foreign minister traveled to Ivory Coast to ensure the participation and commitment of all of the signatories to a cease-fire agreement. These recent experiences raise serious questions about who will play the role of peacekeeper in Africa. In the meantime, new conflicts threaten to become new sources of regional instability. A recent report, the South African Institute of International Affairs (SAIIA) warned that the Zimbabwe crisis could deteriorate to a point where that country could become "another battlefield like the Democratic Republic of Congo."

In "Africa's Security Issues Through 2010," a senior U.S. defense intelligence officer for Africa, William Thom, reviews emerging security trends in sub-Saharan Africa and predicts that interstate warfare will increase even though disparities in military power on the African continent will continue to increase as well. Thom also predicts that transnational criminality and war will become virtually indistinguishable because both are fueled by economic insurgents, warlords, lawless zones harboring criminals, and armies of child soldiers whose political socialization consists of learning how to brutalize and murder civilians. While Thom believes that policing these messy situations will become an international priority, he predicts that some places will remain beyond the reach of Western moral consciousness and continue to experience low-intensity conflict indefinitely.

Despite these gloomy forecasts, relatively peaceful and dramatic political changes in several African countries have occurred. After 40 years of Socialist Party rule in Senegal, the opposition party leader, Abdoulaye Wade of the Senegalese Democratic Party, won a victory with over 60 percent of the vote in a peaceful run-off election in 2000. Relatively peaceful elections also occurred in recent years in the two giants of sub-Saharan Africa—Nigeria and South Africa. After more than 30 years in power, Daniel Moi, the only president that many young Kenyans have known, retired. A long-time political opponent and leader of the united opposition party, Mwai Kibaki won an election at the end of 2002 that was the country's first peaceful change of government since independence in 1963.

Much of the early optimism that accompanied the election of a civilian president, Olusegun Obsanjo, after 14 years of military rule in Nigeria has subsided as a host of serious conflicts in all regions worsened. In recent years thousands have died in riots in northern Nigeria over attempts to implement Shar'ia law. A new round of riots on the eve of the first Miss World contest scheduled to be held in Nigeria led promoters to move the beauty contest to Great Britain. Ethnic conflicts in the oil-rich delta regions add to growing concerns that the presidential elections to be held in 2003 may trigger a new round of violence and possibly another military coup.

In contrast, South Africa's recent experiment with democratic change continues to succeed. The most dramatic evidence was the peaceful government transition after its second majority rule presidential election in 1999. The transfer of power from Nelson Mandela to Thabo Mbeki in a peaceful election and the implementation of a new constitution were considered meaningful successes, but progress towards meaningful economic change or a competitive political party process is proving to be more difficult. In 2001, the dominant political party, the African National Party (ANC), further consolidated its hold on the political system by agreeing to a coalition with the rump party that developed from the ruling National Party during apartheid. A consolidation of power within the ruling party occurred at a time when several senior ANC officials were accused of corruption. South Africa's success to date is remarkable given the fact that it is only a middle-income country struggling with several problems simultaneously.

ROBERT HARRISON—STAFF

PROUD FATHER: In a Pakistan pharmacy, pictured above, Amirul Haq (r.), says he is 'satisfied' that his Muslim son was killed in the Kashmir. He's 'against America, because it doesn't care about those who die in Pakistan.'

'Why do they hate us?'

By Peter Ford
Staff writer

asked President Bush in his speech to Congress last Thursday night. It is a question that has ached in America's heart for the past two weeks. Why did those 19 men choose to wreck the icons of US military and economic power?

Most Arabs and Muslims knew the answer, even before they considered who was responsible. Retired Pakistani Air Commodore Sajad Haider—a friend of the US— understood why. Radical Egyptian-born cleric and US enemy Abu Hamza al-Masri understood. And Jimmy Nur

Zamzamy, a devout Muslim and advertising executive in Indonesia, understood.

They all understood that this assault was more precisely targeted than an attack on "civilization." First and foremost, it was an attack on America.

In the United States, military planners are deciding how to exact retribution. To many people in the Middle East and beyond, where US policy has bred widespread anti-Americanism, the carnage of Sept. 11 *was* retribution.

And voices across the Muslim world are warning that if America doesn't wage its war on terrorism in a way that the Muslim world considers just, America risks creating even greater animosity.

Mr. Haider is a hero of Pakistan's 1965 war against India, and a sworn friend of America. But he and his neighbors in one of Islamabad's toniest districts are clear about why their warm feelings toward the US are not widely shared in Pakistan.

In his dim office in a north London mosque, Abu Hamza al-Masri sympathizes with the goals of Osama bin Laden, fingered by US officials as the prime suspect behind the Sept. 11 attacks. Abu Hamza has himself directed terrorist operations abroad, according to the British police, although for lack of evidence, they have never brought him to trial.

Mr. Zamzamy, a 30-something advertising executive in Jakarta, knew what was behind the attack, too. Trying to give his ads some zip and still stay within the bounds of his Muslim faith, he is keenly aware of the tensions between Islam and American-style global capitalism.

The 19 men—who US officials say hijacked four American passenger jets and flew them on suicide missions that left more than 7,000 people dead or missing—were all from the Middle East. Most of the hijackers have been identified as Muslims.

The vast majority of Muslims in the Middle East were as shocked and horrified as any American by what they saw happening on their TV screens. And they are frightened of being lumped together in the popular American imagination with the perpetrators of the attack.

But from Jakarta to Cairo, Muslims and Arabs say that on reflection, they are not surprised by it. And they do not share Mr. Bush's view that the perpetrators did what they did because "they hate our freedoms."

Rather, they say, a mood of resentment toward America and its behavior around the world has become so commonplace in their countries that it was bound to breed hostility, and even hatred.

And the buttons that Mr. bin Laden pushes in his statements and interviews—the injustice done to the Palestinians, the cruelty of continued sanctions against Iraq, the presence of US troops in Saudi Arabia, the repressive and corrupt nature of US-backed Gulf governments—win a good deal of popular sympathy.

The resentment of the US has spread through societies demoralized by their recent history. In few of the world's 50 or so Muslim countries have governments offered their citizens either prosperity or democracy. Arab nations have lost three wars against their arch-foe—and America's closest ally—Israel. A sense of failure and injustice is rising in the throats of millions.

Three weeks ago, a leading Arabic newspaper, Al-Hayat, published a poem on its front page. A long lament about the plight of the Arabs, addressed to a dead Syrian poet, it ended:

"Children are dying, but no one makes a move.
Houses are demolished, but no one makes a move.
Holy places are desecrated, but no one makes a move....
I am fed up with life in the world of mortals.
Find me a hole near you. For a life of dignity is in those holes."

It sounds as if it could have been written by a desperate and hopeless man, driven by frustration to seek death, perhaps martyrdom. A young Palestinian refugee planning a suicide bomb attack, maybe. In fact, it was written by the Saudi Arabian ambassador to London, a member of one of the wealthiest and most influential families in the kingdom that is Washington's closest Arab ally.

Against the background of that humiliated mood, America's unchallenged military, economic, and cultural might be seen as an affront even if its policies in the Middle East were neutral. And nobody voices that view.

From one end of the region to the other, the perception is that Israel can get away with murder—literally—and that Washington will turn a blind eye. Clearly, the US and Israel have compelling reasons for their actions. But little that US diplomats have done in recent years to broker a peace deal between Israel and the Palestinians has persuaded Arabs that the US is a fair-minded and equitable judge of Middle Eastern affairs.

Over the past year, Arab TV stations have broadcast countless pictures of Israeli soldiers shooting at Palestinian youths, Israeli tanks plowing into Palestinian homes, Israeli helicopters rocketing Palestinian streets. And they know that the US sends more than $3 billion a year in military and economic aid to Israel.

"You see this every day, and what do you feel?" asks Rafiq Hariri, the portly prime minister of Lebanon, who is not an excitable man. "It hurts me a lot. But for hundreds of thousands of Arabs and Muslims, it drives them crazy. They feel humiliated."

RESENTMENT RISES, AND A RADICAL IS BORN

Ask Sheikh Abdul Majeed Atta why Palestinians may not like the United States, and he does not immediately answer. Instead, he pads barefoot across the red swirls of his living room carpet and reaches for three framed photographs on the floor beside a couch.

The black-and-white prints show dusty, rock-strewn hills dotted with tiny tents and cinderblock houses: the early days of Duheisheh refugee camp, south of Bethlehem in the West Bank. It was where Mr. Atta was born, and where his family has lived for more than half a century. Atta's family village was destroyed in the struggle between Palestinian Arabs and Jews after Britain divided Palestine between them in 1948. For 10 years his family of

13 lived in a tent. The year Atta was born, the United Nations gave them a one-room house.

It doesn't matter to Atta that the United States was not directly involved in "the catastrophe," as Palestinians refer to the events of 1948. Washington averted its eyes when it could have helped, he says, and since then has been firmly on Israel's side.

Heavyset, solid, with a neatly trimmed full beard, Atta is the preacher at a nearby mosque. He looks the part of the community leader, always meticulously turned out in crisp shirts and pressed trousers, gold-rimmed reading glasses tucked into a pocket.

In the past year of the Palestinian-Israeli conflict, Atta has joined Hamas, the radical group responsible for recently sending most of the suicide bombers into Israeli towns. Frustration at watching the rising Palestinian death toll at the hands of the Israeli army played a large part in his decision, he says.

His resentment at Israel, though, dates back to his infancy, and the stories he heard of his village, Ras Abu Amar, which he never knew. That village is still alive for him, just as millions of Palestinians in the West Bank and Gaza Strip, and throughout the Middle East cherish photos, house keys, and deeds to homes that no longer exist or which have housed Israelis for generations.

Today he lives in his own house in Duheisheh, a sprawling tangle of densely packed concrete buildings that crowd snaking, narrow alleys. But he still dreams of the home he never knew, and recalls who took it from him, and remembers who they rely on for their strength.

What happened on Sept. 11 "was an awful thing, a tragedy, and since we live a continuous tragedy, we felt like this touched us," he adds. "But when we see something like this in Israel or the US, we feel a contradiction. We see it's a tragedy, but we remember that these are the people behind our tragedy."

"Even small children know that Israel is nothing without America," says Atta. "And here America means F-16, M-16, Apache helicopters, the tools Israelis use to kill us and destroy our homes."

SUPERPOWER SWAGGER

Such weapons are very much the visible face of American policy in the Middle East, where military might has held the balance of power for 50 years. Thousands of US soldiers stationed in the Gulf, and billions of US dollars each year in military aid to Israel, Egypt, and other allies, have shored up Washington's interests in the strategically crucial, oil-rich region.

That military presence and power looks like swagger to some in the Muslim world, even far from the flashpoints. "Now America is ready with its airplanes to bomb this poor nation [Afghanistan], and most people in Indonesia don't like arrogance," says Imam Budi Prasodjo, an Indonesian sociologist and talk-show host.

"You are a superpower, you are a military superpower, and you can do whatever you want. People don't like that, and this is dangerous," he adds.

"America should spread its culture, rather than weapons or tanks," adds Mohammed el-Sayed Said, deputy director of Cairo's influential Al Ahram think tank. "They need to act like any respectable commander or leader of an army. They can't just project an image of contempt for those they wish to lead."

Ten years ago, at the head of a broad coalition of Western and Arab countries, the United States used its superpower status to kick the Iraqi army out of Kuwait. Since then, however, Washington has found itself alone—save for loyal ally Britain—in its determination to keep bombing Iraq, and to keep imposing strict economic sanctions that the United Nations says are partly responsible for the deaths of half a million Iraqi children.

'We wish the American people could see what their governments are doing in the rest of the world.'

—Saniya Ghussein, whose daughter, Raafat, was killed in the 1986 US bombing raid against Libya

---

Those deaths, and those bombs (which US and British planes drop regularly, but without fanfare), are felt keenly among fellow Arabs. And Saniya Ghussein knows all about bombs.

A DAUGHTER DIES, AND PARENTS WAIT FOR US APOLOGY

In the middle of the night of April 16, 1986, the deafening sound of anti-aircraft guns woke Saniya Ghussein with a sudden start. "My God," she thought, "there's a war being fought above my house."

She slipped out of bed and ran into the bedroom where her husband Bassem and their 7-year-old daughter Kinda had fallen asleep earlier in the evening. "Bassem, the Americans are here," she said urgently. "It looks like they're going to hit us."

She checked on her other daughter, Raafat. She had been suffering from her annual bout of hay fever, and the 18-year-old art student was in the television room next to the humidifier so she could breathe easier.

Raafat was still sleeping, completely oblivious of all the commotion going on around her, due to the medication she had taken earlier. There was little Saniya felt she could do. She climbed back into bed and pulled the sheets tight around her.

Bassem lay awake on the bed, listening to the appalling noise in the night sky above.

A Palestinian-born Lebanese national, Bassem had worked in Libya as an engineer for Occidental, the American oil giant, for 20 years, helping exploit the country's

ROBERT HARRISON—STAFF

'When Bush talked of a crusade...it was not a slip of the tongue.'
—Sajad Haider, retired
Pakistani air force officer

massive oil reserves. He and his family lived in the up-market Ben Ashour neighborhood of Tripoli, the Libyan capital, on the ground floor of a two-story apartment block.

Bassem never heard the explosion. Instead, he watched in astonishment as the window frame suddenly flew into the room, and the roof collapsed on top of him and his daughter.

Kinda was screaming in the darkness near him. Bassem tried to move, but was pinned by the rubble. He groped in the blackness for Kinda. "Don't worry," he said, squeezing his daughter's hand. "Daddy's here, don't cry, it will be okay."

The blast had knocked Saniya unconscious. She woke to hear Bassem calling from the next room and Kinda screaming. She stumbled in the darkness, barefoot across the rubble and glass shards, choking on the fumes from the missile blast, as she called her daughter's name "Raafat! Raafat!" for several minutes. But there was no response, and Saniya knew with a terrible certainty that her daughter was dead.

"Bassem," she cried. "Raafat has gone."

Pinned beneath the rubble, Bassem heard his wife's words, and he felt a deep sense of anger and resentment well up inside him. His life and that of his family had been shattered, and nothing would ever be the same again.

It took them eight hours to dig Raafat out from under the ruins of the house. "Our pain and agony, which I cannot describe, started at that moment," Saniya says.

Raafat was one of an estimated 55 victims of an air raid mounted by US warplanes against a series of targets in Tripoli and another Libyan city, Benghazi.

The attacks were in retaliation for the bombing of a disco in Berlin, Germany, 10 days earlier in which 200 people were injured, 63 of them US soldiers; one soldier

and one civilian were killed. The Reagan administration blamed Libyan leader Muammar Qaddafi.

Bassem and Saniya Ghussein are not natural anti-Americans. Bassem studied in the US before going to work for Esso and then Occidental. He sent Raafat to an American Catholic school, and on family trips to the US, Saniya would take Raafat to Disney World in Florida. "We did all the typical American things," she says.

But since that terrible night 16 years ago, neither Bassem nor Saniya have stepped foot in America. They returned to Beirut in 1994 when Bassem retired.

In 1989, the Libyan government enlisted the help of Ramsey Clark, an attorney general during the Carter administration, to file a lawsuit against President Ronald Reagan and British Prime Minister Margaret Thatcher for the civilian deaths during the air raids. "When Clark came to collect our documents and evidence, I asked him if he thought we had a case," Bassem recalls. "He said 'Oh, definitely. This was murder.' "

But US district court judge Thomas Penfield Jackson disagreed. He dismissed the suit, and fined Clark for presenting a "frivolous" case that "offered no hope whatsoever of success."

Twelve years later, the court's decision still rankles with Bassem. "I will only return to America when I know someone will listen to me and say: 'yes, it was our fault your daughter died, and I am sorry.' So long as they think my daughter's death is 'frivolous,' I won't go back," Bassem says.

The Ghusseins have no sympathy for religious extremism and thoroughly condemn the Sept. 11 suicide bombings in New York and Washington. Yet they both maintain that the devastating attack was a result of America's "arrogant" policies in the Middle East and elsewhere. "We wish the American people could see what their governments are doing in the rest of the world," Saniya says.

A FEELING OF BETRAYAL
AMONG FRIENDS

On the other side of Asia, in Pakistan, Air Commodore Haider would sympathize with the Ghusseins' wish. He has always been a friend of the United States, and not just because he enjoyed the 10 years he spent in Washington as his country's military attaché. Like most other members of the ruling elite in Pakistan, in the armed forces, in business, and in the political parties, he sees America as a natural ally.

But not a reliable one. The prevailing mood in Pakistan of anger and suspicion toward the United States springs from a deeply rooted perception that the US has been a fickle friend, Haider says, and not just to Pakistan, but to other nations in the Muslim world.

If there was a moment of betrayal for Haider, it was the 1965 war between India and Pakistan, largely over the future of Kashmir. As Indian tanks advanced on the Pakistani metropolis of Lahore, Haider was head of a squadron of F-86 Sabre jets sent to destroy them. India's Soviet allies helped with money, arms, and diplomatic support. But at a crucial moment, Pakistan's ally, the US, refused to send more weapons. As it turned out, Pakistan was able to defeat the Indian attack on Lahore and elsewhere without US help. Haider's squadron decimated the column of Indian tanks that had reached to within six miles of Lahore. But the lesson lingered: America cannot be trusted.

"There is a feeling of being betrayed, it's a feeling of being let down, and you can only be let down by somebody you care for," says Haider, out for an evening stroll in a tony Islamabad neighborhood.

"They said you will be the bulwark of America and of the free world against Communism. But then they dropped a friend for no good reason."

Today, Haider sees a "convergence of interests" between the United States and Pakistan in the fight against terrorism. But he says that President Bush will need to watch his language when he talks about the Muslim world. "When Bush talked of a Crusade… it was not a slip of the tongue. It was a mindset. When they talk of terrorism, the only thing they have in mind is Islam."

Ultimately, Haider does see a way for America and Muslim nations to become lasting friends, but only if the US begins to give as much weight to the interests of Muslim nations as it does to Israel.

"When you deny justice to people, which you have been doing for several decades in Palestine, and they are intelligent, sensitive people, they are going to find something to do," warns Haider. "They might take shelter in Islam, in fatalism, and some will come to despise you."

AN EGYPTIAN 'INSPIRED' TO JOIN AFGHAN FIGHTERS

Sheikh Abu Hamza al-Masri, the radical Muslim cleric who runs a mosque in a shabby district of north London, has certainly come to despise America.

Abu Hamza says he used to admire the West when he was a young man—so much so that he dropped out of university in his native Alexandria, Egypt, to study in Britain. And he clearly had nothing against the British government when he took a job as a civil engineer at Sandhurst, the British equivalent of West Point, after he graduated.

But as he immersed himself more and more in religious studies, and came into contact with more and more Arab mujahideen, who had travelled from the mountains of Afghanistan to England for medical treatment, he began to change his outlook.

"When you see how happy they are, how anxious to just have a new limb so they can run again and fight again, not thinking of retiring, their main ambition is to get killed in the cause of God… you see another dimension in the verses of the Koran," says Abu Hamza.

How the world views a US military response

In your opinion, once the identity of the terrorists is known, should the American government launch a military attack on the country or countries where the terrorists are based, or should the American government seek to extradite the terrorists to stand trial?

	Launch attack	Try the terrorists	Don't know
Israel	77%	19%	4%
India	72	28	0
United States	54	30	16
Korea	38	54	9
France	29	67	4
Czech Republic	22	64	14
Italy	21	71	8
South Africa	18	75	7
United Kingdom (excluding N. Ireland)	18	75	7
Germany	17	77	6
Bosnia	14	80	6
Colombia	11	85	4
Pakistan	9	69	22
Greece	6	88	6
Mexico	2	94	3

Source: Gallup International surveys Sept. 14 to 17.

Inspired by their example, he took his family to Afghanistan in 1990, to work there as a civil engineer, building roads, tunnels, and "anything I could do." And he also fought with the mujahideen against Afghan President Mohammad Najibullah (seen as a Russian stand-in supported by the Soviets), until he blew both his hands off and lost the sight in his left eye, in a mine explosion.

What transformed him and his comrades-in-arms from anti-Soviet to anti-American militants, he says, was the way Washington abandoned them at the end of the war in Afghanistan, and sought to disarm and disperse them.

"It was when the Americans took the knife out of the Russians and stabbed it in our back, it's as simple as that," says Abu Hamza. "It was a natural turn, not a theoretical one.

"In the meantime, they were bombarding Iraq and occupying the [Arabian] peninsula," he says, referring to the US troops stationed in Saudi Arabia after the Gulf War, "and then with the witch-hunt against the mujahideen, all of it came together, that was a full-scale war, it was very clear."

Abu Hamza would rather see Islamic militants fight corrupt or secular Arab governments before they take on America (indeed, the Yemeni government has sought his extradition from Britain for plotting to overthrow the government in Sana). But he is in no doubt that the American government brought the events of Sept. 11 on its own head.

"The Americans wanted to fight the Russians with Muslim blood, and they could only justify that by triggering the word 'jihad,' " he argues. "Unfortunately for everybody except the Muslims, when that button is pushed, it does not come back that easy. It only keeps going on and on until the Muslim empire swallows every empire existing."

Can he understand the motivation behind the assault on New York and Washington? "The motivation is everywhere," he says, with the current US administration. "When a president stands up before the planet and says an American comes first, he is only preaching hatred. When a president stands up and says we don't honor our missile treaty with the Russians, he is only preaching arrogance. When he refuses to condemn what's happening in Palestine, he is only preaching tyranny.

"American foreign policy has invited everybody, actually, to try to humiliate America, and to give it a bloody nose," he adds.

IN JAKARTA, COUNTERING AMERICAN CULTURE WITHOUT VIOLENCE

You wouldn't catch Rizky "Jimmy" Nur Zamzamy justifying violence that way, though he professes just as deep an attachment to Islam as Abu Hamza.

Mr. Zamzamy, a rangy young Indonesian advertising executive in a pink shirt, is sitting in a Western-style cafe in Jakarta, his cellphone at the ready, and his fried chicken growing cold as he explains how he tries to be a good Muslim by right action, not fighting.

That, he feels, is the best way of countering what he sees as the corrupting influence of American culture and morals on traditional Indonesian ways of life in the largest Muslim country in the world.

Until a few years ago, Zamzamy led a regular secular life, hanging out in bars and dating women. Then he met a Muslim teacher who became his spiritual guide. Now he follows Islamic teachings and donates most of his $1,300 monthly salary to his "guru" to be spent on building mosques and helping the poor.

He says he has made sure that none of the money goes to extremist groups that use violence in the name of Islam, such as the Laskar Jihad group, locked in bloody battle with Christians in the Maluku region of Indonesia.

Two years ago, in line with his growing religious beliefs, he quit the advertising agency he had worked for and set up his own company along Islamic lines: He won't take banks or alcoholic-beverage producers as clients, for example, and he does no business on Friday, the Muslim holy day.

But he is relaxed about those who don't share his beliefs: He does not insist that his wife wear a headscarf, for example, and he is not uncomfortable sitting alongside the rich young Jakartans in the cafe who are flirting and drinking. They must make their own choices, he says.

And though he does not like the sexual overtones of American pop culture, he knows that "you can't hide from American culture." By living his life according to Islamic precepts, he says, "I am fighting America in my own way. But I don't agree with violence."

AMBIVALENCE ABOUT AMERICA

All over the Muslim world, young people like Zamzamy are juggling their sense of Islamic identity with the trappings of a globalized, secular society.

In a classroom of Al Khair University, set in a concrete office park in Islamabad, Nabil Ahmed, a business student, and his classmates are fuming over their president's betrayal of the Pakistani people by pledging to support what they fear will turn into a crusade against Muslims.

Ahmed and his friends are well-dressed, middle-class boys, and represent neither the old-money security of Pakistan's elite nor dirt-poor peasants who make up the bulk of Pakistan's angry conservative masses. They are the silent majority of Pakistan, with their feet firmly planted in both the East and the West. On weekdays, they listen to Whitney Houston and Michael Bolton, wear Dockers and Van Heusen shirts. On weekends, many switch to traditional salwar kameez outfits and go with their fathers to the mosque to pray.

'It is [the] double standard that creates hatred.'

—Nabil Ahmed, a business student at Al Khair University in Islamabad

---

They have much to gain from a Western style of life, and most have plans to move to the United States for a few years to make some money before returning home to Pakistan. Yet despite their attraction to the West, they are wary of it too.

"Most of us here like it both ways, we like American fashion, American music, American movies, but in the end, we are Muslims," says Ahmed. "The Holy Prophet said that all Muslims are like one body, and if one part of the body gets injured, then all parts feel that pain. If one Muslim is injured by non-Muslims in Afghanistan, it is the duty of all Muslims of the world to help him."

Like his friends, Ahmed feels that America has double standards toward its friends and enemies. America attacks Iraq if it invades Kuwait, but allows Israel to bulldoze Palestinian homes in the West Bank and Gaza Strip.

ROBERT HARRISON—STAFF

AFGHAN REFUGEES: These boys are among some 60,000 displaced Afghans at Jalozai refugee camp near Peshawar, Pakistan, along its border with Afghanistan. The camp is crowded, and Pakistan has recently forbidden the UNHCR to register any more refugees.

It ostracizes a Muslim nation like Sudan for oppressing its Christian minority, but allows Russia to bomb its Muslim minority into submission in Chechnya.

And while the US supported many "freedom fighter" movements in the past few decades, including the contra movement in Nicaragua, America labels Pakistan and Afghanistan as terrorist states because they support militant Muslim groups fighting in the Indian state of Kashmir and elsewhere.

'The Americans wanted to fight the Russians with Muslim blood.'

—Sheikh Abu Hamza al-Masri,
a radical Muslim cleric who runs a mosque in London

"There is only one way for America to be a friend of Islam," says Ahmed. "And that is if they consider our lives to be as precious as their own. "If Americans are concerned about the 6,500 deaths in the World Trade Center, let them talk also about the deaths in Kashmir, in Palestine, in Chechnya, in Bosnia. It is this double standard that creates hatred."

Ahmed's ambivalence about America—his desire to live and work there, his admiration for its values, but his anger at its behavior around the world—is broadly shared across the Muslim world and Arab world.

"I think they hate us because of what we do, and it seems to contradict who we say we are," says Bruce Lawrence, a professor of religion at Duke University, referring to people in the Middle East. "The major issue is that our policy seems to contradict our own basic values."

That seems clear enough to Muslims who sympathize with the Palestinians, and who say that Washington should force Israel to abide by United Nations resolutions to withdraw from the occupied territories. "The Americans say September 11th was an attack on civilization," says Mr Hariri, the Lebanese prime minister. "But what does civilized society mean if not a society that lives according to the law?"

It also seems clear to citizens of monarchical states in the Gulf, where elections are unknown and women's rights severely restricted. "Since the Cold War ended, America has talked about promoting democracy," says John Esposito, head of the Center for Muslim-Christian Understanding at Georgetown University in Washington. "But we don't do anything about it in repressive regimes in the Middle East, so you can understand widespread anti-Americanism there."

At the same time, the state-run media—which is all the media there is across much of the Middle East—often fan the flames of anti-American and anti-Israel sentiment because that helps focus citizens' minds on something other than their own government's shortcomings.

In Sana, the Yemeni capital, where queues of visa-seekers line up daily outside the US embassy, the ambivalence about America is clear. "When you go there, you really

50 YEARS OF US POLICY IN THE MIDDLE EAST

1947–48

UN votes to partition Palestine into two states—one for Jews, one for Palestinian Arabs. Arab states invade; 300,000 Palestinians flee Jewish-controlled areas. Jewish forces prevail, declaring Israeli independence. US recognizes Israel.

1953

CIA helps Iran's military stage a coup, deposing elected PM Mohammad Mossadeq, whom US sees as communist threat. US oversees installation of Shah Mohammad Reza Pavlavi as ruler of Iran.

1956

Israel attacks Egypt for control of Suez Canal. Britain and France veto US-sponsored UN resolution calling for halt to military action. British forces attack Egypt.

1960

Iran, Iraq, Kuwait, Saudi Arabia, and Venezuela form Organization of Petroleum Exporting Nations (OPEC).

1966

US sells its firs jet bombers to Israel, breaking with a 1956 decision not to sell arms to the Jewish state.

1967

Six-Day War. Israel launches preemptive strike against Arab neighbors, capturing Jerusalem, the Sinai Peninsula, the Gaza Strip, and the Golan Heights. Kuwait and Iraq cut oil supplies to US, UN adopts Resolution 242, calling on Israel to withdraw from captured territory. Israel refuses.

1968

First major hijacking by Arab militants occurs on El Al flight from Rome to Tel Aviv, marking decades of hostage-takings, hijackings, and assassinations as a strategy by Arab militant groups.

1969

Mummar Qaddafi comes to power in Libyan coup and orders US Air Force to evacuate Tripoli.

1972

Eight Arab commandos of Palestinian group Black September kill 11 Israeli athletes at the Munich Olympic Games.

1973

Egypt and Syria attack Israel over its occupation of the Golan Heights and the Sinai Peninsula. US gives $2.2 billion in emergency aid to Israel, turning tide of battle to Israel's favor. Arab states cut US oil shipments.

1974

UN General Assembly recognizes right of Palestinians to independence.

1976

The UN votes on a resolution accusing Israel of war crimes in occupied Arab territories. US casts lone "no" vote. US Ambassador to Lebanon Francis Meloy and an adviser are shot to death in Beirut. US closes Embassy there.

1978

Egypt and Israel sign US-brokered Camp David peace treaty. Eighteen Arab countries impose an economic boycott on Egypt. Egyptian president Anwar Sadat and Israeli Prime Minister Menachem Begin receive Nobel Peace Prize.

1979

Ayatollah Ruhollah Khomeini leads grass-roots Islamic revolution in Iran, deriding the US as "the great Satan." Iranian students storm US Embassy in Tehran, taking 66 Americans hostage for next 15 months. US imposes sanctions. Protesters attack US Embassies in Libya and Pakistan.

1981

Israel bombs Iraqi nuclear reactor. Muslim militants opposed to Egypt's peace treaty with Israel assassinate Egyptian President Sadat.

1982

Israel invades Lebanon to expel the Palestine Liberation Organization, facilitate election of friendly government, and form 25-mile security zone along Israel's border. Defense Minister Ariel Sharon permits Lebanese Christian militiamen to enter the Sabra and Shatila refugee camps outside Beirut. The ensuing three-day massacre kills 600 or more civilian refugees. US and other nations deploy peacekeeping troops in Lebanon.

1983

A truck bomb explodes in US Marines' barracks in Beirut, Lebanon, killing 241 soldiers. US forces withdraw.

1986

Us bombs Libya in retaliation for the bombing of a Berlin nightclub frequented by US servicemen. The airstrike kills 15 people, including the infant daughter of leader Muammar Qaddafi. All Arab nations condemn the attack.

1987

Start of the Palestinian intifada, or uprising, in the West Bank and Gaza Strip.

1990

Iraq invades Kuwait. Saddam Hussein links pullout to Israel's withdrawal from occupied territories. UN imposes sanctions that continue to hobble Iraq's economy in effort to force Iraqi compliance with weapons resolutions.

(continued)

50 YEARS *(continued)*

1991

US and coalition launch attacks against Iraq from Saudi Arabia. Gulf War ends after some three months, but US deployment continues even now, with 17,000 to 24,000 US troops in region at any time.

1993

World Trade Center in New York is bombed, killing six. US Special Forces, deployed as peacekeepers in Somalia, attempt to capture warlord Mohamed Farah Aidid. Eighteen US servicemen are killed. Israeli PM Yitzhak Rabin and Palestinian leader Yasser Arafat sign historic peace declaration in White House ceremony with President Clinton.

1994

Jordan and Israel sign peace treaty. Yasser Arafat, Yitzhak Rabin, and Foreign Minister Shimon Peres receive Nobel Peace Prize for 1993 agreement.

1995

US announces trade ban against Iran, reinforcing sanctions in effect since 1979. Rabin is assassinated, two years after peace deal with Palestinians. In Riyadh, Saudi Arabia, a car bomb explodes outside an office housing US military personnel. Seven are killed, including five Americans. Three Islamist groups claim responsibility.

1996

A truck bomb explodes outside a US military barracks in Khobar, Saudi Arabia, killing 19 US airmen. UN reports that sanctions cause 4,500 Iraqi children under 5 to die each month.

1997

Egyptian Islamic Group massacres 62 people, mostly foreign tourists, in Luxor, Egypt. The group claims it is retaliation for US imprisonment of Sheikh Omar Abdel al-Rahman, who is later convicted in 1993 World Trade Center bombing.

1998

Bombs explode outside US Embassies in Kenya and Tanzania, killing 224 people. US launches cruise-missile attacks on sites in Sudan and Afghanistan allegedly linked to Osama bin Laden. US indicts bin Laden for committing acts of terrorism against Americans abroad.

1999

Islamic militants, traced to bin Laden, are arrested for plot to bomb tourist sites during millennium celebrations.

2000

Camp David negotiations fail. Sharon visits Temple Mount in Jerusalem, sparking current Palestinian uprising. USS Cole bombing in Yemen's Aden harbor kills 17 American sailors. Bin Laden denies responsibility, but applauds the act.

2001

Hijackers crash two planes into World Trade Center in New York, one into Pentagon, and one in Pennsylvania. More than 7,000 people are dead or missing.

Compiled by Julie Finnin Day

SOURCES: "THE MIDDLE EAST" (CONGRESSIONAL QUARTERLY), NEWS REPORTS.

love the United States," says Murad al-Murayri, a US-trained physicist. "You are treated like a human being, much better than in your own country. But when you go back home, you find the US applies justice and fairness to its own people, but not abroad. In this era of globalization, that cannot stand."

Nor has the mood that has gripped Washington over the past two weeks done much to reassure skeptics, says François Burgat, a French social scientist in Yemen.

"When Bush says 'crusade', or that he wants bin Laden 'dead or alive', that is a *fatwa* (religious edict) without any judicial review," he cautions. "It denies all the principles that America is supposed to be."

A *fatwa* is something Amirul Haq, a Pakistani shopkeeper whose son died two years ago in a jihad in Kashmir, understands better than judicial review. "When I heard that my son died, I was satisfied," he says.

It's a sentiment shared by Azad Khan, too. On a hot Sunday afternoon in Mardan, Pakistan, Mr. Khan and his family have laid out a feast in a small guesthouse next to the local mosque. They are celebrating because they have just heard that Mr. Khan's 20-year-old son, Saeed, has

been killed in a gun battle with Indian troops in the part of Jammu and Kashmir state that is under Indian control. With his death, Saeed has become another *shahid*, a martyr and heroic defender of the Muslims against the enemies of Islam. According to the Koran, *shahideen* are not actually dead; they are still alive, they just can't be seen. And through acts of bravery, a *shahid* guarantees that his whole family will go to heaven.

"It is not a thing to be mourned. We are happy," says Khan, sitting down to a meal of chicken and mutton, rice and bread, along with leaders of the group with which Saeed had fought. "I told him to take part in jihad [holy war] because he is the son of a Muslim," Khan says. "And just as we fight in Kashmir, if we need to fight against the United States in Afghanistan we are ready, because we are Muslims. It is our duty to fight against any infidels who are threatening our Muslim brothers."

It's not likely that many Pakistanis, or other Muslims, will actually go to Afghanistan to fight the Americans—assuming American soldiers land there. Khan's militant views are not shared by most of his countrymen.

But in a broader sense, and in the longer term, many people in the Middle East fear that the coming war against terrorism—unless it is waged with the utmost caution—could unleash new waves of anti-American sentiment.

Jamal al-Adimi, a US-educated Yemeni lawyer, speaks for many when he warns that "if violence escalates, you bring seeds and water for terrorism. You kill someone's brother or mother, and you will just get more crazy people."

Trying to root out terrorism without re-plowing the soil in which it grows—which means rethinking the policies that breed anti-American sentiment—is unlikely to succeed, say ordinary Middle Easterners and some of their leaders.

On the practical level, Hariri points out, "launching a war is in the hands of the Americans, but winning it needs everybody. And that means everybody should see that he has an interest in joining the coalition" that Washington is building.

On a higher level, argues Bassam Tibi, a professor of international relations at Gottingen University in Germany, and an expert on political Islam, "we need value consensus between the West and Islam on democracy and human rights to combat Islamic fundamentalism. We can't do it with bombs and shooting—that will only exacerbate the problem."

Reported by staff writers Scott Baldauf in Islamabad, Pakistan; Cameron W. Barr in Amman, Jordan; Peter Ford in London; Nicole Gaouette in Jerusalem; Robert Marquand in Beijing; Scott Peterson in Sana, Yemen; Ilene R. Prusher in Tokyo; as well as contributors Nicholas Blanford in Beirut, Lebanon; Sarah Gauch in Cairo; and Simon Montlake in Jakarta, Indonesia.

Bin Laden, the Arab "Street," and the Middle East's Democracy Deficit

"Bin Laden speaks in the vivid language of popular Islamic preachers, and builds on a deep and widespread resentment against the West and local ruling elites identified with it. The lack of formal outlets to express opinion on public concerns has created [a] democracy deficit in much of the Arab world, and this makes it easier for terrorists such as bin Laden, asserting that they act in the name of religion, to hijack the Arab street."

DALE F. EICKELMAN

In the years ahead, the role of public diplomacy and open communications will play an increasingly significant role in countering the image that the Al Qaeda terrorist network and Osama bin Laden assert for themselves as guardians of Islamic values. In the fight against terrorism for which bin Laden is the photogenic icon, the first step is to recognize that he is as thoroughly a part of the modern world as was Cambodia's French-educated Pol Pot. Bin Laden's videotaped presentation of self intends to convey a traditional Islamic warrior brought up-to-date, but this sense of the past is a completely invented one. The language and content of his videotaped appeals convey more of his participation in the modern world than his camouflage jacket, Kalashnikov, and Timex watch.

Take the two-hour Al Qaeda recruitment videotape in Arabic that has made its way to many Middle Eastern video shops and Western news media.[1] It is a skillful production, as fast paced and gripping as any Hindu fundamentalist video justifying the destruction in 1992 of the Ayodhya mosque in India, or the political attack videos so heavily used in American presidential campaigning. The 1988 "Willie Horton" campaign video of Republican presidential candidate George H. W. Bush—in which an off-screen announcer portrayed Democratic presidential candidate Michael Dukakis as "soft" on crime while showing a mug shot of a convicted African-American rapist who had committed a second rape during a weekend furlough from a Massachusetts prison—was a propaganda masterpiece that combined an explicit although conventional message with a menacing, underlying one intended to motivate undecided voters. The Al Qaeda video, directed at a different audience—presumably alienated Arab youth, unemployed and often living in desperate conditions—shows an equal mastery of modern propaganda.

The Al Qaeda producers could have graduated from one of the best film schools in the United States or Europe. The fast-moving recruitment video begins with the bombing of the USS *Cole* in Yemen, but then shows a montage implying a seemingly coordinated worldwide aggression against Muslims in Palestine, Jerusalem, Lebanon, Chechnya, Kashmir, and Indonesia (but not Muslim violence against Christians and Chinese in the last). It also shows United States generals received by Saudi princes, intimating the collusion of local regimes with the West and challenging the legitimacy of many regimes, including Saudi Arabia. The sufferings of the Iraqi people are attributed to American brutality against Muslims, but Saddam Hussein is assimilated to the category of infidel ruler.

Osama bin Laden... is thoroughly imbued with the values of the modern world, even if only to reject them.

Many of the images are taken from the daily staple of Western video news—the BBC and CNN logos add to the videos' authenticity, just as Qatar's al-Jazeera satellite television logo rebroadcast by CNN and the BBC has added authenticity to Western coverage of Osama bin Laden.

Alternating with these scenes of devastation and oppression of Muslims are images of Osama bin Laden: posing in front of bookshelves or seated on the ground like a religious scholar, holding the Koran in his hand. Bin Laden radiates charismatic authority and control as he narrates the Prophet Mohammed's flight from Mecca to Medina, when the early Islamic movement was threatened by the idolaters, but returning to conquer them. Bin Laden also stresses the need for jihad, or struggle for the cause of Islam, against the "crusaders" and "Zionists." Later images show military training in Afghanistan (including target practice at a poster of Bill Clinton), and a final sequence—the word "solution" flashes across the screen—captures an Israeli soldier in full riot gear retreating from a Palestinian boy throwing stones, and a reading of the Koran.

THE THOROUGHLY MODERN ISLAMIST

Osama bin Laden, like many of his associates, is imbued with the values of the modern world, even if only to reject them. A 1971 photograph shows him on family holiday in Oxford at the age of 14, posing with two of his half-brothers and Spanish girls their own age. English was their common language of communication. Bin Laden studied English at a private school in Jidda, and English was also useful for his civil engineering courses at Jidda's King Abdul Aziz University. Unlike many of his estranged half-brothers, educated in Saudi Arabia, Europe, and the United States, Osama's education was only in Saudi Arabia, but he was also familiar with Arab and European society.

The organizational skills he learned in Saudi Arabia came in to play when he joined the mujahideen (guerrilla) struggle against the 1979 Soviet invasion of Afghanistan. He may not have directly met United States intelligence officers in the field, but they, like their Saudi and Pakistani counterparts, were delighted to have him participate in their fight against Soviet troops and recruit willing Arab fighters. Likewise, his many business enterprises flourished under highly adverse conditions. Bin Laden skillfully sustained a flexible multinational organization in the face of enemies, especially state authorities, moving cash, people, and supplies almost undetected across international frontiers.

The organizational skills of bin Laden and his associates were never underestimated. Neither should be their skills in conveying a message that appeals to some Muslims. Bin Laden lacks the credentials of an established Islamic scholar, but this does not diminish his appeal. As Sudan's Sorbonne-educated Hasan al-Turabi, the leader of his country's Muslim Brotherhood and its former attorney general and speaker of parliament, explained two decades ago, "Because all knowledge is divine and religious, a chemist, an engineer, an economist, or a jurist" are all men of learning.[2] Civil engineer bin Laden exemplifies Turabi's point. His audience judges him not by his ability to cite authoritative texts, but by his apparent skill in applying generally accepted religious tenets to current political and social issues.

THE MESSAGE ON THE ARAB "STREET"

Bin Laden's lectures circulate in book form in the Arab world, but video is the main vehicle of communication. The use of CNN-like "zippers"—the ribbons of words that stream beneath the images in many newscasts and documentaries—shows that Al Qaeda takes the Arab world's rising levels of education for granted. Increasingly, this audience is also saturated with both conventional media and new media, such as the Internet.[3] The Middle East has entered an era of mass education and this also implies an Arabic lingua franca. In Morocco in the early 1970s, rural people sometimes asked me to "translate" newscasts from the standard transnational Arabic of the state radio into colloquial Arabic. Today this is no longer required. Mass education and new communications technologies enable large numbers of Arabs to hear—and see—Al Qaeda's message directly.

Bin Laden's message does not depend on religious themes alone. Like the Ayatollah Ruhollah Khomeini, his message contains many secular elements. Khomeini often alluded to the "wretched of the earth." At least for a time, his language appealed equally to Iran's religiously minded and to the secular left. For bin Laden, the equivalent themes are the oppression and corruption of many Arab governments, and he lays the blame for the violence and oppression in Palestine, Kashmir, Chechnya, and elsewhere at the door of the West. One need not be religious to rally to some of these themes. A poll taken in Morocco in late September 2001 showed that a majority of Moroccans condemned the September 11 bombings, but 41 percent sympathized with bin Laden's message. A British poll taken at about the same time showed similar results.

Osama bin Laden and the Al Qaeda terrorist movement are thus reaching at least part of the Arab "street." Earlier this year, before the September terrorist attacks, United States policymakers considered this "street" a "new phenomenon of public accountability, which we have seldom had to factor into our projections of Arab behavior in the past. The information revolution, and particularly the daily dose of uncensored television coming out of local TV stations like al-Jazeera and international coverage by CNN and others, is shaping public opinion, which, in turn, is pushing Arab governments to respond. We don't know, and the leaders themselves don't know, how that pressure will impact on Arab policy in the future."[4]

Director of Central Intelligence George J. Tenet was even more cautionary on the nature of the "Arab street." In testimony before the Senate Select Committee on Intelligence in February 2001, he explained that the "right catalyst—such as the outbreak of Israeli-Palestinian violence—can move people to act. Through access to the Internet and other means of communication, a restive public is increasingly capable of taking action without any identifiable leadership or organizational structure."

Because many governments in the Middle East are deeply suspicious of an open press, nongovernmental organizations, and open expression, it is no surprise that the "restive" public, increasingly educated and influenced by hard-to-censor new media, can take action "without any identifiable leadership or organized structure." The Middle East in general has a democracy deficit, in which "unauthorized" leaders or critics, such as Egyptian academic Saad Eddin Ibrahim—founder and director of the Ibn Khaldun Center for Development Studies, a nongovernmental organization that promotes democracy in Egypt—suffer harassment or prison terms.

One consequence of this democracy deficit is to magnify the power of the street in the Arab world. Bin Laden speaks in the vivid language of popular Islamic preachers, and builds on a deep and widespread resentment against the West and local ruling elites identified with it. The lack of formal outlets to express opinion on public concerns has created the democracy deficit in much of the Arab world, and this makes it easier for terrorists such as bin Ladin, asserting that they act in the name of religion, to hijack the Arab street.

The immediate response is to learn to speak directly to this street. This task has already begun. Obscure to all except specialists until September 11, Qatar's al-Jazeera satellite television is a premier source in the Arab world for uncensored news and opinion. It is more, however, than the Arab equivalent of CNN. Uncensored news and opinions increasingly shape "public opinion"—a term without the pejorative overtones of "the street"—even in places like Damascus and Algiers. This public opinion in turn pushes Arab governments to be more responsive to their citizens, or at least to say that they are.

Rather than seek to censor al-Jazeera or limit Al Qaeda's access to the Western media—an unfortunate first response of the United States government after the September terror attacks— we should avoid censorship. Al Qaeda statements should be treated with the same caution as any other news source. Replacing Sinn Fein leader Gerry Adams' voice and image in the British media in the 1980s with an Irish-accented actor appearing in silhouette only highlighted what he had to say, and it is unlikely that the British public would tolerate the same restrictions on the media today.

Ironically, at almost the same time that national security adviser Condoleezza Rice asked the American television networks not to air Al Qaeda videos unedited, a former senior CIA officer, Graham Fuller, was explaining in Arabic on al-Jazeera how United States policymaking works. His appearance on al-Jazeera made a significant impact, as did Secretary of State Colin Powell's presence on a later al-Jazeera program and former United States Ambassador Christopher Ross, who speaks fluent Arabic. Likewise, the timing and content of British Prime Minister Tony Blair's response to an earlier bin Laden tape suggests how to take the emerging Arab public seriously. The day after al-Jazeera broadcast the bin Laden tape, Blair asked for and received an opportunity to respond. In his reply, Blair—in a first for a Western leader—directly addressed the Arab public through the Arab media, explaining coalition goals in attacking Al Qaeda and the Taliban and challenging bin Laden's claim to speak in the name of Islam.

PUTTING PUBLIC DIPLOMACY TO WORK

Such appearances enhance the West's ability to communicate a primary message: that the war against terrorism is not that of one civilization against another, but against terrorism and fanaticism in all societies. Western policies and actions are subject to public scrutiny and will often be misunderstood. Public diplomacy can significantly diminish this misapprehension. It may, however, involve some uncomfortable policy decisions. For instance, America may be forced to exert more diplomatic pressure on Israel to alter its methods of dealing with Palestinians.

Western public diplomacy in the Middle East also involves uncharted waters. As Oxford University social linguist Clive Holes has noted, the linguistic genius who thought up the first name for the campaign to oust the Taliban, "Operation Infinite Justice," did a major disservice to the Western goal. The expression was literally and accurately translated into Arabic as *adala ghayr mutanahiya,* implying that an earthly power arrogated to itself the task of divine retribution. Likewise, President George W. Bush's inadvertent and unscripted use of the word "crusade" gave Al Qaeda spokesmen an opportunity to attack Bush and Western intentions.

Mistakes will be made, but information and arguments that reach the Arab street, including on al-Jazeera, will eventually have an impact. Some Westerners might condemn al-Jazeera as biased, and it may well be in terms of making assumptions about its audience. However, it has broken a taboo by regularly inviting official Israeli spokespersons to comment live on current issues. Muslim religious scholars, both in the Middle East and in the West, have already spoken out against Al Qaeda's claim to act in the name of Islam. Other courageous voices, such as Egyptian playwright Ali Salem, have even employed humor for the same purpose.[5]

We must recognize that the best way to mitigate the continuing threat of terrorism is to encourage Middle Eastern states to be more responsive to participatory demands, and to aid local nongovernmental organizations working toward this goal. As with the case of Egypt's Saad Eddin Ibrahim, some countries may see such activities as subversive. Whether Arab states like it or not, increasing levels of education, greater ease of travel, and the rise of new communications media are turning the Arab street into a public sphere in which greater numbers of people, and not just a political and economic elite, will have a say in governance and public issues.

NOTES

1. It is now available on-line with explanatory notes in English. See <http://www.ciaonet.org/cbr/cbr00/video/excerpts_index.html>.

2. Hasan al-Turabi, "The Islamic State," in *Voices of Resurgent Islam,* John L. Esposito, ed. (New York: Oxford University Press, 1983), p. 245.

3. On the importance of rising levels of education and the new media, see Dale F. Eickelman, "The Coming Transformation in the Muslim World," *Current History,* January 2000.

4. Edward S. Walker, "The New US Administration's Middle East Policy Speech," *Middle East Economic Survey,* vol. 44, no. 26 (June 25, 2001). Available at <http://www.mees.com/news/a44n26d01.htm>.

5. See his article in Arabic, "I Want to Start a Kindergarten for Extremism," *Al-Hayat* (London), November 5, 2001. This is translated into English by the Middle East Media Research Institute as Special Dispatch no. 298, Jihad and Terrorism Studies, November 8, 2001, at <http://www.memri.org>.

DALE F. EICKELMAN *is Ralph and Richard Lazarus Professor of Anthropology and Human Relations at Dartmouth College. His most recent book is* The Middle East and Central Asia: An Anthropological Approach, *4th ed. (Englewood Cliffs, N. J.: Prentice Hall, 2002). An earlier version of this article appeared as "The West Should Speak to the Arab in the Street," Daily* Telegraph *(London), October 27, 2001.*

The Costs of Chaos in Palestine

Israel has launched a comprehensive war of attrition in the Occupied Territories, whose objective is a decisive military victory leading to prolonged interim arrangements dictated by Israel. Facing these overwhelming odds, the Palestinians remain plagued by a crisis of leadership that has already exacted a high price indeed.

Mouin Rabbani

As the Palestinian uprising against Israeli occupation enters its third year, the sense of crisis is pervasive. Strategic and tactical in nature, the crisis is military, political and socio-economic in character, and domestic, regional and international in scope. The depth of the Palestinian predicament suggests that the uprising could suffer a defeat that will take with it not only the limited achievements of the Palestinian Authority(PA)—circumscribed self-government in the West Bank and Gaza—but the more substantial ones of the Palestine Liberation Organization (PLO) as well. Even the prospect of mass expulsions of Palestinians is again a regular topic of discussion.

Such dire predictions are at present premature, but the continued absence of change will in due course make them all too real. The Palestinian leadership has failed to organize, mobilize and deploy the resources at its disposal in a way that could meaningfully influence the actions and policies of Israel, the Arab states and the international community. The Palestinians, decidedly the weaker party in the continued confrontation with Israel, can ill afford this crisis of leadership, which is extracting an increasingly—high and visible—price.

Strategic Predicament

From the outset, the uprising has been plagued by the absence of unified leadership, strategic clarity or tactical consistency. Instead, a variety of autonomous political forces simultaneously pursued differing tactics serving contradictory strategies, undermining each other in the process.

The PA, determined to achieve Palestinian statehood as part of a negotiated permanent settlement with Israel, simply shunned the command role normally exercised by political leadership in times of crisis. Unwilling to lead a rebellion it is unable to control, and seeking to avoid the domestic and international consequences of asserting formal command, it related to the uprising as an autonomous phenomenon of primarily tactical significance. At times the PA condones or encourages an escalation in fighting to improve its bargaining position, and at others it counsels or enforces restraint to demonstrate its authority. The PA is consistent only in claiming credit for any achievements and shirking responsibility for failures. The PA security forces, under strict instructions not to participate as an organized force except for purposes of territorial self-defense (and often not even that), covertly assist and overtly obstruct Palestinian paramilitary formations as the occasion demands. Ambivalent in attitude toward both the political factions and the population it rules, the PA has for all intents and purposes voluntarily disqualified itself from providing genuine leadership.

In visible contrast to the PA, the emerging generation of Fatah cadres seized upon the uprising as a strategy. By assuming leadership of the rebellion during its early stages, sustaining and militarizing it, Fatah transformed it into a war of attrition with which to vanquish the Israeli occupation and eclipse the discredited PA/PLO elite. By presenting itself as the main agent of Palestinian independence, Fatah also sought to reclaim from the Islamist movement its role as leader of Palestinian resistance to Israel. But Fatah could not subordinate either the PA or the other political factions to its agenda, nor did it sufficiently mobilize the civilian population—deficiencies that reflect the party's notorious fractiousness. Hence Fatah has neither the advantage of strategic hegemony nor the popular base required to prosecute a successful guerrilla war.

Hamas did not stand idly by and applaud Fatah's renewed militancy. Rather, the Islamist movement used the freedom created by the uprising to rebuild its infrastructure and launch an armed campaign that consciously went well beyond that espoused by Fatah. Where Fatah confined its attacks to the Occupied Territories, Hamas upstaged its nationalist rivals by retaliating against Israeli killings within the West Bank and Gaza Strip with bombings in Israel.[1] It was a tactic guaranteed to score with a Palestinian public enraged by the actions of the Israeli military, devastated by the tightening siege and left in the lurch by both its own leaders and the international community. Popular support emboldened Hamas to periodically derail diplomatic initiatives which would have strengthened the PA at its expense and, eventually, saw Fatah expand its own operations across the pre-1967 boundaries.

The timing of major Hamas attacks has led a number of Palestinian commentators to suggest that its military decision-making structures have been infiltrated by Israeli security. A more logical explanation is that escalations that systematically weaken the PA and steal Fatah's thunder are consistent with the agenda of the radical faction within the Hamas leadership, which believes the movement is prepared for power. More generally, Hamas is convinced that the systematic disruption of normal life within Israel—with its attendant socio-economic consequences—is the most effective method of forcing an end to the occupation.

War of Attrition

Such convictions notwithstanding, the June 1, 2001 Hamas bombing of a Tel Aviv discotheque turned the tables on both the PA and Fatah. Capitalizing on international outrage at an attack which killed 12 teenagers and 9 others—and confident in the knowledge that the world condemns Israeli casualties but only regrets Palestinian ones—the coalition government led by Ariel Sharon and Shimon Peres shed any remaining pretense of seeking a negotiated political settlement.[2] Instead, Israel launched a comprehensive war of attrition against the Palestinians, whose objective is a decisive military victory leading to prolonged interim arrangements dictated by Israel. During the following year, Israel's campaign of creeping escalation, assisted at critical junctures by deliberate foot-dragging and calculated provocations—and throughout by increasingly unqualified US support—transformed Sharon's agenda of eliminating the PA from an implicit threat to an explicit reality.

By July 2002, Israel had effectively destroyed the PA's civil and military infrastructure and reoccupied all West Bank territory under PA jurisdiction with the exception of Jericho. Washington had thwarted regional and international efforts to revive Middle East diplomacy, and adopted Israel's agenda of promoting "a new and different Palestinian leadership" whose primary function would be to ensure Israeli security. Then, as if the negotiations initiated in Oslo a decade earlier had never happened, Israeli officials proposed an agreement in which Palestinians would exercise their authority in Gaza and Bethlehem first.

Some presumed that the political fallout of the September 11 attacks in the US would seal the occupation's fate, but the Palestinians' incoherent response to the changed geopolitical reality frustrated their efforts to transform the Bush administration's unprecedented yet vague "vision" of a Palestinian state into a concrete program. Simultaneously, the crisis of Palestinian leadership facilitated Israel's efforts to fold its colonial war into the US "war on terrorism" and obtain explicit American support for the further consolidation of its rule. The Bush administration's approval of the prolonged occupation of West Bank PA enclaves was outflanked on the right by House Majority Leader Dick Armey's suggestion in May that Israel could expel Palestinians from the West Bank and Defense Secretary Donald Rumsfeld's repeated references in August to the "so-called occupied territories." Indeed, US and Israeli interests have become so intimately intertwined during the past year that increased political pressure on the Palestinians and the isolation of Syria are considered important motives for a US invasion of Iraq, while traditional US client states which have otherwise supported its efforts to destroy the al-Qaeda network are recast as adversaries on account of their antagonism to Israel.[3]

The Search for Consensus

Clearly, effective coordination between the PA, Fatah and Hamas is a precondition for any Palestinian attempt to overcome such overwhelming odds. To date, the most concrete attempt to achieve a Palestinian consensus has been the formation of the National and Islamic Forces (NIF). Established at Fatah's initiative during the early stages of the uprising and led by it since, this broad coalition of 14 political factions and several civic organizations coordinates the uprising in the spirit of the United National Leadership of the Uprising (UNLU) during the 1987–1993 *intifada.* In practice, the NIF has sought to exert strategic pressure upon the PA to reject partial or transitional agreements and respect the right of the factions to continue the uprising, tactical pressure upon the opposition to refrain from actions that undermine or endanger the PA and popular pressure through the organized mobilization of key sectors of the civilian population. However, because the PA does not accept the NIF as a supervisory authority, the factions do not consider themselves bound by its decisions, and the civic organizations represent institutional rather than popular interests, the NIF has achieved only limited success.[4]

Confronted with this dilemma, an increasing number of Palestinian activists, including Haidar Abd al-Shafi and the imprisoned Marwan Barghouthi, secretary-general of Fatah in the West Bank, have since early 2001 been calling for the formation of a national unity government. Such a government would combine the PA and NIF, and be empowered to adopt binding resolutions on political strategy and resistance tactics. Accepted in principle by its envisaged members, the national unity government has nevertheless failed to materialize. Yasser Arafat has proposed coopting the NIF factions into the PA cabinet by providing their leaders with ministerial posts, but the NIF has insisted that a common political program be formulated first.

Against this background, and with the assistance of European diplomats, in July 2002, Fatah brokered a preliminary understanding with Hamas, which among other items called for a termination of all attacks within Israel and a continuation of the uprising within the Occupied Territories. The initiative, which was more promising than the PA's unilateral ceasefires because it "originated in the [Fatah] *tanzim* itself,"[5] was nonetheless sabotaged by the Sharon-Ben Eliezer government, a mere 90 minutes before its scheduled proclamation on July 22, when an F-16 fighter dropped a one-ton bomb upon a residential building in Gaza's al-Daraj neighborhood, killing Hamas military wing founder Salah Shehada, 11 children and five others.

Efforts to achieve a national consensus on Palestinian objectives, strategy and tactics have since resumed, most visibly in Gaza under NIF auspices.[6] Given the depth of the current crisis, such efforts are likely eventually to bear fruit. If Hamas refuses to explicitly ratify the outcome, it will be presented the option of supporting it tacitly in order to keep it on board. Similarly, the factions will not be expected to indefinitely adhere to Israeli-Palestinian security arrangements independently negotiated by the PA that do not result in timely and visible progress. Given the recalcitrance of both the PA and Hamas toward a formal partnership, the available alternatives for a *modus vivendi*, and the growing field alliance between the various NIF factions, the establishment of a Palestinian coalition government, however, remains unlikely.

Yet, achieving consensus is only one element of the Palestinian challenge. Another is maintaining it in an environment of severe pressure upon the PA to either dismantled militant paramilitary formations or be dismantled itself, and extreme provocation by an Israeli government determined to crush the prospect for a coherent and purposeful Palestinian strategy— not least one facilitated by European diplomats that endorses continued attacks upon Israeli occupation forces.

The threat of such a collapse is all the more real in view of the current momentum in Israeli-Palestinian relations. Since its advent, the Sharon government has consistently followed a "strategy of blocking political negotiations until the fulfillment of a long and—in the view of many international observers— deliberately unrealistic list of demands."[7] Because the success of such incremental security plans depends upon a decisive confrontation between the PA and the factions, their failure can be reduced to "Arafat's refusal to confront terror" and thus serve as the launching pad for renewed Israeli escalation.[8] Either scenario leaves Israel strengthened and the Palestinians weakened. Indeed, from the moment the PA accepted the precedence of Israeli security interests through its endorsement of the May 2001 Mitchell Report,[9] the downward spiral of increasingly lopsided understandings facilitating increasingly comprehensive Israeli military offensives was all but inevitable. Today Palestinians are confronting the calamitous consequences of the PA's inability to withstand international pressure, and the NIF's failure to prosecute the uprising in a manner which does not place the PA on the defensive at key political junctures.

Consequences

With only the briefest of intervals, since March 2002 all West Bank PA enclaves except the town of Jericho have been fully reoccupied by the Israeli military. Israeli conduct during this period has been designed to reduce the PA to a formality, wipe out the factions and break the will of the Palestinian civilian population.

Although Israel has succeeded in thoroughly immobilizing the PA as a political institution and administrative apparatus, it has withheld the *coup de grace* for a number of reasons. Concerns about the international response are only secondary in this respect. More importantly, neither Israel nor the US has succeeded in cultivating a viable Palestinian alternative to Arafat. Faced with the prospect of a leadership vacuum being filled by a decentralized, cross-factional alliance of militant paramilitaries, they have opted to neuter the PA from within through a series of "reforms." The transparently political motives of Washington's sudden obsession with Palestinian democracy and good governance was most recently demonstrated when it pressed the PA to revise its system of government from the current presidential one, which it concluded only preserves Arafat's authority and legitimacy, to a parliamentary one leading to the indirect selection of a powerful executive prime minister.

Of even greater relevance from Israel's point of view is that it can exploit the continued existence of a Palestinian authority to avoid direct responsibility for the civil affairs of the Palestinian population. Thus, the Israeli military can pursue policies of prolonged, comprehensive siege and curfew that, according to the World Bank, USAID and others, have resulted in critical and unprecedented levels of unemployment, poverty and child malnutrition,[10] and put forward the claim that dealing with the consequences is the responsibility of the PA and international donors.

Given that residents of Nablus have been permitted out of their homes for less than 40 hours in a period of 65 consecutive days, that villagers remain cut off from urban centers where most vital services are located and that Palestinians are being arrested by the thousands while their homes are demolished by the dozens, it would be no exaggeration to state that in the summer of 2002 Palestinians are subject to a level of collective punishment unseen since Amram Mitzna, a new candidate for leadership of the Labor Party, was military commander of the West Bank during the first years of the 1987–1993 *intifada*.

Other practices that prompted strong international criticism during Operation Defensive Shield in the spring of 2002 have also persisted during the ongoing Operation Determined Path, though the critics are now more muted. On a regular basis, militants are hunted down and killed by Israeli forces in apparent extrajudicial executions. The Israeli army has continued to use Palestinian civilians as human shields, as exposed in Tubas on August 14 when Nidal Daraghma, 17, was killed during his enforced participation in an effort to eliminate his neighbor, Hamas activist Nasir Jarrar. Jarrar, missing three of four limbs and bound to a wheelchair, was killed either by artillery fire directed at his home, or when an armored Israeli bulldozer reduced its remnants to rubble with him still inside.

Israel has followed each of its failures to "end" the uprising by military force with the deployment of greater and more indiscriminate levels of violence. Palestinians, for their part, appear to be concluding that their own resort to systematic armed violence—particularly attacks which target civilians within Israeli cities—has blurred the character of their anti-colonial revolt in the court of international public opinion, and deprived them of crucial support in their confrontation with an otherwise more powerful adversary.

The future of the current uprising, and perhaps of the contemporary Palestinian national movement as a whole, is more likely than not to be determined by the ability of Palestinian leadership to formulate and implement a coherent strategy of resistance, based upon the premises that not all actions in support of a just cause are necessarily legitimate, and that not all legitimate actions are necessarily shrewd.

Endnotes

1. The first Hamas attack within Israel during the current uprising occurred on March 28, 2001, by which time Israel had killed 398 Palestinians (almost a third of them children, including 16 pre-teens). For Palestinian casualties see www.palestinercs.org; for Israeli ones consult www.mfa.gov.il.

2. Since September 28, 2000, the US government has not once explicitly condemned an Israeli killing of a Palestinian. This double standard extends, in varying degrees, to the UN Security Council, the European Union and even major international human rights organizations like Human Rights Watch.

3. See, for example, Jason Vest, "The Men from JINSA and CSP," *The Nation,* September 2, 2002; Brian Whitaker, "Selective Memri," *Guardian,* August 12, 2002; Brian Whitaker, "US Think Tanks Give Lessons in Foreign Policy," *Guardian,* August 19, 2002. In the words of Wesley Clark, former NATO commander in Europe: "Those who favor this attack now will tell you candidly, and privately, that it is probably true that Saddam Hussein is no threat to the United States. But they are afraid at some point he might decide, if he had a nuclear weapon, to use it against Israel." *The Independent,* August 21, 2002. The Nuclear Posture Review leaked in March characterized an attack upon Israel—a nuclear power with which the US has no formal defense treaty—as justification for a US nuclear response. *Los Angeles Times,* March 10, 2002.

4. A further complication is the occasional inability of the factions' political leadership—especially Fatah—to exercise authority over affiliated paramilitary formations.

5. *Al-Ahram Weekly,* August 1–7, 2002.

6. *Washington Post,* August 14, 2002; *Ha'aretz,* August 12, 2002; *al-Ahram Weekly,* August 15–21, 2002. While the Gaza deliberations are formally being undertaken by an NIF "Higher Follow-Up Committee," the discussions have included a number of prominent legislative council members and individuals associated with the PA and PLO as well.

7. *The Independent,* August 26, 2002.

8. In the words of one senior Israeli official, "Either they [the PA] fight against Hamas, or against us. There's no other way." *Ha'aretz,* August 25, 2002.

9. Mouin Rabbani, "The Mitchell Report: Oslo's Last Gasp?" MERIP Press Information Note 59, June 1, 2001. http://www.merip.org/pins/pin59.html

10. *Washington Post,* August 6, 2002.

Mouin Rabbani *is director of the Palestinian American Research Center in Ramallah.*

Kenya's Porous Border Lies Open to Arms Smugglers

By DEXTER FILKINS with MARC LACEY

KIUNGA, Kenya, Dec. 3—The smugglers who ply the empty expanse that spans the boundary with Somalia are thought to deal in a wide array of goods: lobsters, drugs, people, even missiles.

Kenyan authorities say they believe the surface-to-air rockets that were fired at an Israeli passenger jet in Mombasa last week, and possibly the bomb that killed 16 people in a nearby hotel, were smuggled across the border from the lawless state of Somalia. A senior Kenyan official said today that an intelligence warning had been sent to agents posted on the border in October, advising them to be on the lookout for something sensitive.

The official, who spoke on the condition of anonymity, said the border guards were not told what to look for, who the smugglers might be or whether the goods might be coming overland or down the coast. The border guards, many of whom Kenyan officials say work closely with area smugglers, made no intercepts.

"The border with Somalia is so porous, everything comes through," the Kenyan official said. "We think the missiles came across there."

In the days since the attacks, Kenyan officials have detained 10 men, including four Somalis, who were captured in a boat off the Kenyan coast. American and Kenyan officials have pointed to a Somali group called Al Itihaad, which is believed to have links to Al Qaeda. Both groups have been named as possible suspects in the attacks.

The attention on the illegal traffic across the Kenyan-Somali border casts a new light on the historic smuggling routes that form a 2,000-mile arc from Pakistan down the eastern coast of Africa to the Comoros Islands, between Mozambique and Madagascar. Western officials suspect that in recent months Al Qaeda operatives have used the routes to move guns and people around the region, taking advantage of the predominantly Muslim areas along the coastline.

If so, Al Qaeda would be continuing a pattern it established years ago. Kenyan officials say the bombs used in the 1998 attacks on the American embassies in Nairobi and Tanzania, which killed 224 people and injured 4,000, were smuggled in from Somalia.

The American military is stepping up its efforts to cut off the traffic in East Africa. It has begun outfitting the border post here with modern equipment like satellite telephones, and agreed to provide speed boats and four-wheel-drive vehicles as well. As one of its contributions to the war on terror, the German government regularly dispatches low-flying planes to track the movements of dhows, the wooden boats that meander up and down the coast here.

People who know the region well express skepticism that the American-led effort will penetrate the age-old smuggling network, which is sustained by tribal links and religious sympathies.

"They have been using these routes for hundreds of years, and they know every dip and cut in the coastline," said a Western aid worker, who has worked in

East Africa for 15 years. "Every one of them is a Muslim, and they only trust each other."

Of all the region's smuggling centers, the Kenyan-Somali border seems to highlight the difficulties in policing the illegal traffic in people and guns. The boundary, formed in 1925 by an agreement between what was then British East Africa and Italian-controlled Somaliland, cuts across 424 miles of trees and savannah to the shores of the Indian Ocean.

From the air, there is not a hint of activity by either government for more than 100 miles. On the land, there is the seemingly infinite expanse of the African bush; and by sea, the wide vistas of ocean. Up close, the boundary seems lost amid dense stands of palms and acacia trees.

"There is no border here, it is not even demarcated," said Capt. Felix Ndiga, a Kenyan naval officer stationed at Kiunga, about 15 miles from the Somali border.

Indeed, as Captain Ndiga drove his four-wheel-drive vehicle down a sandy road in search of the border, he drove right past it and into Somalia—a potentially hazardous mistake. Since the early 1990's, Somalia has been without a central government, with feuding warlords carving up chunks of the country. Just across the border here, the area is controlled by a band of fighters known as the Akhwan, an Islamic fundamentalist group believed by some Kenyan officials to have links to Al Qaeda.

At one point today, a truck full of Somali men rumbled through the Kenyan border town of Kiunga and rolled on into Somalia without stopping. Some of the men appeared to be wearing desert-tan American military uniforms, which may have been left behind in the ill-fated effort to restore order to Somalia in the early 1990's.

At the Kenyan border post at Kiunga, the local officials lack the tools to combat the smugglers, who, they say, are often armed with rocket-propelled grenades and heavy machine guns.

And it's unclear whether they would stop the smuggling if they could. The Kenyan official said many of the border guards in Kiunga appear to cooperate with the smugglers in return for cash.

"They have family on both sides of the border, and they let things come through," the official said.

Off the coast, the small wooden boats stream back and forth, some south towards Mombasa, the site of last week's

bombing, and others northward to the Somali border. No one but the sailors themselves seem to know what they are carrying.

Though many of the centuries-old sailing vessels look barely seaworthy, experienced sailors can travel from Kenya to Mogadishu in three days, darting in and out of the bays and waterways along the coast. And that is just with a sail. A small outboard can shorten the trip to 12 hours, said fisherman who have made the trip.

For months, the German Navy has been conducting surveillance flights out of Mombasa to track suspicious vessels. The spy planes have found dhows 50 miles or more off the coast, sharing the water with huge container ships. The 10 men in Kenyan custody on suspicion of being involved in last week's attacks were detained while on board a dhow that had come all the way from Pakistan.

The centrality of the dhow in the East African smuggling network has not been lost on American officials. This afternoon, two American investigators were spotted poking around Mombasa's old port, the entry point for many of the wooden boats.

Further north, on the Somali border, American officials are supposed to arrive in the next several weeks with speedboats, and perhaps four wheel-drive trucks, for the border guards there.

Yet for all the help that is on the way, even observers who want the effort to succeed are skeptical that it will make a difference.

"When people talk about a porous border, this is it," said a Western diplomat, who has visited the region. "You don't even know where the border is. It's as remote as can be."

"It's nowhere."

Africa's Security Issues Through 2010

by William Thom

Prospects for Africa over the next 10 years hinge on the continent's severe security problems. Peace is the foundation for Africa's future because all goals for development, plans for good governance and alleviation of human suffering depend on a secure and stable environment. South of the Sahara, Africa suffers from a vicious cycle of poverty, which contributes to criminal and political violence that inhibits investment and discourages economic development. One in three sub-Saharan states is currently experiencing some form of military conflict.

Abject poverty is at the root of many African conflicts, and the number of risk takers willing to take up arms to claim their piece of the meager economic pie is growing. The global communications revolution fuels rising expectations, and as Africans realize the depths of their poverty for the first time, they are losing patience with ineffective political leaders and traditional rulers—opportunities for economic advancement are painfully beyond their grasp. Poorly governed states with weak or uncontrollable armies face collapse.

Concern for basic safety is another factor. When a state can no longer protect its citizens, its primary reason to exist ceases; individuals will seek protection elsewhere. Insecurity fans ethnic, religious and regional animosities, even where differences have long been beneath the surface. When all else fails, individuals fall back on their tribal unit, encouraging the rise of warlords, often based on ethnic affiliations.

Another major change in Africa's security calculus has occurred in the aftermath of the Cold War: African countries are now setting their own security agendas. After more than 100 years of colonial domination and Cold War distortion, Africans are taking charge of events around the continent. Africans sense a waning security commitment from traditional external powers—their former colonial rulers and Cold War partners.

France's more constrained role recently as the self-styled "gendarme of Africa" is instructive. Paris's unilateral intervention in Rwanda in 1994 brought accusations that France had sided with the Hutu against the Tutsi. Two years later, when longtime French ally Zairian President Mobutu Sese Seko faced a serious rebellion supported by an alliance of regional states, Paris demurred. The inaction sent a message that there were new, more restrictive limits to French intervention in Africa.

Today's African leaders see a new freedom to act militarily. On the positive side, African states are more inclined to take responsibility for solving African security problems. In the post-Cold War era, some 20 countries have participated in peacekeeping and peacemaking operations on the continent, mostly on their own. On the negative side, this new freedom has also fostered military adventures that have complicated regional security problems.

Sub-Saharan Africa's position in the post-Cold War global security constellation is emerging. The continent has unfinished business from the Cold War and even the colonial period. In this land of mostly small internal wars, a limited military investment can potentially yield immense profits. Among the numerous weak states with poor armies and fragile institutions, even a small war can generate great destruction, as in Somalia and Sierra Leone. In 10 years Africa will likely still be at war with itself, continuing the process of nation-building, as relatively strong, stable states survive, and weak, hopelessly fractured ones do not. What follows are some key military themes that will help shape African realities over the next 10 years.

Warfare in the Era of Independence

Since the end of World War II, there have been three identifiable periods of warfare in sub-Saharan Africa. They span the spectrum of combat from guerrilla wars to coalition warfare, but with insurgency as a constant. During this period, an estimated 3.5 million soldiers and civilians have perished in African conflicts. The first period involved wars of liberation against the colonial powers, which extended well into the 1970s. These armed insurgencies against the remaining colonial powers were essentially low-budget, small-scale conflicts backed by communist powers. But, other revolts against colonialism did not align with the communist cause and—at least initially—did not receive significant support from Moscow. Examples from the 1950s and 60s include the Mau Mau revolt in Kenya, the early uprising in Angola and the Eritrean independence struggle. In Southern Africa there were wars of national liberation to end white-dominated settler regimes.

The second period involved the appearance of a few interstate wars and large-scale civil wars that were militarily significant, mostly conventional and politically galvanizing. By the 1970s a number of African states had developed armies capable of projecting power across their borders. The two best examples of African interstate conflict during this period were the Ogaden War between Ethiopia and Somalia (1977–78), and the Tanzania-Uganda War (1978–79). White-ruled South Africa pursued a forward-defense strategy during the 1970s and 1980s, which resulted in episodic combat with black-ruled states to the north. In Angola, however, Pretoria's apartheid government deployed conventional forces in strength to fight Angolan and Cuban forces. Two pivotal states where communist regimes had come to power in the 1970s—Ethiopia and Angola—faced large-scale civil wars in the 1980s. Communist powers poured in troops, advisors and billions of dollars of conventional weaponry in a vain attempt to preserve their perceived strategic gains in these two anchor countries. To balance the ledger, the West provided military assistance to professed anticommunist "freedom fighters" in Angola, and such anti-Marxist bulwarks as Zairian President Mobutu.

By the post-Cold War 1990s, however, a third period had emerged, one that points toward the next decade. The significant wars have once again become mainly internal contests fought at the unconventional or semiconventional level, leading to state collapse and wars of intervention. Easy to finance and difficult to defend against, guerrilla warfare—long the bane of Africa—remains its most prevalent form of conflict. Today's vicious insurgencies differ from yesterday's armed liberation movements in motivation: current struggles are based on power and economics, not a political cause or ideology. In weak states with unprofessional, underpaid armies, armed bandits become armed insurgents as they fill the power vacuum.

War in the 1990s became more destructive as internecine conflicts destroyed already fragile infrastructures. Today's African insurgents tend to be better armed and out number their 1960s predecessors. As the distinctions between guerrilla warfare and organized banditry blur, the targets often become the people themselves. Prolonged internal wars can destroy the fabric of the state and the society. On a continent where the majority of the population is no more than 15 years old, the communications revolution has highlighted the enormous gap between rich and poor. Youth without hope in dysfunctional nation states provide a ready manpower pool for local warlords; elsewhere, children are kidnapped out of villages by roving insurgent bands. The result can be young combatants socialized by an intensely violent right of passage, who begin to see banditry, murder and pillaging as normal behavior.

For African states the present is a time of experimentation with the uses and limits of applying military force. The next 10 to 20 years will bring polarized military power on the subcontinent and a small but growing number of strong states increasingly willing to use military force. Conventional wars will be fought over resources such as oil, other minerals, water and arable land, and to determine regional dominance. Armed insurgency will prevail in many of the weaker states, much as it does now, with regional powers or power blocs selectively intervening to protect their vital interests, often merely the capital and valuable resources in the interior. Eventually, power blocs will give way to dominant subregional military powers willing to engage in conflict, which will frequently take the form of peace enforcement and counterinsurgency.

An Uneven Balance

Nearly all postcolonial African armies began as colonial adjuncts to European armies and served primarily as tripwire forces in the colonies. As such, they were lightly armed and dependent on their colonial power for training, logistics and leadership. For example, the Kenyan African Rifles descended from the King's African Rifles. Over the past 40 years these armies grew to resemble, on a smaller scale, the forces of their colonial rulers or Cold War patrons.

Throughout this period, there have been great inequities in the military capabilities of African states. Until the mid-1990s, power imbalances have been held in check by the threat of intervention by powers external to Africa. During the Cold War in particular, these external powers intervened militarily to reverse adverse security trends or at least level the playing field. Soviets and Cubans intervened in Angola to balance South African intervention in 1975, and France worked to form a posse of African states to save the Mobutu regime in Zaire in 1977 and 1978.

By Western standards, today's African armies are still lightly armed, poorly equipped and trained, and dependent on external military aid. Nevertheless, a growing number of states, notably Nigeria, Angola, South Africa, Uganda, Rwanda, Ethiopia and Zimbabwe, are capable of using military force to pursue their own interests on the continent because of the gross inequities in raw military power. In a conventional scenario a country with a few operational jet fighters or attack helicopters and 30 armored vehicles backed by artillery has an immense advantage over a country that can oppose it with only light infantry units. Without an external or effective regional brake on their activities, emergent local powers can and will take the military option when they believe their vital interests are at stake.

Angola, for example, used its experienced army to intervene once in Congo-Brazzaville and twice in Congo-Kinshasa in the late 1990s to effect outcomes that it perceived as beneficial relative to its struggle with the insurgent Union for the Total Independence of Angola (UNITA). Nigeria managed to field a force up to division size in Liberia and then in Sierra Leone to pursue regional peace enforcement and its own hegemony in West Africa. Zimbabwe also deployed a division-sized force into the Democratic Republic of Congo (DRC), and South Africa (along with Botswana) sent troops into Lesotho to quell disturbances there. Uganda's army fought in three neighboring states in the 1990s—Rwanda, Sudan and the DRC. Rwanda has launched its forces into the DRC twice in recent years, and Ethiopia mobilized a force of 250,000 for its border war with Eritrea and continues to pursue hostile elements into the former Somalia.

The next few years promise little change in this military inequity. In 10 to 20 years the gap between the few dominant military powers and the rest of the countries will likely grow exponentially. Among the stronger states, large infantry forces will give way to smaller, more mobile forces with greater reach and firepower. The most capable states will maintain a variety of forces tailored for specific missions such as power projection, peacekeeping, peace enforcement and counterinsurgency. While the best sub-Saharan armies will grow more impressive, they will remain several generations behind the global leaders.

Regional Powers and Power Blocs

The original continental organization—the Organization of African Unity (OAU)—organized around the principle of decolonizing Africa. But it did not have a mandate to intervene as a regional military organization or adjudicate military disputes. Thus, in the post-Cold War period continental power blocs have begun to develop and act in conjunction with the OAU. They stem mostly from economic unions, the best example being the Economic Community of West African States and its military arm, the Economic Community of West African States Cease Fire Monitoring Group (ECOMOG). Dominated by regional power Nigeria, ECOMOG has served in Liberia, Sierra Leone and Guinea-Bissau, earning both respect and ridicule. Elsewhere, the Southern Africa Development Community, bolstered by South Africa, has assumed a regional security role, but its unity has been strained by sharp disagreement over Zimbabwe and Namibia's involvement in the DRC. On the Horn of Africa, the Inter-Governmental Authority on Development has engaged in diplomatic conflict resolution in Sudan but lacks any military cooperation among its members. The East African Cooperation—composed of Kenya, Tanzania and Uganda—has conducted joint military exercises. Some groupings appear to be ad hoc and temporary, such as the "Frontline States of East Africa" (Uganda, Ethiopia and Eritrea) which foundered when the Ethiopia-Eritrea border war erupted in 1998. The "Great Lakes Powers" of Uganda, Rwanda and Burundi have acted as an informal bloc in the DRC war, although tensions between Kigali and Kampala resulted in a shout-out at Kisangani in 1999.

Other groupings will likely emerge and some extant groupings rearrange themselves to accommodate changing national interests among members. Power blocs attempt to deal with collective regional security concerns as Africans see themselves increasingly on their own. They see viral forms of economic insurgency and highly destructive internal wars that disregard borders and appear out of control. Responsible leaders band together fearing that these conflicts, left unchecked, could destroy states and create pockets of complete lawlessness. The OAU, by its inaction, encourages the development of such subregional groupings. The OAU has only a token military mechanism, and prefers to endorse military interventions by others rather than take the lead itself. Recently, however, the OAU has shown signs of becoming more active by playing a prominent role in helping negotiate an end to the Ethiopia-Eritrea dispute and by sponsoring a joint military commission in the DRC.

Regional power blocs are only as solvent and effective as the powers that lead them. In sub-Saharan Africa few states are powerful enough to lead now. South Africa and Nigeria are the two best-known military leaders in the sub-Saharan region. Both face severe internal challenges but should maintain their roles as regional powers, and in the long run they have the potential to become continent-wide powers. Such a development could lead to recolonization by African powers although the context would be different from the European experience. Pretoria and Abuja, for example, could develop hegemonic tendencies; one could argue that Nigeria already has. Beyond these two countries, predicting other major developing powers is difficult. Among those that could emerge over the next decade or so are Kenya, Angola, Zimbabwe, Ethiopia and perhaps Senegal. Even small countries such as Rwanda and Eritrea have already shown an ability to project force and influence the local military balance.

A proving ground for budding regional powers will be peace enforcement missions and other military interventions in failed states. Peacekeeping may become a lost art in Africa in this century. Namibia and Mozambique have been relative UN successes, but Sierra Leone, Angola, Somalia and Liberia have shown limited returns for expensive peacekeeping ventures. Military interventions in collapsed states will continue, but they are apt to be police actions to ward off insurgents or multinational struggles for resources. The DRC case applies here. The imbalance in military capabilities will not be redressed over the next decade and will likely become more pronounced.

Arms Trade Trends

Arms acquisition is occurring on three levels—light arms, heavy stock-in-trade items and more sophisticated weapon systems. The extremely active trade in small arms and other light infantry weapons has captured international attention since the Cold War because they help fuel local wars around the continent. These light weapons include small arms, machine guns, rocket-propelled grenade launchers and small-caliber mortars—all man-portable.

These weapons have three principal origins. During the Cold War millions of assault rifles and other firearms were pumped into Africa, mostly by communist powers equipping "allies," notably Angola, Ethiopia, Mozambique, Somalia and Sudan. Rifles such as the AK-47 have become so numerous that they are regarded as a form of currency in some places. Second, in the post-Cold War era a brisk trade has developed, through middlemen, to acquire light arms from the former Soviet Union and other East European countries where such weapons are now cheap and plentiful. Third, a half-dozen or so sub-Saharan states manufacture light arms, and their production far exceeds their own needs.

Small arms are difficult to track, yet one commercial airliner can carry enough of them and their ammunition to start a guerrilla war. That is precisely why the trade in light weapons is so dangerous. The current glut of small arms in Africa should gradually contract over the next 10 to 20 years as the millions of small arms delivered in the 1970s and 80s age, become unserviceable and are not replaced in such quantities. Nevertheless, light arms will remain relatively easy to acquire and a major concern.

The trade in heavy weapons and large pieces of military equipment increased in the late 1990s with the growing number of conflicts on the continent and the unprecedented number of countries participating in military operations. Throughout 1998 and 1999 African armies deployed to other African nations 19 times, while 17 countries experienced significant combat on their territory. These deployments included armored vehicles, artillery, surface-to-air missiles, and combat and transport aircraft. These weapon systems, although not new to the sub-Saharan scene, are now frequently upgraded versions of old classics. The T-55 tank, for example, is now available with reactive armor, night vision equipment and the ability to fire antitank missiles from its main gun. MiG-21 and MiG-23 fighter-bombers are now frequently upgraded with better avionics, power plants, weapon suites and other performance enhancements. Other popular items of equipment in the 1990s include infantry fighting vehicles, hand-held surface-to-air missiles, multiple rocket launchers, and combat and transport helicopters—most of Soviet design. The next decade will likely see modest growth in the delivery of heavy weapons to sub-Saharan Africa. Although some observers consider armor and combat aircraft inappropriate for African wars, countries that have recently acquired them are shopping for more. For example, the T-55 is now a prime player in wars from the Horn to Angola, from Rwanda to Guinea. Mi-24 HIND attack helicopters are popular as a

counterinsurgency and close-air-support platform, and are used by a dozen African countries.

In the late 1990s a new generation of military equipment began to appear in the sub-Saharan region—much of it aviation. The Ethiopia-Eritrea border war has brought Su-27 and MiG-29 fighters, a first for the region. At least a few other countries, such as Angola and Nigeria, will probably acquire these and other new-generation aircraft over next the two to three years. Ethiopia has also received the 2S19 152mm self-propelled artillery system, a quantum leap in sophistication over the post-World War II designed artillery commonly found in Africa. With no Cold War restraints, African countries can successfully seek the next level of sophisticated weaponry.

How can African states afford these arms? The Cold War's military equipment grant aid and easy credit terms are over. The few large or wealthy African countries are understandably in the market for major equipment acquisition. But smaller, poorer countries, driven by perceived threats or the fact that they are already embroiled in a conflict, are also in the arms market. Imaginative financing, such as barter agreements and concessions, makes predictions about who can afford future arms highly speculative.

Black and gray market arms dealers further complicate the scenario as they increasingly replace the classic state-to-state arms deals. Most big-ticket purchases still happen through government agencies, and the dollar costs still overwhelmingly favor state-to-state transactions, but the business going to arms peddlers is increasing. This is a troubling development because the independent dealers are motivated strictly by profit, will sell to anyone—insurgents or governments—and care little about the consequences.

The Question of Privatization

The longstanding reliance on mercenaries will likely continue as African state and substate actors contract out military services to dramatically improve their capabilities. The privatization of state security functions provides African countries with a force multiplier—a cheaper, quicker, albeit controversial, solution for a flagging military. Contractors can be more responsive than states in helping a government. The South African firm Executive Outcomes (EO) was employed effectively in the mid-1990s in both Angola and Sierra Leone and is generally credited with helping to reverse the poor military postures of both governments. EO strayed into operations, however, bringing charges that it was merely a thinly disguised mercenary outfit. The difference between legitimate security contractor and illegal mercenary has blurred. In Africa, mercenaries are a loaded issue, yet many states see contractors as alternatives to Cold War security assistance programs.

State security functions are generally outsourced in the areas of training, advisory assistance and logistics (maintenance is a key deficiency in African militaries). Air transportation has become an especially critical area for privatization. Without contract air transport, many of the recent African engagements would not have been possible. In the current DRC war, air transport is considered the most costly expense for each side.

Security contractors cross the line and become mercenaries when they act as operators and fighters and not just as maintainers and teachers. They cross another line when they begin dealing with substate actors and not recognized governments. Security entrepreneurs may be increasingly willing to sell their services to insurgent movements, tribal militias, local warlords and even nongovernment organizations. While better-known security firms—such as MPRI and Sandline International—strive to foster a legitimate business image, other lesser-known, spin-off or free-lance groups are concerned only with the bottom line and will deal with just about anyone. It seems likely that private security activities will expand both above board and below. Security vendors selling to substate actors will further destabilize the region.

The new wave of interest in contracting and mercenary services stemmed primarily from arms dealers. When items are sold, package deals include trainers, technicians and advisors. From there it is a short leap to providing people to fight. While mercenary combat troops continue to show up occasionally in Africa, the next decade would seem to prize "technomercenaries," technicians who can keep equipment running and train the locals on how to use it, without actually pulling the trigger.

Prospects for Intrastate Wars

African military conflicts since the Cold War have again become almost exclusively internal affairs far more damaging to economic and social underpinnings than traditional interstate wars. The most prevalent forms of conflict in Africa are armed insurgency and civil war, with the latter often growing out of the former. Such unrest seems all but certain to persist over the next 10 years. The conditions that foster the development of economic insurgencies (extreme poverty, large pool of disenfranchised and disaffected youth, ethnic tensions and easy availability of arms) are likely to persist and may intensify. Dissident groups evolve from simple banditry to insurgent warfare as they become larger and more successful. Credos and manifestos are quickly manufactured to provide a fig leaf of political legitimacy. Eventually, insurgencies may become recognized as civil wars as the rebel chiefs acquire respectability as legitimate political leaders.

Almost all internal wars in Africa attract, or in some cases are created by, the meddling of outside powers. Every insurgency depends at some level on outside assistance, so internal struggles can be viewed as proxy wars disguised as internal conflicts. Weak states are vulnerable to collapse, and internal wars hasten the process. State collapse as defined here is not merely the failure of the machinery of government to work, as in Zaire under Mobutu; it is the complete breakdown of national government authority, as in Somalia under a gaggle of feuding warlords. National control disappears when the rot from within erodes the military to the point that it can no longer serve as the guardian of the state. Ironically, either unwise military downsizing, or worse, unwise rapid military mobilization, can exacerbate internal security problems. Armed groups opposing the government, or merely oriented toward self defense, fill the void left by receding state power and create ethnic, regional or social networks. In this regard, the expanding number of paramilitaries (armed militias, political factions and ethnic self-defense forces) contributes to instability by increasing the number of armed substate actors with their own agendas. Further, these groups are susceptible to foreign manipulation. This dangerous form of internal warfare, characteristic of the 1990s, will likely be a major problem in Africa throughout the next decade.

It also seems that solvent, functioning African states will selectively intervene militarily to control insurgencies that either threaten neighboring countries or harbor dangerous elements, such as terrorist groups and radical fundamentalist movements. Strong African states and the subregional bodies they dominate will increasingly recognize danger signs such as the subdivision of insurgent forces into warlord gangs, the manipulation of rebel groups by outside interests seeking to capitalize on conflict and the emergence of a criminal empire in a lawless environment. Over the next decade Western powers will recognize that Africa's internal wars, which destabilize some states and cause others to collapse, ultimately threaten their strategic interests as well. This lesson is not likely to be driven home, however, until some environmental or criminal disaster strikes that directly threatens Western interests.

Prospects for Interstate Wars

Wars between sovereign states in sub-Saharan Africa have taken place throughout the era of independence, but they have rarely been more than a regional concern. The Ogaden war between Ethiopia and Somalia gained notice because of the involvement of Cuban troops and Soviet advisors, but most interstate conflicts, like the five-day 1985 Christmas war between Mali and Burkina Faso, have been mere footnotes to modern African history. That may well change over the next 10 to 20 years as the militarily strong states attempt to stake out their areas of interest unintimidated by external powers.

A legitimate question is whether African states can afford to participate in interstate military contests. Countries in the Great Lakes region and on the Horn of Africa have shown a surprising and sobering ability to finance current military campaigns. Even in areas where oil, diamonds or other high-priced natural resources are not ev-

Current Conflicts

The DRC Civil War 1998–?

Status: Peace agreement signed, being violated by most signatories.
Type: Coalition civil war with extensive participation by foreign powers and substate actors.
Number of combatants: 120–140,000.
Displaced persons: 290,000
Significant formations: Battalion, company.
Casualties: 20–27,000 (mostly civilian).
Tactics: Semiconventional (a mix).
Foreign involvement: Zimbabwe, Angola, Namibia, Chad and Sudan for the government; Rwanda, Uganda and Burundi for the rebels.

Angolan Civil War 1998–?

Status: Lusaka Protocol violated by both sides, ongoing conflict.
Type: Latest phase of long-running civil war.
Number of combatants: 150–180,000.
Displaced persons: over 1.4 million.
Significant formations: Brigade, regiment and battalion.
Casualties: Unknown (mostly civilian).
Tactics: Primarily conventional.
Foreign involvement: Private contract military assistance (both sides).

Sierra Leone Civil War 1991–?

Status: Lusaka Protocol violated by both sides, ongoing conflict.
Type: Coalition civil war with extensive participation by foreign powers and substate actors.
Number of combatants: 30–40,000.
Displaced persons: 600,000+.
Significant formations: Battalion and company.
Casualties: over 10,000 (mostly civilian).
Tactics: Semiconventional (a mix).
Foreign involvement: West African force (headed by Nigeria) transitioning to a UN peacekeeping force for the government; Liberia and private contractors for rebels.

Ethiopia-Eritrea War 1998–?

Status: OAU/UN peace negotiations stalled, temporary lull in fighting.
Type: Large-scale border war.
Number of combatants: 400,000.
Displaced persons: over 400,000.
Significant formations: Division, brigade and battalion.
Casualties: 30–45,000 killed (military).
Tactics: Conventional.
Foreign involvement: Contract personnel on both sides but primarily in Ethiopia.

ident, countries find ways to pay for heavy, modern weapons. Financing African conflicts, especially conventional interstate wars, remains problematic, but the lack of resources is no reason to rule out future interstate wars.

In the sub-Saharan environment, a growing number of states have the raw military capability to engage in interstate wars, even when they do not involve an adjacent country. Contract air transport has revolutionized warfare in Africa by giving countries strategic reach. Further, many of Africa's new dynamic leaders, such as Ugandan president Yoweri Museveni and Rwandan president Paul Kagame, who came to power by force of arms, tend to view military power as a legitimate—even preferred— tool of statecraft. Additionally, some old-line rulers, such as Angolan president Jos Eduardo dos Santos and Zimbabwe's president Robert Mugabe, also see flexing military muscle as an acceptable way to do business.

As regional powers become more active in the next decade, and their strategic interests become well defined, occasional interstate wars loom. While intrastate conflicts will remain the principal form of warfare, interstate warfare will be more likely than in the past 40 years. Some conflict may take the form of coalition warfare, such as that now under way in the DRC. Others will be more tra-

ditional one-on-one contests such as the Ethiopia-Eritrea war. The battle of wills and principles driving that dispute serve as a reminder that many wars are fought for symbolic and moral purposes. More future interstate wars in Africa are, however, likely to be fought over scarce or vanishing resources—and not just high-value commodities such as oil and diamonds. Water, fisheries, arable land and ethnic solidarity will be among the root causes of interstate wars. Borders established by the colonials will continue to become less relevant and more easily altered by Africa's emerging power structure.

Wars in Africa will stem from acute poverty and a sense of hopelessness among its burgeoning population, especially alienated young men. Fed by rising expectations stemming from increased media exposure, these wars will be primarily internal and unconventional. They will exact a high price on the people, the fragile infrastructures and the foundering states themselves. More states will collapse, be propped up by external powers from within Africa or be patrolled by international peacekeepers.

The disparities in military power on the African continent will become even greater. Emergent local powers and power blocs will be the significant military actors on the continent. As great powers limit their involvement, these emerging powers will pursue their own agendas that by 2010 will change Africa's political map.

The current scope of African military conflict is unprecedented. In the late 1990s sub-Saharan Africa may have entered into a "Thirty Years War," a metamorphic process that will profoundly change the continent. In some corners of Africa, the fires of war will remain difficult to extinguish for another reason: they have gone on for so long that they have attained a sense of normalcy. Entire generations in places such as Angola, Eritrea, Liberia and Somalia have grown up knowing nothing other than war.

In Africa, as elsewhere, transnational criminality and war will become virtually indistinguishable. Economic insurgents, warlords for profit, lawless zones harboring criminals, armies of child soldiers and brutalized civilians will all offend the moral senses of Western nations and seemingly demand a response. Policing these messy situations will become an international priority. Nevertheless, some places will remain beyond the reach of Western moral consciousness and continue to experience low intensity conflict indefinitely.

The next two to three years do not portend much change in African security, but by 2010 Africa's political relief map will likely show stark differences. Islands of stability may be built around relatively strong and prosperous states such as South Africa, Kenya and perhaps Nigeria. In countries riven by insurgency and facing collapse, international forces protecting the capital may in effect create city-states. Elsewhere, local powers will demonstrate hegemonic interests, and geographic boundaries will reflect the continent's new political order.

William G. Thom is the Defense Intelligence Officer for Africa, Defense Information Agency, Washington DC. He received a B.S. from the State University of New York at New Paltz and an M.A. from American University. He is a graduate of the US Army War College and the National Security Leadership Course. He has held a variety of positions in the DIA, including chief, Africa Military Capabilities Branch; senior analyst, Directorate for Estimates; and Africa analyst, J2. He has also served as senior analyst for Africa, J2, Headquarters, US European Command, Vaihingen, Germany.

From *Military Review*, July/August 2000. © 2000 by *Military Review*.

UNIT 10

International Organizations and Global Issues

Unit Selections

Key Points to Consider

- What specific actions would you recommend that the United Nations take to diffuse the current conflict in Iraq and to ensure that the government in power in Iraq in the future does not attempt to resume covert research on weapons of mass destruction?

- Explain why you agree or disagree with Henry Kissinger's negative assessment and lack of support for the new International Criminal Court.

- Which one of the networks described by Moisés Naím related to terrorism, drugs, arms, intellectual property, people, and money is the most destabilizing for world stability?

 **Links: www.dushkin.com/online/**
These sites are annotated in the World Wide Web pages.

Commission on Global Governance
http://www.sovereignty.net/p/gov/gganalysis.htm

Global Trends 2005 Project
http://www.csis.org/gt2005/sumreport.html

InterAction
http://www.interaction.org

IRIN
http://www.irinnews.org

The North-South Institute
http://www.nsi-ins.ca/ensi/index.html

Uniited Nations Home Page
http://www.un.org

The most visible international institution since World War II has been the United Nations. Membership grew from the original 50 in 1945 to 185 in 1995. The UN, across a variety of fronts, achieved noteworthy results—eradication of disease, immunization, provision of food and shelter to refugees and victims of natural disasters, and help to dozens of countries that have moved from colonial status to self-rule.

After the Gulf War in the early 1990s, the United Nations guided enforcement of economic sanctions against Iraq, sent peacekeeping forces to the former Yugoslavia and to Somalia, monitored an unprecedented number of elections and cease-fire agreements, and played an active peacekeeping role in almost every region of the world. However, the withdrawal of the UN mission in Somalia, the near-collapse of the UN peacekeeping mission in Bosnia prior to the intervention of NATO-sponsored troops, and the delayed response of the UN in sending troops to monitor cease-fire agreements in East Timor and Sierra Leone in 1999 raised doubts about the ability of the organization to continue to be involved in peacekeeping worldwide. Some observers now call for the United Nations to scale back its current level of peacekeeping in order to focus more effectively on global problems that nobody else can or will tackle.

Pressures to reduce UN peacekeeping efforts are also fueled by the realities of scarce resources created largely by the refusal of many member states, including the United States, to pay back dues. The United States withheld payments of back dues for several years as part of a campaign led by conservative members of the U.S. Congress. They demanded that the United Nations undertake extensive internal reforms and reduce the amount of the U.S. contribution before the United States would pay its back dues. The United States narrowly averted losing the right to vote in the General Assembly by agreeing at the end of 1999 to pay a portion of its $1.02 billion debt over a 3-year period. At the end of 2000, UN and U.S. negotiators reached agreement on a deal that includes a large reduction in the dues that the United States pays. How much the terrorist attacks in September 2001 changed the political landscape in the United States was in evidence at the end of 2001. As the United States made the second of three installments to pay off the U.S. debt to the United Nations, there was no criticism of the $582 million outlay. Congressional opposition to paying for the United Nations melted away in the face of the country's need for international support in its war on terrorism.

The importance of having the support of major donor nations was illustrated once again as the United Nations met to discuss reconstruction aid for Afghanistan. Japan, still suffering from a prolonged recession, reduced its contribution to all UN programs except Afghani reconstruction. Donor nations met in Tokyo at the beginning of 2002 to work out the final details of a 5-year reconstruction plan. The primary priority of the plan was to ensure that economic and political aid go hand in hand. The long-range coordinated plan calls for additional reforms to ensure that the links between the United Nations' pre- and post-conflict work, peacekeeping activities, and development programs are strengthened.

Increased pressure by many nation-states contributed to the Bush administration's decision to seek a mandate from the United Nations Security Council that demanded that Saddam Hussein readmit and cooperate with UN weapons-of-mass-destruction inspectors or risk a U.S.–led military invasion backed by the support of the international community. In "Who Needs the U.N. Security Council?" James Traub discusses how useful the recent Security Council resolution that requires Iraq to accept a team of UN weapons inspectors and disarm was for all parties. For the Bush administration the UN resolution provided needed international cover while everybody else supported the resolution as a means to rein in the United States.

Tensions among member states over dues allocation and the proper role of the United Nations reflects more fundamental disagreements over the meaning of international security. For many, the United Nations has not lived up to expectations in securing a disarmed and peaceful world. Especially since the genocide in Rwanda in 1994, there have been several efforts to obtain justice at the international level for the worst war crimes. The lack of a global consensus on what a broader definition of security would entail and who should be responsible for guaranteeing universal human rights remains a core dispute in international relations and law. As Henry Kissinger notes in "The Pitfalls of Universal Jurisdiction," the recent establishment of a permanent international criminal court is one of the most controversial developments in the field of international law. Kissinger warns that its creation substitutes the tyranny of judges for that of governments. This view mirrors the view of President Bush and most of his senior advisers, and may help to explain why the United States vigorously opposes the new international institution.

Disagreements also extend to other issue areas where analysts, policy makers of nation-states, and international and nongovernmental organizations disagree about who is the most appropriate actor to tackle a variety of broader transnational problems often called "human security" issues. In "The Five Wars of Globalization" Moisés Naím describes how in addition to terrorism, governments today are fighting several other networks involved in drugs, arms, intellectual property, people, and money. Naim argues that governments will lose these wars until they adopt new strategies to deal with a larger, unprecedented struggle that now shapes the world as much as confrontations between nation-states once did.

Several societies in the world today are also struggling to cope with the extreme social dislocations caused by continued poverty and an unprecedented death rate caused by the AIDS epidemic among adults. After years of ignoring the problem, the United Nations finally started to play a lead role in 2000 in fighting AIDS at the global level. In 2001 the secretary general of the United Nations, Kofi Annan, won the Nobel Peace Prize for his work on human rights, poverty, AIDS, and terrorism. Annan helped to raise the profile of several new global scourges including AIDS and international terrorism as part of a broader restructuring of the 56-year-old institution.

Kofi Annan also headed the first UN General Assembly special session on AIDS. This meeting resulted in the establishment of a new Global Fund to Fight AIDS, Tuberculosis, and Malaria. The Bush administration was the first nation-state to pledge funds to this account. The U.S. pledge of $500 million over 2 years is the largest one to date. But when measured by the size of the U.S. economy, it means that the United States has committed to giving about half as much as Europe. Most health experts agree that the United States' contribution would have to be at least $2.5 billion a year to make a difference to the current HIV/AIDS pandemic. The pledges of other countries to the new fund are also small compared with the magnitude of the problems. Nevertheless, many observers view the fund's establishment as an important start on a long-overdue recognition that the emergence of infectious diseases is a global problem.

The HIV/AIDS pandemic and the resulting dislocating effects it has caused on citizens' health, food production, the economy, and the stability of society is painfully evident in Africa today. Even though the HIV/AIDS pandemic has not yet peaked in Southern Africa, some countries such as Swaziland and Botswana are report devastating rates of infection of up to a quarter of their adult population. The devastation currently being experienced by several countries in southern Africa reflects the fact that the HIV/AIDs pandemic is the worst infectious disease pandemic since the black plague swept through feudal Europe. Both created a series of interrelated and unfamiliar problems.

Who Needs the U.N. Security Council?

1) The Bush administration, seeking international cover to do what it wants, and 2) everybody else, seeking to rein in the United States. Welcome to the New World Order.

By James Traub

The Mission of France to the United Nations is located on the 44th floor of a high-rise building on East 47th Street, just a few blocks from the U.N. itself. When I went to visit Jean-David Levitte, France's ambassador to the U.N., in the midst of the tortured and enormously protracted negotiations over a resolution requiring Iraq to accept a team of U.N. weapons inspectors and disarm, Levitte drew me to the south-facing plate-glass window of his office and delivered an ever-so-slightly defensive speech on Franco-American amity. "I watched the World Trade Center buildings come down from here," he said. "At that time, France was president of the Security Council, and the very next day, Sept. 12, we introduced Resolution 1368, which, for the first time in U.N. history, described a terrorist attack as a threat to international peace and security"—and thus gave the United States an unequivocal right to retaliate against Al Qaeda and against the Taliban regime in Afghanistan.

The campaign against Al Qaeda represented one of those rare moments when the Security Council swings quickly behind American aims. The U.N. itself felt implicated in the terrorist attack: its headquarters was evacuated both that day and the next, and there was brief talk of holding a Security Council meeting in a local coffee shop. But the moment of solidarity couldn't last. For the Security Council, Afghanistan was a momentary departure from a tradition of conflict resolution; for the Bush Administration, it was the first battle in a global war.

It is not only the United States but also the United Nations that has become a different place after 9/11. Only yesterday, it seems, the great issue was getting an increasingly disengaged United States to pay its back dues and pay attention; now the problem is keeping an aroused America from sallying off on what virtually every other member of the Security Council considers a reckless crusade. The Security Council needs the United States in order for it to play a meaningful role in world affairs, but it appears as though the United States doesn't need the Security Council—or at least that many of the leading members of the Bush administration think that it doesn't. Secretary of State George Marshall had predicted in 1948 that should there be "a complete lack of power equilibrium in the world, the United Nations cannot function successfully." And now, for the first time since the U.N.'s establishment, that state of affairs has come to pass.

As the Brazilian ambassador to the U.N. put it archly, 'You have a situation of dual containment: you have to contain the United States; you have to contain Iraq.'

And so the resolution on Iraq has been the first test case of the new world of American supersupremacy. As Gelson Fonseca, the Brazilian ambassador to the U.N., put it archly, "You have a situation of dual containment: you have to contain the United States; you have to contain Iraq." Containing the Bush administration has meant finding a middle ground between rubber-stamping American policy—and thus making the council superfluous—and blocking American policy, and thus provoking America to unilateral action, which of course would make the council irrelevant. Fonseca seemed to feel that containing the U.S. is a harder job than containing Iraq, and possibly a more important one.

As I prepared to leave after talking to Levitte, he said, "You know, I told you this story about Resolution 1368 because I want you to understand something: we want the United States in." Of course they do. But what neither Levitte nor his colleagues know for sure is what it will take to keep the Bush administration in. Last week, the Security Council unanimously passed a resolution demanding that Iraq allow the weapons inspectors who left four years ago not only to return but also to conduct more rigorous inspections for signs of biological, chemical and nuclear weapons production. If Saddam rebuffs the inspectors, there will almost surely be war. If he meekly complies, there probably won't. But Saddam may decide to test the fragile entente now reigning in the council by complying well enough for the French and the Russians but not well enough for the Americans. And then… who knows?

The central question posed by the debate over Iraq remains: Is the blessing of the international community so valuable a good that even this administration, at this moment of American power, is prepared to sacrifice something of its freedom of action in order to secure it? And if it is not, what, exactly, is the Security Council for?

The Security Council has for many years been a dim shadow of what it was intended to be by the architects of the U.N. system. In "F.D.R. and the Creation of the U.N.," Townsend Hoopes and Douglas Brinkley argue that from the earliest days of World War II, President Roosevelt foresaw a new world order governed by what he called "the Four Policemen"—the United States, Great Britain, Russia and China. The failure of the post-World War I League of Nations had made Roosevelt skeptical of the merits of a world body, but by 1942, Sumner Welles, his under secretary of state, had drawn up a proposal for a "United Nations Authority" with a "security commission" of the four policemen, who would provide the forces needed to quash threats to world peace. At Teheran in November 1943, Roosevelt persuaded Stalin to accept a single, centralized body consisting of all the world's nations and governed by a council dominated by the big four. (France was added later.) Secretary of State Cordell Hull addressed a joint session of Congress and magnificently asserted, "There will no longer be need for spheres of influence, for alliances, for balances of power."

The Security Council, then, was a new system, designed to prevent another 1914 or 1939, in which the most powerful nations would exercise an effective monopoly on force. Sir Brian Urquhart, one of the first employees of the United Nations and later one of its most important chroniclers, says: "We got into World War I owing to a kind of ludicrous diplomatic folk dance that didn't pan out, and there was no international delay mechanism, no breakwater to stop this rush to war; and that's what they set up the League of Nations to prevent. Then in 1939, you had a war caused by unchecked aggression. And so the new side of the U.N. as opposed to the league is that it provides a mechanism for taking action." The Security Council, which would consist of the five "permanent members" as well as 10 other members who would rotate on and off, was intended to serve both as a "delay mechanism" and, should deliberations fail, as an enforcement body.

The United Nations Charter, drawn up in San Francisco in the summer of 1945, makes amazing reading today, when American conservatives talk about signing on to U.N. treaties as a surrender of national sovereignty. Chapter VII deals with "Action with respect to threats to the peace, breaches of the peace and acts of aggression." Article 42 of Chapter VII empowers the Security Council to "take such action by air, sea or land forces as may be necessary to maintain or restore international peace and security." Article 43 requires U.N. members to make armed forces and "facilities" available to the body "on its call." And Article 47 establishes a "Military Staff Committee," consisting of the five permanent members' chiefs of staff, which would be responsible for the "strategic direction of any armed forces placed at the disposal of the Security Council." What is perhaps even more amazing, from our own perspective, is that Congress passed the U.N. Charter almost without debate.

Nevertheless, it was clear even at the time that the five permanent nations, soon to be abbreviated as the P5, might not permit their foreign policies to be directed by the Security Council. Stalin insisted on veto power for each of the P5 members, as did, if less vehemently, the United States. And then the cold war settled in, and each side used its veto, or the threat of one, to check initiatives dear to the other. Urquhart recalls that the U.S. had agreed to make a "very substantial force" available for Chapter VII enforcement actions, but that the entire arrangement was scuttled in the late 40's when Stalin balked at the idea. Roosevelt's dream of a global police force led by the great powers died before it could even be tested.

But something else was happening at the same time. The process of "decolonization" was unfolding much faster than anyone had expected, and new nations were pouring into the United Nations. Chapter VII was already a memory by the time most of these third-world nations joined, and in any case, the new members were more concerned with economic development than with peacekeeping. Starting in the 1950's, the U.N. began to spawn a whole range of agencies largely directed at the needs of the new members—bodies dealing with health, food, education, relief, refugees and so on. And the culture of the institution drifted further and further from the muscular principles of Chapter VII. This is the domain most of us associate with the U.N.—high-minded confabulations on intractable global problems, solar-powered cookers, declarations on the rights of historically oppressed communities, etc. This sense of an organization preoccupied with terribly important things it can't actually do very much about has not done much for the U.N.'s reputation, at least in the United States.

But the U.N. does not, of course, belong to the United States. David Malone, a former Canadian diplomat who runs the International Peace Academy, a research group far more hardheaded than its fuzzy-sounding name implies, notes: "The U.N. in its own mind is largely about a positive agenda that the agencies deliver. Many members aren't comfortable with anything beyond the positive agenda. They view the fight against bad guys and evil as incompatible with the ethos of the organization and being conducted at the behest of a few big powers they don't trust." By the end of the cold war, 90 percent of the U.N.'s resources were being devoted to the agencies; peacekeeping had become a vestigial activity, carried out largely in quiet places like Cyprus.

AND THEN CAME the U.N.'s very own Prague Spring. The ideological deadlock of the cold war was melting away, and with it the constraints on the Security Council. Saddam Hussein came along just when the Security Council was ready to deal with him. When Iraq overran Kuwait in 1990, the council passed a resolution of condemnation. Once the first President Bush decided on a war, he sought a Security Council resolution before he went to Congress, calculating, correctly, that the authority of the council would pressure the Congress into approval—the exact opposite of the thinking that has governed the current Bush administration. And the council, after a very brief debate, invoked Chapter VII to authorize a coalition force to dislodge the Iraqi forces.

The war was, of course, a great success, and an excited President Bush became the tribune of a new U.N. Veterans of the organization still recall with wonder the speech in which Bush offered to turn over Fort Dix for the training of U.N. peacekeepers. In June 1992, Secretary General Boutros Boutros-Ghali, acting at the behest of the permanent five, submitted "An Agenda for Peace," which argued that the time had come to rejuvenate U.N. enforcement and to make the Security Council the genuine global peacekeeper envisioned by Chapter VII. It was a moment of euphoria throughout the U.N. community. There was a feeling, as Shashi Tharoor, the U.N.'s head of public information and one of its chief institutional voices, recalls, that "every problem can come to us, and we can prescribe a solution." Boutros-Ghali, Tharoor told me, "even spoke rashly of 'a problem of too much credibility.' Within two and a half years he was eating those words, because we had a crisis of too little credibility."

What happened, of course, was Srebenica—a byword for moral failure. This fiasco virtually discredited U.N. peacekeeping. Perhaps the fault lay less with the U.N. than with the nations of the Security Council, which were unwilling to send the kind of troops needed to counter the savagery of the Bosnian conflict. Perhaps the fault lay with the expectations themselves. William Shawcross wrote an account of peacekeeping efforts titled "Deliver Us From Evil." That was the job description; but peacekeeping could not deliver the world from evil.

No one knows the answer to the peacekeeping problem. If you ask people around the U.N. what will happen the next time genocide threatens—say, in Burundi—the answer will be, "We'll authorize a regional organization to go in." Few people expect a U.N. force of "blue helmets" to venture into mayhem. And if you point out that no "regional organizations" happen to exist in the neighborhood of Burundi—well, that's true.

THROUGHOUT THE LATE 90's, conservatives like John Ashcroft accused the Clinton administration of seeking to "subcontract" American foreign policy to the U.N. The Bush administration took office vowing never to fall into that trap. While still an adviser to the campaign, Condoleezza Rice, later Bush's national security adviser, criticized the belief that "the support of many states—or even better, of institutions like the United Nations—is essential to the legitimate exercise of power." Legitimacy, for at least most members of the Bush foreign-policy team, arose from a clearheaded assessment of national interest; little could

be expected from the U.N., a moralistic body squeamish about the exercise of power and largely hostile to American interests. Only a few in the administration, most notably Secretary of State Colin Powell, believed that acting in concert with others was such a good in itself that the U.S. should seek to do so whenever possible.

Describing the Security Council, the French U.N. ambassador said, 'The Americans tend to consider that there is somewhere a supreme power imposing its will on America.'

Exhibit A in the indictment of U.N. pusillanimity handed up by the hawkish wing of the Bush team was policy toward Iraq. In the aftermath of the gulf war, the Security Council established the combination of harsh economic sanctions, "no fly" zones and weapons inspections that more or less kept Saddam Hussein in his cage. But by 1998, inspectors were complaining of constant interference, and in December of that year, the United States and the British insisted—over the objections of the French and the Russians—that the inspectors be withdrawn, and then mounted a bombing campaign directed at Iraqi weapons sites. Since that time, the Security Council had seemed to be much more disturbed by the consequences of confronting Iraq than by the consequences of failing to do so. The big issue in the Security Council last year was not Saddam Hussein's weapons of mass destruction but the harm done by the sanctions. Russia, Iraq's chief ally among the P5, argued that Iraq had fully disarmed itself of nuclear weapons and began agitating to have the "nuclear file" closed and Iraq be given some benefit for its compliance.

Until Sept. 11 of last year, the Bush administration officially viewed Iraq less as a threat to world peace than as a hopeless mess; Secretary of State Colin Powell had even taken up the call for less burdensome "smart sanctions." After the attacks on the World Trade Center and the Pentagon, the Bush administration, worried about Saddam making chemical and biological and nuclear weapons and perhaps allowing them to slip into the hands of terrorists, was prepared to disarm him by any means. When Bush addressed the General Assembly two months ago, he asserted provocatively that the U.N. would turn into a League of Nations if it continued to permit Saddam Hussein to keep inspectors out in defiance of U.N. resolutions.

When I suggested to Levitte, the French ambassador, that the Security Council would never have acted to restore inspectors to Iraq without the American threat of war, he said: "This is absolutely true. But let's be fair. What is the Security Council? The Americans tend to consider that there is somewhere a supreme power imposing its will on America. It is 15 members; it is not a kind of supreme power. And the United States is first among equals."

And that is why the president's bristling speech provoked nervous applause, a few snide jokes and, as David Malone of the International Peace Academy says, "an instant refocusing on Iraqi noncompliance." The threat of war, of course, concentrated many minds, but it is also true that the United States can determine the agenda of the Security Council if it wants to, even if it cannot quite dictate outcomes. That has not always been true, but the era of ideological opposition to the United States, whether from the Soviet bloc or the third world, has largely come to an end; it has been a long time since anyone called for a "new world economic order" to redistribute wealth to poor nations. There is a widespread acceptance in the U.N. of the fact of American supremacy, even if high-handed American behavior has left a deep residue of resentment.

It is, in fact, precisely because the U.N. system cares so much about preserving its own relevance that the members have generally resigned themselves to the new American hegemony. Even the Security Council's most zealous defenders recognize that it must be a "mirror" of world power in order to remain effective. And so council members have tended to swallow their anger over American arrogance. "With the amount of hair we tear out of our heads over the United States, we should all be bald," a representative of one of America's allies put it. You often hear expressions of gratitude over even minimal American gestures of respect for the U.N., like Shashi Tharoor's comment that "in a case where the United States is acting in the name of the international community, the fact that it has come to seek the blessings of the international community is welcome."

Diplomats are trained to accept reality and work with it—to play a weak hand as strongly as they can. Whenever I think of the U.N., I picture Gelson Fonseca, the Brazilian ambassador, a charming, bushy-browed, multilingual gentleman possessed of a strikingly sinuous mind. "Diplomacy is about an illusion," Fonseca told me recently, "the illusion that I am equal to you. When I sit across the table from America, I have the illusion that I can convert you by the force of my argument, and you, in order to sit across the table from me, must share that illusion." And yet because it is an illusion, Fonseca accepts the fact that the United States cannot be successfully confronted; it must rather, he says, be "seduced"—a very Brazilian approach to diplomacy. What Fonseca means is that diplomacy in the unipolar world must be devoted to coaxing the United States inside the system of rules and institutions. Like many of his colleagues, he has a tremendous fear of a United States freed from those constraints; he would almost rather surrender to American wishes than see the U.S. dismiss the Security Council. "Order," Fonseca says, "is better than justice."

AND SO, as the first debate in this new era of American dominance got under way in September, the Security Council was prepared to play a weak hand strongly; but no one could say for sure whether the Bush administration was willing to play at all. The administration had come grudgingly to the negotiating ta-

ble, and the initial draft resolution circulated by the Americans included elements so manifestly intolerable to the Iraqis—like the idea of a military force supporting U.N. inspectors—that it appeared to be designed to be repudiated, and thus to bring about the war that Vice President Dick Cheney, Defense Secretary Donald Rumsfeld and other leading figures in the administration were strongly suggesting was the only alternative. This was too great a price for the U.N. to pay. At a gathering in mid-September, a group of diplomats and top U.N. officials asked themselves whether they should simply accept the invitation to rubber-stamp American policy. "The preponderant view was that this was not better," says Paul Heinbecker, the U.N. ambassador from Canada, which is not currently a Security Council member. "The U.N. has to stand for something; it has to have some principles."

It was left to the French to state those principles. No major nation cherishes the illusory equality of the U.N. more than the French, whose sense of great-power standing depends almost entirely on their membership in the P5. The French were infuriated in 1998 when the British and the Americans bombed Iraq without seeking council authorization. The French would not permit the council to be sidelined once again. In late September, the French proposed that in the event of Iraqi noncompliance with the inspections regime, the United States be compelled to seek a second Security Council resolution authorizing the use of force. The Russians publicly, and other members of the council privately, backed the French position. After several weeks of deadlocked talks, the Americans, working with the British, proposed compromise language in which the Security Council would "convene immediately" for deliberations in the case of Iraqi noncompliance, though the administration would not have to seek further authority from the council to begin a military assault. This was a two-*step* process, not the two-*resolution* process the French wanted, but it took only a small sophistry to blur the distinction. The French accepted.

I happened to visit Levitte two days after the new American language arrived. Levitte is a rising star in French diplomacy, the former foreign-policy advisor to President Jacques Chirac and a new kind of French diplomat, informal and affable. He insisted that the French position had much less to do with a sense of national prerogative than with the need for a united front. "In France, we have five to six million Muslims," he said, "between 8 and 10 percent of the population." In case of war, he went on, "we have to do whatever is possible to limit the bitterness. It is in our common interest to move step by step. If Saddam Hussein violates the rules set by the international community, then it is not a unilateral act of aggression against Saddam Hussein. It is Saddam Hussein who is creating an attack against the international community."

This was not only a French position; I heard some variant of it from other members of the council and even more from U.N. professionals who feared they would have to live with a Middle Eastern world aflame with anger. For them, "international legitimacy" was not a diplomatic nicety or an abstraction; it was the means by which war became a collective good rather than an exercise in American self-interest. Beyond Iraq, there was anxiety that American action could loose the dogs of unilateralism;

Russia could well cite the Iraqi precedent to take its own war on Chechen terrorism into neighboring Georgia. And so the other council members were imploring the United States to stay inside the international system as the price for their acquiescence to a process that might well lead to war—a war that virtually none of them believed in.

But the French also understood that simply by agreeing to engage in diplomacy, the Bush administration had already accepted some limits on its freedom of action in order to gain the imprimatur of a Security Council resolution. At this critical moment, according to administration sources, Bush had come down on the side of Powell and the multilateralists. What happened? First of all, polls showed that a clear majority of Americans preferred the inspections-first route and would be much more comfortable with a war conducted with United Nations support—a source of tremendous comfort within the United Nations itself. Moreover, the U.N. had hardly been unresponsive as regards the war on terrorism. As Richard Haass, a senior State Department official (and a strong advocate of the multilateralist view), puts it, "This is the second major issue we have brought to the U.N., the first being Afghanistan, and there we did just fine." Haass adds, "You've got to ask yourself what's the price in terms of delay and constraint, and what the benefits are."

Those benefits were overwhelming: a military threat backed by a united Security Council would be far likelier to persuade Saddam Hussein to disarm, and in the event of war, the United States would be much likelier to obtain overflight rights and access to Middle East bases. (There would also, in all likelihood, be financial support and international help in administering postwar Iraq.)

And so with Secretary Powell conducting the negotiations (the American ambassador to the U.N., John Negroponte, served largely as a message carrier), the Bush administration ultimately came a long way toward the French position, leaving more latitude to the inspectors and modifying language that appeared to provide a pretext for war no matter what the Iraqis did—what the French called "automaticity." At the dramatic Security Council meeting at which all 15 nations endorsed the resolution, Negroponte explicitly confirmed that the language permitted no automaticity.

But the session also exposed the continuing deep fears among council members of the administration's unilateralist impulses. When Ireland's ambassador to the United Nations said at the meeting, "It is for the *council* to decide on any ensuing action," no one could mistake his meaning. Bush administration officials, on the other hand, fear that council members will try to block a military response. The era of good feelings could dissipate fast.

EVERY OTHER FRIDAY, the military attachés of the five permanent members of the Security Council meet in a conference room at the U.N., read the minutes of their last meeting—which consist only of the reading of the minutes of their previous meeting—and then adjourn. This is all that remains of the Military Staff Committee established by Article 47 of the U.N. Charter. And it is virtually all that remains of the founding vision of the Security Council as an institutionalized version of the World War II alliance. And yet as the debate over the Iraq resolution demonstrates, the Security Council has not been consigned to irrelevance. As Edward Luck, the author of "Mixed Messages," a history of America's turbulent relationship with the U.N., observes, "While the Security Council may have failed as a military tool, as a political tool it's more important than ever." Luck predicts that "the demand of the American public for international authorization is going to become greater and greater."

The Security Council need not be wholly reduced to its legitimacy-granting function. The U.N. and its various agencies have built up enormous expertise in the thankless task of nation building; one U.N. official says that the Bush administration even approached the organization to take over the civil administration of Afghanistan after the Taliban had been dislodged, a role that it played in East Timor and elsewhere. (The U.N. had the good sense to decline.) The American resolve to engage in strategic nation wrecking may increasingly force the Security Council, which will be asked to authorize these missions, into the role of picking up the pieces.

But in a world defined by the fight against terrorism, which is to say a world shaped by a single power overwhelmingly preoccupied with the fight against terrorism, the Security Council's central role will be to shape the terms and establish the conditions under which that fight becomes broadly acceptable. Its job is both to restrain and to license the superpower.

Conservative critics of the U.N., some of whom now occupy important posts in the Bush administration, have long argued that the Security Council is useful only when it accepts American leadership and embraces American interests—which, they would add, is virtually never. And yet what has become obvious in recent weeks is that with only the most modest gestures toward multilateralism on the part of the U.S., the Security Council is prepared to offer that embrace. And the Bush administration is likely to hug back when it suits its needs. It's a relationship of convenience. But it's a relationship.

One Security Council diplomat who finds this prospect both professionally pleasing and deeply gratifying to his sense of irony points out that the final draft of the Iraq resolution essentially gives Hans Blix, the head of the U.N. inspection team, the power to decide whether or not the Iraqis are in compliance (though Blix has told the Council he doesn't want to bear that burden). Blix, he points out, "is a nice, soft-spoken, grandfatherly Swede, not an American, not a warmonger"—the perfect legitimator of the American war effort. "If the war comes," he adds, "I see Bush making an 8 o'clock speech to the nation, with Hans Blix mentioned at least 10 times."

The Pitfalls of Universal Jurisdiction

by Henry A. Kissinger

RISKING JUDICIAL TYRANNY

IN LESS THAN a decade, an unprecedented movement has emerged to submit international politics to judicial procedures. It has spread with extraordinary speed and has not been subjected to systematic debate, partly because of the intimidating passion of its advocates. To be sure, human rights violations, war crimes, genocide, and torture have so disgraced the modern age and in such a variety of places that the effort to interpose legal norms to prevent or punish such outrages does credit to its advocates. The danger lies in pushing the effort to extremes that risk substituting the tyranny of judges for that of governments; historically, the dictatorship of the virtuous has often led to inquisitions and even witch-hunts.

The doctrine of universal jurisdiction asserts that some crimes are so heinous that their perpetrators should not escape justice by invoking doctrines of sovereign immunity or the sacrosanct nature of national frontiers. Two specific approaches to achieve this goal have emerged recently. The first seeks to apply the procedures of domestic criminal justice to violations of universal standards, some of which are embodied in United Nations conventions, by authorizing national prosecutors to bring offenders into their jurisdictions through extradition from third countries. The second approach is the International Criminal Court (ICC), the founding treaty for which was created by a conference in Rome in July 1998 and signed by 95 states, including most European countries. It has already been ratified by 30 nations and will go into effect when the total reaches 60. On December 31, 2000, President Bill Clinton signed the ICC treaty with only hours to spare before the cutoff date. But he indicated that he would neither submit it for Senate approval nor recommend that his successor do so while the treaty remains in its present form.

The very concept of universal jurisdiction is of recent vintage. The sixth edition of *Black's Law Dictionary*, published in 1990, does not contain even an entry for the term. The closest analogous concept listed is *hostes humani generis* ("enemies of the human race"). Until recently, the latter term has been applied to pirates, hijackers, and similar outlaws whose crimes were typically committed outside the territory of any state. The notion that heads of state and senior public officials should have the same standing as outlaws before the bar of justice is quite new.

In the aftermath of the Holocaust and the many atrocities committed since, major efforts have been made to find a judicial standard to deal with such catastrophes: the Nuremberg trials of 1945–46, the Universal Declaration of Human Rights of 1948, the genocide convention of 1948, and the anti-torture convention of 1988. The Final Act of the Conference on Security and Cooperation in Europe, signed in Helsinki in 1975 by President Gerald Ford on behalf of the United States, obligated the 35 signatory nations to observe certain stated human rights, subjecting violators to the pressures by which foreign policy commitments are generally sustained. In the hands of courageous groups in Eastern Europe, the Final Act became one of several weapons by which communist rule was delegitimized and eventually undermined. In the 1990s, international tribunals to punish crimes committed in the former Yugoslavia and Rwanda, established ad hoc by the U.N. Security Council, have sought to provide a system of accountability for specific regions ravaged by arbitrary violence.

But none of these steps was conceived at the time as instituting a "universal jurisdiction." It is unlikely that any of the signatories of either the U.N. conventions or the Helsinki Final Act thought it possible that national judges would use them as a basis for extradition requests regarding alleged crimes committed outside their jurisdictions. The drafters almost certainly believed that they were stating general principles, not laws that would be enforced by national courts. For example, Eleanor Roosevelt, one of the drafters of the Universal Declaration of Human Rights, referred to it as a "common standard." As one of the negotiators of the Final Act of the Helsinki conference, I can affirm that the administration I represented considered it primarily a diplomatic weapon to use to thwart the communists' attempts to pressure the Soviet and captive peoples. Even with respect to binding undertakings such as the genocide convention, it was never thought that they would subject past and future leaders of one nation to prosecution by the national magistrates of another state where the violations had not occurred. Nor, until recently, was it argued that the various U.N. declarations subjected past and future leaders to the possibility of prosecution by national magistrates of third countries without either due process safeguards or institutional restraints.

Yet this is in essence the precedent that was set by the 1998 British detention of former Chilean President Augusto Pinochet as the result of an extradition request by a Spanish judge seeking to try Pinochet for crimes committed against Spaniards on Chilean soil. For advocates of universal jurisdiction, that detention—lasting more than 16 months—was a landmark establishing a just principle. But any universal system should contain procedures not only to punish the wicked but also to constrain the righteous. It must not allow legal principles to be used as weapons to settle political scores. Questions such as these must therefore be answered: What legal norms are being applied? What are the rules of evidence? What safeguards exist for the defendant? And how will prosecutions affect other fundamental foreign policy objectives and interests?

A DANGEROUS PRECEDENT

It is decidedly unfashionable to express any degree of skepticism about the way the Pinochet case was handled. For almost all the parties of the European left, Augusto Pinochet is the incarnation of a right-wing assault on democracy because he led a coup d'état against an elected leader. At the time, others, including the leaders of Chile's democratic parties, viewed Salvador Allende as a radical Marxist ideologue bent on imposing a Castro-style dictatorship with the aid of Cuban-trained militias and Cuban weapons. This was why the leaders of Chile's democratic parties publicly welcomed—yes, welcomed—Allende's overthrow. (They changed their attitude only after the junta brutally maintained its autocratic rule far longer than was warranted by the invocation of an emergency.)

The world must respect Chile's own attempt to come to terms with its brutal past.

Disapproval of the Allende regime does not exonerate those who perpetrated systematic human rights abuses after it was overthrown. But neither should the applicability of universal jurisdiction as a policy be determined by one's view of the political history of Chile. The appropriate solution was arrived at in August 2000 when the Chilean Supreme Court withdrew Pinochet's senatorial immunity, making it possible to deal with

the charges against him in the courts of the country most competent to judge this history and to relate its decisions to the stability and vitality of its democratic institutions.

On November 25, 1998, the judiciary committee of the British House of Lords (the United Kingdom's supreme court) concluded that "international law has made it plain that certain types of conduct… are not acceptable conduct on the part of anyone." But that principle did not oblige the lords to endow a Spanish magistrate—and presumably other magistrates elsewhere in the world—with the authority to enforce it in a country where the accused had committed no crime, and then to cause the restraint of the accused for 16 months in yet another country in which he was equally a stranger. It could have held that Chile, or an international tribunal specifically established for crimes committed in Chile on the model of the courts set up for heinous crimes in the former Yugoslavia and Rwanda, was the appropriate forum.

The unprecedented and sweeping interpretation of international law in *Exparte Pinochet* would arm any magistrate anywhere in the world with the power to demand extradition, substituting the magistrate's own judgment for the reconciliation procedures of even incontestably democratic societies where alleged violations of human rights may have occurred. It would also subject the accused to the criminal procedures of the magistrate's country, with a legal system that may be unfamiliar to the defendant and that would force the defendant to bring evidence and witnesses from long distances. Such a system goes far beyond the explicit and limited mandates established by the U.N. Security Council for the tribunals covering war crimes in the former Yugoslavia and Rwanda as well as the one being negotiated for Cambodia.

Perhaps the most important issue is the relationship of universal jurisdiction to national reconciliation procedures set up by new democratic governments to deal with their countries' questionable pasts. One would have thought that a Spanish magistrate would have been sensitive to the incongruity of a request by Spain, itself haunted by transgressions committed during the Spanish Civil War and the regime of General Francisco Franco, to try in Spanish courts alleged crimes against humanity committed elsewhere.

The decision of post-Franco Spain to avoid wholesale criminal trials for the human rights violations of the recent past was designed explicitly to foster a process of national reconciliation that undoubtedly contributed much to the present vigor of Spanish democracy. Why should Chile's attempt at national reconciliation not have been given the same opportunity? Should any outside group dissatisfied with the reconciliation procedures of, say, South Africa be free to challenge them in their own national courts or those of third countries?

It is an important principle that those who commit war crimes or systematically violate human rights should be held accountable. But the consolidation of law, domestic peace, and representative government in a nation struggling to come to terms with a brutal past has a claim as well. The instinct to punish must be related, as in every constitutional democratic political structure, to a system of checks and balances that includes other elements critical to the survival and expansion of democracy.

Another grave issue is the use in such cases of extradition procedures designed for ordinary criminals. If the Pinochet case becomes a precedent, magistrates anywhere will be in a position to put forward an extradition request without warning to the accused and regardless of the policies the accused's country might already have in place for dealing with the charges. The country from which extradition is requested then faces a seemingly technical legal decision that, in fact, amounts to the exercise of political discretion—whether to entertain the claim or not.

Once extradition procedures are in train, they develop a momentum of their own. The accused is not allowed to challenge the substantive merit of the case and instead is confined to procedural issues: that there was, say, some technical flaw in the extradition request, that the judicial system of the requesting country is incapable of providing a fair hearing, or that the crime for which the extradition is sought is not treated as a crime in the country from which extradition has been requested—thereby conceding much of the merit of the charge. Meanwhile, while these claims are being considered by the judicial system of the country from which extradition is sought, the accused remains in some form of detention, possibly for years. Such procedures provide an opportunity for political harassment long before the accused is in a position to present any defense. It would be ironic if a doctrine designed to transcend the political process turns into a means to pursue political enemies rather than universal justice.

The Pinochet precedent, if literally applied, would permit the two sides in the Arab-Israeli conflict, or those in any other passionate international controversy, to project their battles into the various national courts by pursuing adversaries with extradition requests. When discretion on what crimes are subject to universal jurisdiction and whom to prosecute is left to national prosecutors, the scope for arbitrariness is wide indeed. So far, universal jurisdiction has involved the prosecution of one fashionably reviled man of the right while scores of East European communist leaders—not to speak of Caribbean, Middle Eastern, or African leaders who inflicted their own full measures of torture and suffering—have not had to face similar prosecutions.

Some will argue that a double standard does not excuse violations of international law and that it is better to bring one malefactor to justice than to grant immunity to all. This is not an argument permitted in the domestic jurisdictions of many democracies—in Canada, for example, a charge can be thrown out of court merely by showing that a prosecution has been selective enough to amount to an abuse of process. In any case, a universal standard of justice should not be based on the proposition that a just end warrants unjust means, or that political fashion trumps fair judicial procedures.

AN INDISCRIMINATE COURT

T HE IDEOLOGICAL supporters of universal jurisdiction also provide much of the intellectual compass for the emerging In-

ternational Criminal Court. Their goal is to criminalize certain types of military and political actions and thereby bring about a more humane conduct of international relations. To the extent that the ICC replaces the claim of national judges to universal jurisdiction, it greatly improves the state of international law. And, in time, it may be possible to negotiate modifications of the present statute to make the ICC more compatible with U.S. constitutional practice. But in its present form of assigning the ultimate dilemmas of international politics to unelected jurists—and to an international judiciary at that—it represents such a fundamental change in U.S. constitutional practice that a full national debate and the full participation of Congress are imperative. Such a momentous revolution should not come about by tacit acquiescence in the decision of the House of Lords or by dealing with the ICC issue through a strategy of improving specific clauses rather than as a fundamental issue of principle.

> At any future time, U.S. officials involved in the NATO air campaign in Kosovo could face international prosecution.

The doctrine of universal jurisdiction is based on the proposition that the individuals or cases subject to it have been clearly identified. In some instances, especially those based on Nuremberg precedents, the definition of who can be prosecuted in an international court and in what circumstances is self-evident. But many issues are much more vague and depend on an understanding of the historical and political context. It is this fuzziness that risks arbitrariness on the part of prosecutors and judges years after the event and that became apparent with respect to existing tribunals.

For example, can any leader of the United States or of another country be hauled before international tribunals established for other purposes? This is precisely what Amnesty International implied when, in the summer of 1999, it supported a "complaint" by a group of European and Canadian law professors to Louise Arbour, then the prosecutor of the International Criminal Tribunal for the Former Yugoslavia (ICTY). The complaint alleged that crimes against humanity had been committed during the NATO air campaign in Kosovo. Arbour ordered an internal staff review, thereby implying that she did have jurisdiction if such violations could, in fact, be demonstrated. Her successor, Carla Del Ponte, in the end declined to indict any NATO official because of a general inability "to pinpoint individual responsibilities," thereby implying anew that the court had jurisdiction over NATO and American leaders in the Balkans and would have issued an indictment had it been able to identify the particular leaders allegedly involved.

Most Americans would be amazed to learn that the ICTY, created at U.S. behest in 1993 to deal with Balkan war criminals, had asserted a right to investigate U.S. political and military leaders for allegedly criminal conduct—and for the indefinite future, since no statute of limitations applies. Though the ICTY prosecutor chose not to pursue the charge—on the ambiguous ground of an inability to collect evidence—some national prosecutor may wish later to take up the matter as a valid subject for universal jurisdiction.

The pressures to achieve the widest scope for the doctrine of universal jurisdiction were demonstrated as well by a suit before the European Court of Human Rights in June 2000 by families of Argentine sailors who died in the sinking of the Argentine cruiser *General Belgano* during the Falklands War. The concept of universal jurisdiction has moved from judging alleged political crimes against humanity to second-guessing, 18 years after the event, military operations in which neither civilians nor civilian targets were involved.

Distrusting national governments, many of the advocates of universal jurisdiction seek to place politicians under the supervision of magistrates and the judicial system. But prosecutorial discretion without accountability is precisely one of the flaws of the International Criminal Court. Definitions of the relevant crimes are vague and highly susceptible to politicized application. Defendants will not enjoy due process as understood in the United States. Any signatory state has the right to trigger an investigation. As the U.S. experience with the special prosecutors investigating the executive branch shows, such a procedure is likely to develop its own momentum without time limits and can turn into an instrument of political warfare. And the extraordinary attempt of the ICC to assert jurisdiction over Americans even in the absence of U.S. accession to the treaty has already triggered legislation in Congress to resist it.

The independent prosecutor of the ICC has the power to issue indictments, subject to review only by a panel of three judges. According to the Rome statute, the Security Council has the right to quash any indictment. But since revoking an indictment is subject to the veto of any permanent Security Council member, and since the prosecutor is unlikely to issue an indictment without the backing of at least one permanent member of the Security Council, he or she has virtually unlimited discretion in practice. Another provision permits the country whose citizen is accused to take over the investigation and trial. But the ICC retains the ultimate authority on whether that function has been adequately exercised and, if it finds it has not, the ICC can reassert jurisdiction. While these procedures are taking place, which may take years, the accused will be under some restraint and certainly under grave public shadow.

The advocates of universal jurisdiction argue that the state is the basic cause of war and cannot be trusted to deliver justice.

If law replaced politics, peace and justice would prevail. But even a cursory examination of history shows that there is no evidence to support such a theory. The role of the statesman is to choose the best option when seeking to advance peace and justice, realizing that there is frequently a tension between the two and that any reconciliation is likely to be partial. The choice, however, is not simply between universal and national jurisdictions.

MODEST PROPOSALS

THE PRECEDENT SET by international tribunals established to deal with situations where the enormity of the crime is evident and the local judicial system is clearly incapable of administering justice, as in the former Yugoslavia and Rwanda, have shown that it is possible to punish without removing from the process all political judgment and experience. In time, it may be possible to renegotiate the ICC statute to avoid its shortcomings and dangers. Until then, the United States should go no further toward a more formal system than one containing the following three provisions. First, the U.N. Security Council would create a Human Rights Commission or a special subcommittee to report whenever systematic human rights violations seem to warrant judicial action. Second, when the government under which the alleged crime occurred is not authentically representative, or where the domestic judicial system is incapable of sitting in judgment on the crime, the Security Council would set up an ad hoc international tribunal on the model of those of the former Yugoslavia or Rwanda. And third, the procedures for these international tribunals as well as the scope of the prosecution should be precisely defined by the Security Council, and the accused should be entitled to the due process safeguards accorded in common jurisdictions.

In this manner, internationally agreed procedures to deal with war crimes, genocide, or other crimes against humanity could become institutionalized. Furthermore, the one-sidedness of the current pursuit of universal jurisdiction would be avoided. This pursuit could threaten the very purpose for which the concept has been developed. In the end, an excessive reliance on universal jurisdiction may undermine the political will to sustain the humane norms of international behavior so necessary to temper the violent times in which we live.

HENRY A. KISSINGER, Chairman of Kissinger Associates, Inc., is a former Secretary of State and National Security Adviser. This essay is adapted from his latest book, *Does America Need a Foreign Policy? Toward a Diplomacy for the 21st Century.*

THE
Five Wars
OF GLOBALIZATION

The illegal trade in drugs, arms, intellectual property, people, and money is booming. Like the war on terrorism, the fight to control these illicit markets pits governments against agile, stateless, and resourceful networks empowered by globalization. Governments will continue to lose these wars until they adopt new strategies to deal with a larger, unprecedented struggle that now shapes the world as much as confrontations between nation–states once did.

By Moisés Naím

The persistence of al Qaeda underscores how hard it is for governments to stamp out stateless, decentralized networks that move freely, quickly, and stealthily across national borders to engage in terror. The intense media coverage devoted to the war on terrorism, however, obscures five other similar global wars that pit governments against agile, well-financed networks of highly dedicated individuals. These are the fights against the illegal international trade in drugs, arms, intellectual property, people, and money. Religious zeal or political goals drive terrorists, but the promise of enormous financial gain motivates those who battle governments in these five wars. Tragically, profit is no less a motivator for murder, mayhem, and global insecurity than religious fanaticism.

In one form or another, governments have been fighting these five wars for centuries. And losing them. Indeed, thanks to the changes spurred by globalization over the last decade, their losing streak has become even more pronounced. To be sure, nation-states have benefited from the information revolution, stronger political and economic linkages, and the shrinking importance of geographic distance. Unfortunately,

criminal networks have benefited even more. Never fettered by the niceties of sovereignty, they are now increasingly free of geographic constraints. Moreover, globalization has not only expanded illegal markets and boosted the size and the resources of criminal networks, it has also imposed more burdens on governments: Tighter public budgets, decentralization, privatization, deregulation, and a more open environment for international trade and investment all make the task of fighting global criminals more difficult. Governments are made up of cumbersome bureaucracies that generally cooperate with difficulty, but drug traffickers, arms dealers, alien smugglers, counterfeiters, and money launderers have refined networking to a high science, entering into complex and improbable strategic alliances that span cultures and continents.

Defeating these foes may prove impossible. But the first steps to reversing their recent dramatic gains must be to recognize the fundamental similarities among the five wars and to treat these conflicts not as law enforcement problems but as a new global trend that shapes the world as much as confrontations between nation-states did in the past. Customs officials, police officers, lawyers, and judges alone will never win these wars. Governments must recruit and deploy more spies, soldiers, diplomats, and economists who understand how to use incentives and regulations to steer markets away from bad social outcomes. But changing the skill set of government combatants alone will not end these wars. Their doctrines and institutions also need a major overhaul.

THE FIVE WARS

Pick up any newspaper anywhere in the world, any day, and you will find news about illegal migrants, drug busts, smuggled weapons, laundered money, or counterfeit goods. The global nature of these five wars was unimaginable just a decade ago. The resources—financial, human, institutional, technological—deployed by the combatants have reached unfathomable orders of magnitude. So have the numbers of victims. The tactics and tricks of both sides boggle the mind. Yet if you cut through the fog of daily headlines and orchestrated photo ops, one inescapable truth emerges: The world's governments are fighting a qualitatively new phenomenon with obsolete tools, inadequate laws, inefficient bureaucratic arrangements, and ineffective strategies. Not surprisingly, the evidence shows that governments are losing.

Drugs

The best known of the five wars is, of course, the war on drugs. In 1999, the United Nations' "Human Development Report" calculated the annual trade in illicit drugs at $400 billion, roughly the size of the Spanish economy and about 8 percent of world trade. Many countries are reporting an increase in drug use. Feeding this habit is a global supply chain that uses everything from passenger jets that can carry shipments of cocaine worth $500 million in a single trip to custom-built submarines that ply the waters between Colombia and Puerto Rico. To foil eavesdroppers, drug smugglers use "cloned" cell phones and broadband radio receivers while also relying on complex financial structures that blend legitimate and illegitimate enterprises with elaborate fronts and structures of cross-ownership.

The United States spends between $35 billion and $40 billion each year on the war on drugs; most of this money is spent on interdiction and intelligence. But the creativity and boldness of drug cartels has routinely outstripped steady increases in government resources. Responding to tighter security at the U.S.-Mexican border, drug smugglers built a tunnel to move tons of drugs and billions of dollars in cash until authorities discovered it in March 2002. Over the last decade, the success of the Bolivian and Peruvian governments in eradicating coca plantations has shifted production to Colombia. Now, the U.S.-supported Plan Colombia is displacing coca production and processing labs back to other Andean countries. Despite the heroic efforts of these Andean countries and the massive financial and technical support of the United States, the total acreage of coca plantations in Peru, Colombia, and Bolivia has increased in the last decade from 206,200 hectares in 1991 to 210,939 in 2001. Between 1990 and 2000, according to economist Jeff

DeSimone, the median price of a gram of cocaine in the United States fell from $152 to $112.

Even when top leaders of drug cartels are captured or killed, former rivals take their place. Authorities have acknowledged, for example, that the recent arrest of Benjamin Arellano Felix, accused of running Mexico's most ruthless drug cartel, has done little to stop the flow of drugs to the United States. As Arellano said in a recent interview from jail, "They talk about a war against the Arellano brothers. They haven't won. I'm here, and nothing has changed."

Arms Trafficking

Drugs and arms often go together. In 1999, the Peruvian military parachuted 10,000 AK-47s to the Revolutionary Armed Forces of Colombia, a guerrilla group closely allied to drug growers and traffickers. The group purchased the weapons in Jordan. Most of the roughly 80 million AK-47s in circulation today are in the wrong hands. According to the United Nations, only 18 million (or about 3 percent) of the 550 million small arms and light weapons in circulation today are used by government, military, or police forces. Illicit trade accounts for almost 20 percent of the total small arms trade and generates more than $1 billion a year. Small arms helped fuel 46 of the 49 largest conflicts of the last decade and in 2001 were estimated to be responsible for 1,000 deaths a day; more than 80 percent of those victims were women and children.

Small arms are just a small part of the problem. The illegal market for munitions encompasses top-of-the-line tanks, radar systems that detect Stealth aircraft, and the makings of the deadliest weapons of mass destruction. The International Atomic Energy Agency has confirmed more than a dozen cases of smuggled nuclear-weapons-usable material, and hundreds more cases have been reported and investigated over the last decade. The actual supply of stolen nuclear-, biological-, or chemical-weapons materials and technology may still be small. But the potential demand is strong and growing from both would-be nuclear powers and terrorists. Constrained supply and increasing demand cause prices to rise and create enormous incentives for illegal activities. More than one fifth of the 120,000 workers in Russia's former "nuclear cities"—where more than half of all employees earn less than $50 a month—say they would be willing to work in the military complex of another country.

Governments have been largely ineffective in curbing either supply or demand. In recent years, two countries, Pakistan and India, joined the declared nuclear power club. A U.N. arms embargo failed to prevent the reported sale to Iraq of jet fighter engine parts from Yugoslavia and the Kolchuga anti-Stealth radar system from Ukraine. Multilateral efforts to curb the manufacture and distribution of weapons are faltering, not least because some powers are unwilling to accept curbs on their own activities. In 2001, for example, the United States blocked a legally binding global treaty to control small arms in part because it worried about restrictions on its own citizens' rights to own guns. In the absence of effective international legislation and enforcement, the laws of economics dictate the sale of more weapons at cheaper prices: In 1986, an AK-47 in Kolowa, Kenya, cost 15 cows. Today, it costs just four.

Other Fronts

Drugs, arms, intellectual property, people, and money are not the only commodities traded illegally for huge profits by international networks. They also trade in human organs, endangered species, stolen art, and toxic waste. The illegal global trades in all these goods share several fundamental characteristics: Technological innovations and political changes open new markets, globalization is increasing both the geographical reach and the profit opportunities for criminal networks, and governments are on the losing end of the fight to stop them. Some examples:

Human organs: Corneas, kidneys, and livers are the most commonly traded human parts in a market that has boomed thanks to technology, which has improved preservation techniques and made transplants less risky. In the United States, 70,000 patients are on the waiting list for major organ transplants while only 20,000 of them succeed in getting the organ they need. Unscrupulous "organ brokers" partly meet this demand by providing, for a fee, organs and transplant services.

Some of the donors, especially of kidneys, are desperately poor. In India, an estimated 2,000 people a year sell their organs. Many organs, however, come from nonconsenting donors forced to undergo operations or from cadavers in police morgues. For example, medical centers in Germany and Austria were recently found to have used human heart valves taken without consent from the cadavers of poor South Africans.

Endangered species: From sturgeon for caviar in gourmet delicatessens to tigers or elephants for private zoos, the trade in endangered animals and plants is worth billions of dollars and includes hundreds of millions of plant and animal types. This trade ranges from live animals and plants to all kinds of wildlife products derived from them, including food products, exotic leather goods, wooden musical instruments, timber, tourist curiosities, and medicines.

Stolen art: Paintings and sculptures taken from museums, galleries, and private homes, from Holocaust victims, or from "cultural artifacts" poached from archeological digs and other ancient ruins are also illegally traded internationally in a market worth an estimated $2 billion to $6 billion each year. The growing use of art-based transactions in money laundering has spurred demand over the last decade. The supply has boomed because the Soviet Union's collapse flooded the world's market with art that had been under state control. The Czech Republic, Poland, and Russia are three of the five countries most affected by art crime worldwide.

Toxic waste: Innovations in maritime transport, tighter environmental regulations in industrialized countries coupled with increased integration of poor countries to the global economy and better telecommunications have created a market where waste is traded internationally. Greenpeace estimates that during the 20 years prior to 1989, just 3.6 million tons of hazardous waste were exported; in the five years after 1989, the trade soared to about 6.7 billion tons. The environmental organization also reckons that 86 to 90 percent of all hazardous waste shipments destined for developing countries—purportedly for recycling, reuse, recovery, or humanitarian uses—are toxic waste.

—*M.N.*

Intellectual Property

In 2001, two days after recording the voice track of a movie in Hollywood, actor Dennis Hopper was in Shanghai where a street vendor sold him an excellent pirated copy of the movie with his voice already on it. "I don't know how they got my voice into the country before I got here," he wondered. Hopper's experience is one tiny slice of an illicit trade that cost the United States an estimated $9.4 billion in 2001. The piracy rate of business software in Japan and France is 40 percent, in Greece and South Korea it is about 60 percent, and in Germany and Britain it hovers around 30 percent. Forty percent of Procter & Gamble shampoos and 60 percent of Honda motorbikes sold in China in 2001 were pirated. Up to 50 percent of medical drugs in Nigeria and Thailand are bootleg copies. This problem is not limited to consumer products: Italian makers of industrial valves worry that their $2 billion a year export market is eroded by counterfeit Chinese valves sold in world markets at prices that are 40 percent cheaper.

The drivers of this bootlegging boom are complex. Technology is obviously boosting both the demand and the supply of illegally copied products. Users of Napster, the now defunct Internet company that allowed anyone, anywhere to download and reproduce copyrighted music for free, grew from zero to 20 million in just one year. Some 500,000 film files are traded daily through file-sharing services such as Kazaa and Morpheus; and in late 2002, some 900 million music files could be downloaded for free on the Internet—that is, almost two and a half times more files than those available when Napster reached its peak in February 2001.

Global marketing and branding are also playing a part, as more people are attracted to products bearing a well-known brand like Prada or Cartier. And thanks to the rapid growth and integration into the global economy of countries, such as China, with weak central governments and ineffective laws, producing and exporting near perfect knockoffs are both less expensive and less risky. In the words of the CEO of one of the best known Swiss watchmakers: "We now compete with a product manufactured by Chinese prisoners. The business is run by the Chinese military, their families and friends, using roughly the same machines we have, which they purchased at the same industrial fairs we go to. The way we have rationalized this problem is by assuming that their customers and ours are different. The person that buys a pirated copy of one of our $5,000 watches for less than $100 is not a client we are losing. Perhaps it is a future client that some day will want to own the real thing instead of a fake. We may be wrong and we do spend money to fight the piracy of our products. But given that our efforts do not seem to protect us much, we close our eyes and hope for the better." This posture stands in contrast to that of companies that sell cheaper products such as garments, music, or videos, whose revenues are directly affected by piracy.

Governments have attempted to protect intellectual property rights through various means, most notably the World Trade Organization's Agreement on Trade-Related Aspects of Intellectual Property Rights (TRIPS). Several other organizations such as the World Intellectual Property Organization, the World Customs Union, and Interpol are also involved. Yet the large and growing volume of this trade, or a simple stroll in the streets of Manhattan or Madrid, show that governments are far from winning this fight.

Alien Smuggling

The man or woman who sells a bogus Hermes scarf or a Rolex watch in the streets of Milan is likely to be an illegal alien. Just as likely, he or she was transported across several continents by a trafficking network allied with another network that specializes in the illegal copying, manufacturing, and distributing of high-end, brand-name products.

Alien smuggling is a $7 billion a year enterprise and according to the United Nations is the fastest growing business of organized crime. Roughly 500,000 people enter the United States illegally each year—about the same number as illegally enter the European Union, and part of the approximately 150 million who live outside their countries of origin. Many of these backdoor travelers are voluntary migrants who pay smugglers up to $35,000, the top-dollar fee for passage from China to New York. Others, instead, are trafficked—that is, bought and sold internationally—as commodities. The U.S. Congressional Research Service reckons that each year between 1 million and 2 million people are trafficked across borders, the majority of whom are women and children. A woman can be "bought" in Timisoara, Romania, for between $50 and $200 and "resold" in Western Europe for 10 times that price. The United Nations Children's Fund estimates that cross-border smugglers in Central and Western Africa enslave 200,000 children a year. Traffickers initially tempt victims with job offers or, in the case of children, with offers of adoption in wealthier countries, and then keep the victims in subservience through physical violence, debt bondage, passport confiscation, and threats of arrest, deportation, or violence against their families back home.

Governments everywhere are enacting tougher immigration laws and devoting more time, money, and technology to fight the flow of illegal aliens. But the plight of the United Kingdom's government illustrates how tough that fight is. The British government throws money at the problem, plans to use the Royal Navy and Royal Air Force to intercept illegal immigrants, and imposes large fines on truck drivers who (generally unwittingly) transport stowaways. Still, 42,000 of the 50,000 refugees who have passed through the Sangatte camp (a main entry point for illegal immigration to the United Kingdom) over the last three years have made it to Britain. At current rates, it

will take 43 years for Britain to clear its asylum backlog. And that country is an island. Continental nations such as Spain, Italy, or the United States face an even greater challenge as immigration pressures overwhelm their ability to control the inflow of illegal aliens.

Money Laundering

The Cayman Islands has a population of 36,000. It also has more than 2,200 mutual funds, 500 insurance companies, 60,000 businesses, and 600 banks and trust companies with almost $800 billion in assets. Not surprisingly, it figures prominently in any discussion of money laundering. So does the United States, several of whose major banks have been caught up in investigations of money laundering, tax evasion, and fraud. Few, if any, countries can claim to be free of the practice of helping individuals and companies hide funds from governments, creditors, business partners, or even family members, including the proceeds of tax evasion, gambling, and other crimes. Estimates of the volume of global money laundering range between 2 and 5 percent of the world's annual gross national product, or between $800 billion and $2 trillion.

Smuggling money, gold coins, and other valuables is an ancient trade. Yet in the last two decades, new political and economic trends coincided with technological changes to make this ancient trade easier, cheaper, and less risky. Political changes led to the deregulation of financial markets that now facilitate cross-border money transfers, and technological changes made distance less of a factor and money less "physical." Suitcases full of banknotes are still a key tool for money launderers, but computers, the Internet, and complex financial schemes that combine legal and illegal practices and institutions are more common. The sophistication of technology, the complex web of financial institutions that crisscross the globe, and the ease with which "dirty" funds can be electronically morphed into legitimate assets make the regulation of international flows of money a daunting task. In Russia, for example, it is estimated that by the mid-1990s organized crime groups had set up 700 legal and financial institutions to launder their money.

Faced with this growing tide, governments have stepped up their efforts to clamp down on rogue international banking, tax havens, and money laundering. The imminent, large-scale introduction of e-money—cards with microchips that can store large amounts of money and thus can be easily transported outside regular channels or simply exchanged among individuals—will only magnify this challenge.

WHY GOVERNMENTS CAN'T WIN

The fundamental changes that have given the five wars new intensity over the last decade are likely to persist. Technology will continue to spread widely; criminal networks will be able to exploit these technologies more quickly than governments that must cope with tight budgets, bureaucracies, media scrutiny, and electorates. International trade will continue to grow, providing more cover for the expansion of illicit trade. International migration will likewise grow, with much the same effect, offering ethnically based gangs an ever growing supply of recruits and victims. The spread of democracy may also help criminal cartels, which can manipulate weak public institutions by corrupting police officers or tempting politicians with offers of cash for their increasingly expensive election campaigns. And ironically, even the spread of international law—with its growing web of embargoes, sanctions, and conventions—will offer criminals new opportunities for providing forbidden goods to those on the wrong side of the international community.

Even the spread of international law will offer criminals new opportunities for providing forbidden goods to those on the wrong side of the international community.

These changes may affect each of the five wars in different ways, but these conflicts will continue to share four common characteristics:

They are not bound by geography.

Some forms of crime have always had an international component: The Mafia was born in Sicily and exported to the United States, and smuggling has always been by definition international. But the five wars are truly global. Where is the theater or front line of the war on drugs? Is it Colombia or Miami? Myanmar (Burma) or Milan? Where are the battles against money launderers being fought? In Nauru or in London? Is China the main theater in the war against the infringement of intellectual property, or are the trenches of that war on the Internet?

They defy traditional notions of sovereignty.

Al Qaeda's members have passports and nationalities—and often more than one—but they are truly stateless. Their allegiance is to their cause, not to any nation. The same is also true of the criminal networks engaged in the five wars. The same, however, is patently *not* true of government employees—police officers, customs agents, and judges—who fight them. This asymmetry is a crippling disadvantage for governments waging these wars. Highly paid, hypermotivated, and resource-rich combatants on one side of the wars (the criminal gangs) can seek refuge in and take advantage of national borders, but combatants of the other side (the governments) have fewer resources and are hampered by traditional notions of sovereignty. A former senior CIA official reported that international criminal gangs are able to move people, money, and weapons globally faster than he can move resources inside his own agency, let alone worldwide. Coordination and information sharing among government agencies in different countries has certainly improved, especially after September 11. Yet these tactics fall short of what is needed to combat agile organizations that can exploit every nook and cranny of an evolving but imperfect body of international law and multilateral treaties.

They pit governments against market forces.

In each of the five wars, one or more government bureaucracies fight to contain the disparate, uncoordinated actions of thousands of independent, stateless organizations. These groups are motivated by large profits obtained by exploiting international price differentials, an unsatisfied demand, or the cost advantages produced by theft. Hourly wages for a Chinese cook are far higher in Manhattan than in Fujian. A gram of cocaine in Kansas City is 17,000 percent more expensive than in Bogotá. Fake Italian valves are 40 percent cheaper because counterfeiters don't have to cover the costs of developing the product. A well-funded guerrilla group will pay anything to get the weapons it needs. In each of these five wars, the incentives to successfully overcome government-imposed limits to trade are simply enormous.

They pit bureaucracies against networks.

The same network that smuggles East European women to Berlin may be involved in distributing opium there. The proceeds of the latter fund the purchase of counterfeit Bulgari watches made in China and often sold on the streets of Manhattan by illegal African immigrants. Colombian drug cartels make deals with Ukrainian arms traffickers, while Wall Street brokers controlled by the U.S.-based Mafia have been known to front for Russian money launderers. These highly decentralized groups and individuals are bound by strong ties of loyalty and common purpose and organized around semiautonomous clusters or "nodes" capable of operating swiftly and flexibly. John Arquilla and David Ronfeldt, two of the best known experts on these types of organizations, observe that networks often lack central leadership, command, or headquarters, thus "no precise heart or head that can be targeted. The network as a whole (but not necessarily each node) has little to no hierarchy; there may be multiple leaders.... Thus the [organization's] design may sometimes appear acephalous (headless), and at other times polycephalous (Hydra-headed)." Typically, governments respond to these challenges by forming interagency task forces or creating new bureaucracies. Consider the creation of the new Department of Homeland Security in the United States, which encompasses 22 former federal agencies and their 170,000 employees and is responsible for, among other things, fighting the war on drugs.

RETHINKING THE PROBLEM

Governments may never be able to completely eradicate the kind of international trade involved in the five wars. But they can and should do better. There are at least four areas where efforts can yield better ideas on how to tackle the problems posed by these wars:

Develop more flexible notions of sovereignty.

Governments need to recognize that restricting the scope of multilateral action for the sake of protecting their sovereignty is often a moot point. Their sovereignty is compromised daily, not by nation-states but by stateless networks that break laws and cross borders in pursuit of trade. In May 1999, for example, the Venezuelan government denied U.S. planes authorization to fly over Venezuelan territory to monitor air routes commonly used by narcotraffickers. Venezuelan authorities placed more importance on the symbolic value of asserting sovereignty over air space than on the fact that drug traffickers' planes regularly violate Venezuelan territory. Without new forms of codifying and "managing" sovereignty, governments will continue to face a large disadvantage while fighting the five wars.

Strengthen existing multilateral institutions.

The global nature of these wars means no government, regardless of its economic, political, or military power, will make much progress acting alone. If this seems obvious, then why does Interpol, the multilateral agency in charge of fighting international crime, have a staff of 384, only 112 of whom are police officers, and an annual budget of $28 million, less than the price of some boats or planes used by drug traffickers? Similarly, Europol, Europe's Interpol equivalent, has a staff of 240 and a budget of $51 million.

One reason Interpol is poorly funded and staffed is because its 181 member governments don't trust each other. Many assume, and perhaps rightly so, that the criminal networks they are fighting have penetrated the police departments of other countries and that sharing information with such compromised officials would not be prudent. Others fear today's allies will become tomorrow's enemies. Still others face legal impediments to sharing intelligence with fellow nation-states or have intelligence services and law enforcement agencies with organizational cultures that make effective collaboration almost impossible. Progress will only be made if the world's governments unite behind stronger, more effective multilateral organizations.

Devise new mechanisms and institutions.

These five wars stretch and even render obsolete many of the existing institutions, legal frameworks, military doctrines, weapons systems, and law enforcement techniques on which governments have relied for years. Analysts need to rethink the concept of war "fronts" defined by geography and the definition of "combatants" according to the Geneva Convention. The functions of intelligence agents, soldiers, police officers, customs agents, or immigration officers need rethinking and adaptation to the new realities. Policymakers also need to reconsider the notion that ownership is essentially a physical reality and not a "virtual" one or that only sovereign nations can issue money when thinking about ways to fight the five wars.

Move from repression to regulation.

Beating market forces is next to impossible. In some cases, this reality may force governments to move from repressing the market to regulating it. In others, creating market incentives may be better than using bureaucracies to curb the excesses of these markets. Technology can often accomplish more than government policies can. For example, powerful encryption techniques can better protect software or CDs from being copied in Ukraine than would making the country enforce patents and copyrights and trademarks.

In all of the five wars, government agencies fight against networks motivated by the enormous profit opportunities created by other government agencies. In all cases, these profits can be traced to some form of government intervention that creates a major imbalance between demand and supply and makes prices and profit margins skyrocket. In some cases, these government interventions are often justified and it would be imprudent to eliminate them—governments can't simply walk away from the fight against trafficking in heroin, human beings, or weapons of mass destruction. But society can better deal with other segments of these kinds of illegal trade through regulation, not prohibition. Policymakers must focus on opportunities where market regulation can ameliorate problems that have defied approaches based on prohibition and armed interdiction of international trade.

Ultimately, governments, politicians, and voters need to realize that the way in which the world is conducting these five wars is doomed to fail—not for lack of effort, resources, or political will but because the collective thinking that guides government strategies in the five wars is rooted in wrong ideas, false assumptions, and obsolete institutions. Recognizing that governments have no chance of winning unless they change the ways they wage these wars is an indispensable first step in the search for solutions.

Moisés Naím is editor of FOREIGN POLICY magazine.

From *Foreign Policy*, January/February 2003, pp. 28-37. © 2002 by the Carnegie Endowment for International Peace. Reprinted by permission.

Index

Test Your Knowledge Form

We encourage you to photocopy and use this page as a tool to assess how the articles in *Annual Editions* expand on the information in your textbook. By reflecting on the articles you will gain enhanced text information. You can also access this useful form on a product's book support Web site at *http://www.dushkin.com/online/*.

NAME: DATE:

TITLE AND NUMBER OF ARTICLE:

BRIEFLY STATE THE MAIN IDEA OF THIS ARTICLE:

LIST THREE IMPORTANT FACTS THAT THE AUTHOR USES TO SUPPORT THE MAIN IDEA:

WHAT INFORMATION OR IDEAS DISCUSSED IN THIS ARTICLE ARE ALSO DISCUSSED IN YOUR TEXTBOOK OR OTHER READINGS THAT YOU HAVE DONE? LIST THE TEXTBOOK CHAPTERS AND PAGE NUMBERS:

LIST ANY EXAMPLES OF BIAS OR FAULTY REASONING THAT YOU FOUND IN THE ARTICLE:

LIST ANY NEW TERMS/CONCEPTS THAT WERE DISCUSSED IN THE ARTICLE, AND WRITE A SHORT DEFINITION:

We Want Your Advice

ANNUAL EDITIONS revisions depend on two major opinion sources: one is our Advisory Board, listed in the front of this volume, which works with us in scanning the thousands of articles published in the public press each year; the other is you—the person actually using the book. Please help us and the users of the next edition by completing the prepaid article rating form on this page and returning it to us. Thank you for your help!

ANNUAL EDITIONS: World Politics 03/04

ARTICLE RATING FORM

Here is an opportunity for you to have direct input into the next revision of this volume.
We would like you to rate each of the articles listed below, using the following scale:

1. **Excellent: should definitely be retained**
2. **Above average: should probably be retained**
3. **Below average: should probably be deleted**
4. **Poor: should definitely be deleted**

Your ratings will play a vital part in the next revision.
Please mail this prepaid form to us as soon as possible.
Thanks for your help!

RATING	ARTICLE
_____	1. Clash of Globalizations
_____	2. Transnational Terrorism and the al Qaeda Model: Confronting New Realities
_____	3. Sovereignty
_____	4. Reconciling Non-Intervention and Human Rights
_____	5. Terrorism's Financial Lifeline: Can It Be Severed?
_____	6. Measuring Globalization
_____	7. The Rich Should Not Forget the ROW (Rest of the World)
_____	8. Prisoners of Geography
_____	9. Nuclear Nightmares
_____	10. Return of the Nuclear Debate
_____	11. In North Korea and Pakistan, Deep Roots of Nuclear Barter
_____	12. Towards an Internet Civil Defence Against Bioterrorism
_____	13. A Grand Strategy of Transformation
_____	14. The Eagle Has Crash Landed
_____	15. Economic Crossroads on the Line
_____	16. Canada's Military 'Needs Overhaul'
_____	17. Fragile Democracies
_____	18. Argentina May Be Down But I Don't Plan to Get Out
_____	19. Two Ways to Go Global
_____	20. The Balts and the European Union: Welcome Aboard!
_____	21. Reforging the Atlantic Alliance
_____	22. The Real Trans-Atlantic Gap
_____	23. Europe's Eastern Enlargement: Who Benefits?
_____	24. George W. Bush and Russia
_____	25. Where Does Europe End?
_____	26. Japan's Slow-Moving Economic Avalanche
_____	27. India, Pakistan, and the Prospect of War
_____	28. Great Leap Backwards
_____	29. "Why Do They Hate Us?"
_____	30. Bin Laden, the Arab "Street," and the Middle East's Democracy Deficit
_____	31. The Costs of Chaos in Palestine
_____	32. Kenya's Porous Border Lies Open to Arms Smugglers

RATING	ARTICLE
_____	33. Africa's Security Issues Through 2010
_____	34. Who Needs the U.N. Security Council?
_____	35. The Pitfalls of Universal Jurisdiction
_____	36. The Five Wars of Globalization

(Continued on next page)

NO POSTAGE
NECESSARY
IF MAILED
IN THE
UNITED STATES

BUSINESS REPLY MAIL
FIRST-CLASS MAIL PERMIT NO. 84 GUILFORD CT

POSTAGE WILL BE PAID BY ADDRESSEE

McGraw-Hill/Dushkin
530 Old Whitfield Street
Guilford, Ct 06437-9989

ABOUT YOU

Name   Date
_____

Are you a teacher? ☐ A student? ☐
Your school's name
_____

Department
_____

Address   City   State   Zip
_____

School telephone #
_____

YOUR COMMENTS ARE IMPORTANT TO US!

Please fill in the following information:
For which course did you use this book?
_____

Did you use a text with this ANNUAL EDITION? ☐ yes ☐ no
What was the title of the text?
_____

What are your general reactions to the *Annual Editions* concept?
_____

Have you read any pertinent articles recently that you think should be included in the next edition? Explain.
_____

Are there any articles that you feel should be replaced in the next edition? Why?
_____

Are there any World Wide Web sites that you feel should be included in the next edition? Please annotate.
_____

May we contact you for editorial input? ☐ yes ☐ no
May we quote your comments? ☐ yes ☐ no